Smart Farming, Smarter Solutions

Revolutionizing Agriculture with Artificial Intelligence

Editors

Manojit Chowdhury
Krishi Vigyan Kendra
ICAR-Central Institute of Agricultural Engineering
Bhopal, India

Nand Lal Kushwaha
Division of Agricultural Engineering
ICAR-Indian Agricultural Research Institute
New Delhi, India

Gourav Dhar Bhowmick
Department of Agricultural and Food Engineering
Indian Institute of Technology Kharagpur, India

CRC Press
Taylor & Francis Group
Boca Raton London New York

CRC Press is an imprint of the
Taylor & Francis Group, an **informa** business
A SCIENCE PUBLISHERS BOOK

First edition published 2026
by CRC Press
2385 NW Executive Center Drive, Suite 320, Boca Raton FL 33431

and by CRC Press
4 Park Square, Milton Park, Abingdon, Oxon, OX14 4RN

CRC Press is an imprint of Taylor & Francis Group, LLC

Library of Congress Cataloging-in-Publication Data (applied for)

ISBN: 978-1-032-86360-3 (hbk)
ISBN: 978-1-032-88259-8 (pbk)
ISBN: 978-1-003-53693-2 (ebk)

DOI: 10.1201/9781003536932

Typeset in Times New Roman
by Prime Publishing Services

Preface

Agriculture has always been a sector of innovation, adapting to changing environmental, economic, and technological landscapes. In recent years, artificial intelligence (AI) has emerged as a game-changer, offering data-driven solutions to optimize farming practices, reduce waste, and enhance sustainability. The motivation behind this book, *Smart Farming, Smarter Solutions: Revolutionizing Agriculture with Artificial Intelligence*, was to provide a comprehensive reference that encapsulates the latest advancements and practical applications of AI in agriculture.

This edited volume brings together leading researchers and practitioners from around the world, offering insights into the transformative role of AI across various agricultural domains. The chapters explore critical aspects such as big data processing, IoT sensors, drones, robotics, disease detection, weed management, smart irrigation, and supply chain optimization. By providing both theoretical foundations and real-world case studies, this book aims to serve as a practical guide for researchers, students, policymakers, and agribusiness professionals looking to leverage AI for enhanced agricultural productivity.

We extend our sincere gratitude to all the authors who contributed their expertise to this volume. Their commitment to advancing knowledge in this field has been instrumental in shaping the content of this book. We also appreciate the support of our publisher who has played a crucial role in refining and improving this work. We hope this book will serve as a valuable resource, sparking new research, innovations, and practical implementations in the field of AI-driven agriculture.

Manojit Chowdhury
Nand Lal Kushwaha
Gourav Dhar Bhowmick

Contents

CHAPTER
1

The Role of Artificial Intelligence in Agriculture

Subrata Mandal[1], Akshay Kinge[1], Shrijita Bid[2*], Saikat Karmakar[3] and Manpreet Singh[4]

[1] Division of Agricultural Engineering, Indian Agricultural Research Institute, New Delhi - 110012, India
[2] Department of Soil Science, Punjab Agricultural University, Ludhiana, India
[3] Department of Agricultural and Food Engineering, Indian Institute of Technology Kharagpur, Kharagpur - 721302, India
[4] Department of Plant and Soil Science, Texas Tech University, Texas, USA

Introduction

Agriculture is vital as it serves as the main source of food, raw materials, and livelihoods for billions of people globally. Artificial intelligence (AI) is revolutionizing agriculture by improving precision, efficiency, and sustainability. The term "artificial intelligence" was first introduced by John McCarthy in 1955 during the Dartmouth Conference. The initial application of AI in agriculture began in 1985 with McKinion and Lemmon, who created GOSSYM (Gossypium Simulation Model). This Expert System helped optimize cotton production by considering factors like climate, irrigation, fertilization, weed control, and cultivation. The development of AI in agriculture began with early technologies like Global Positioning System (GPS) and remote sensors, which helped farmers improve efficiency through precision farming in the early 20th century (Zha 2020). AI is developing intelligence in machines so machines behave like humans or develop decision-making abilities in machines.

AI-powered tools help in precision farming, enabling real-time monitoring of crops, soil, and water through sensors and drones (Jha et al. 2023). Machine learning algorithms provide predictive insights on weather, crop diseases, and yields, helping farmers make informed decisions. AI-driven autonomous machinery, such as tractors and robots, automate planting, harvesting, and weed control tasks, reducing labour costs (Subeesh and Mehta 2021; Sharma et al. 2023). AI optimizes resource use,

*Corresponding author: shrijita2568@gmail.com

DOI: 10.1201/9781003536932-1

improves water and fertilizer management, and contributes to more sustainable agricultural practices. Better decision-making is made possible by predictions such as forecast yields, market trends, and climate change (Kisliuk et al. 2024).

AI in agriculture faces several challenges, such as investment costs, internet connectivity in remote locations, and the requirement for quality data. Anything that has advantages also has some disadvantages for example if any mistake is made by machines who will be responsible for that (Verma 2018). Artificial intelligence is complex to apply in agriculture, and agriculture's surrounding environment also limits AI in agriculture. For precise application of AI technology requires precise and high-tech resources, and these are very expensive (Javaid et al. 2023). For example, object detection in agriculture is difficult as compared to industry because the variation, environmental conditions, complex backgrounds, and non-uniformity between objects make it difficult. AI-based robotics also has less efficiency and precision in agriculture than industrial robotics (Wakchaure et al. 2023; Pradhan et al. 2024). Standard norms or regulations of AI in agriculture are not available and experts or skilled technicians are also minimal (Gardezi et al. 2023).

Aerial imaging and drone technology have also become indispensable tools in modern agriculture. Drones equipped with high-resolution cameras and AI-driven imaging analysis can capture large volumes of data across expansive fields in a short time (Drohan 2023). AI algorithms analyze these images to identify plant health indicators, such as chlorophyll levels, canopy cover, and leaf coloration, which help assess crop vigor, growth, and overall health (Patra et al. 2023; Chowdhury et al. 2024a; 2024b). Early detection of issues, such as nutrient deficiencies or water stress, enables timely intervention and prevents potential yield loss (Su et al. 2023). Moreover, drones offer a cost-effective and time-efficient solution for crop monitoring, especially in large and remote fields where manual inspection would be impractical or too labor-intensive. Farmers using drone technology have reported an increase in yield by up to 20% due to timely, data-informed interventions (Kumar et al. 2020).

The world population continuously increases but the available resources are limited so to meet the demand of the rising population new technological advancements help to achieve it. Soil and water are the major agricultural inputs AI-driven sensors help to apply water to the crop according to their needs and apply fertilizer precisely to prevent the degradation of land by the overuse of fertilizers. Climate change now a days adversely affects agriculture here AI provides information to farmers by predicting weather, drought, and floods it helps farmers select proper crop varieties and management plans. Automation by AI system helps reduce labour requirements, helps in transportation and storage, postharvest losses, and food safety and quality control (OECD 2019).

This chapter aims to explore how AI technologies are transforming modern agriculture, offering innovative solutions to critical challenges. By examining applications such as precision farming, crop monitoring, soil management, and pest control, the chapter illustrates how AI enhances productivity and resource efficiency while minimizing waste. It covers advanced technologies like IoT, robotics, deep learning, and satellite imaging, showing how data-driven insights enable farmers to optimize decisions, improve yields, and reduce environmental impact. Economic, environmental, and social benefits of AI adoption are highlighted, alongside real-world case studies that demonstrate its success in various agricultural settings. The chapter also addresses challenges in AI adoption, including infrastructure costs and data privacy concerns, and provides an outlook on future advancements. This comprehensive overview offers insights for farmers, researchers, and policymakers on leveraging AI for a sustainable, resilient agricultural sector.

Importance of Artificial Intelligence in Agriculture

AI is transforming agriculture, offering tools that boost efficiency, optimize resource use, and enhance productivity, all of which are critical as global food demand rises. One of AI's primary contributions is in precision agriculture, where data-driven insights help farmers make informed decisions on irrigation, fertilizer use, and pest control. By analyzing real-time data from soil sensors, weather patterns, and satellite imagery, AI systems allow for precise management, reducing input costs and minimizing environmental impact. AI also excels in monitoring crops through computer vision and machine learning, enabling early detection of diseases, nutrient deficiencies, and pest infestations, allowing for targeted interventions that protect yields (Shaikh et al. 2022).

In addition to crop management, AI strengthens supply chains, improving efficiency from farm to market. Predictive analytics forecast crop yields and market trends, helping farmers adjust planting and harvesting schedules to meet demand. Blockchain technology, combined with AI, enhances supply chain transparency, which ensures product quality and bolsters consumer trust (Anitha and Rai 2022). For post-harvest management, AI maintains optimal storage conditions, reducing losses and extending the shelf life of perishables, which is vital for food security and sustainability (Fadiji et al. 2023).

AI-powered automation, like autonomous tractors and robotic harvesters, addresses labor shortages and enhances efficiency, enabling timely planting and harvesting (Kushwah et al. 2024btractor). Resource optimization is another significant area; AI-driven irrigation systems monitor soil moisture to deliver just the right amount of water, conserving this vital resource. AI also plays a crucial role in adapting agriculture to climate change by providing early warnings for extreme weather events, which helps farmers mitigate risks. In livestock management, wearable AI devices monitor health indicators, optimizing feeding and breeding strategies to increase productivity while ensuring animal welfare (Sood et al. 2022).

Overall, AI's impact on agriculture is profound, empowering farmers with data-driven insights and scalable, sustainable practices that address food security and environmental challenges. With ongoing advancements, AI will continue to modernize agriculture, making it more resilient and capable of meeting global food needs.

AI enables precision farming, crop and soil health monitoring, early disease detection, weed detection, weather forecasting, yield prediction and mapping, smart irrigation system management, operating drones and autonomous machinery, livestock, fisheries and poultry monitoring (Fig. 1) (Thakur and Singh 2021).

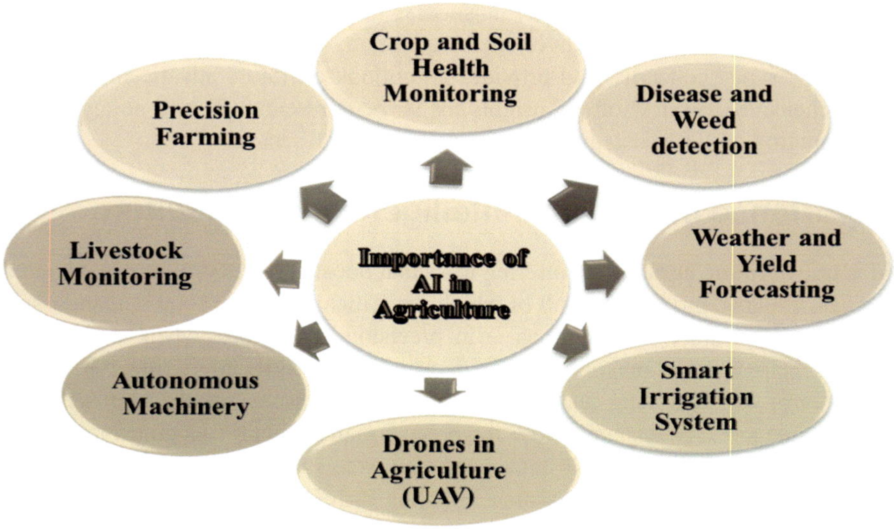

Fig. 1: Importance of AI in agriculture.

Key AI Technologies in Agriculture

Machine Learning and Predictive Analytics

Machine Learning is a subset of AI that employs methods, such as deep learning, to enable machines to perform tasks like classifying and predicting outcomes (Fig. 2). Machine learning and predictive analytics enable farmers to analyze both historical and real-time data on crop growth, weather patterns, and soil health to predict outcomes and optimize farm operations. By making use of these insights, farmers can anticipate crop yields, manage pest infestations, and improve soil conditions. Machine learning applications also help optimize planting and harvesting schedules, offering proactive guidance that reduces risk and improves yield quality. With real-time data analysis, machine learning empowers farmers to make precise decisions that adapt to changing weather and environmental conditions. As AI models continue to advance, predictive analytics will play a vital role in managing agricultural challenges, leading to higher yields and more efficient farming practices (Farooqui et al. 2024).

Deep Learning and Neural Networks

Deep learning and neural networks are advanced AI tools that perform complex analyses, allowing for high-accuracy predictions and decision-making in agriculture. These models are particularly effective in pattern recognition tasks, such as identifying crop diseases or predicting yield outcomes based on multiple data points. By processing vast amounts of data, neural networks uncover patterns and correlations that may not be evident through conventional analysis, helping farmers make informed decisions. For example, deep learning models can analyze

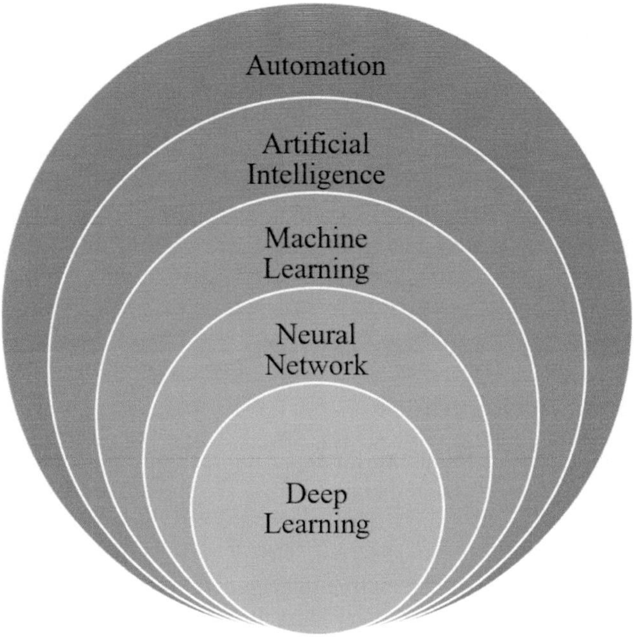

Fig. 2: Relationship between AI and its subsets.

images to detect early signs of disease or nutrient deficiency, allowing for timely intervention. These technologies are valuable in adapting to unpredictable factors like weather variations and environmental stressors. As neural networks become more sophisticated, their ability to provide precise, data-driven insights will play an essential role in optimizing agricultural productivity, enabling farmers to navigate complex challenges with enhanced accuracy and proactive strategies for crop management (Kashyap et al. 2021).

Natural Language Processing

Natural Language Processing (NLP) enables communication and knowledge sharing between farmers and AI systems. Through virtual assistants and chatbots, NLP provides agricultural advice and support in local languages, making expert information more accessible. For instance, chatbots can guide farmers on crop management, pest control, and weather updates, providing quick responses to their questions. NLP-driven tools are particularly beneficial for smallholder farmers and those in remote areas who may lack direct access to agronomists. By offering instant, language-based solutions, NLP supports farmers in making informed decisions that improve yields and reduce risks. These tools also serve as an educational platform, helping farmers gain a deeper understanding of best practices. As NLP technology evolves, it will increasingly bridge knowledge gaps in agriculture, enabling farmers to make data-driven decisions with the support of AI-driven conversational agents, thus fostering innovation and sustainable farming practices (Niranjan et al. 2019).

Computer Vision and Image Recognition

Computer vision and image recognition technologies are equally impactful, using images from drones, satellites, and mobile devices to monitor crop health and identify early signs of disease, pest infestations, or nutrient deficiencies. By enabling early detection, computer vision helps farmers address issues before they lead to significant crop losses, promoting timely and targeted interventions. This technology is valuable for enhancing productivity by ensuring crops receive optimal care (Liu and Bruch 2020).

Applications of AI in Agriculture

AI in Precision Agriculture

With the growing global population, the need for agricultural products is going to increase by 70% by 2050 (Bhat and Huang 2021). To meet this demand with limited resources, agricultural sector is now adopting data-driven approaches to optimize the inputs like water, fertiliser, pesticides, etc. This is termed as precision agriculture, which enables more responsive, adaptive management strategies. Coupled with AI, precision agriculture leverages advanced data analytics, machine learning, and IoT (Gawande et al. 2023; Satpute et al. 2024). AI helps in providing real-time monitoring of crop health, weather conditions, and soil parameters to obtain precise information and plan management strategies for each farm (Fig. 3). Thus, AI helps in improving yield with sustainable use of resources and cutting down the labour cost. In this section the role of AI in different sectors of precision agriculture is discussed.

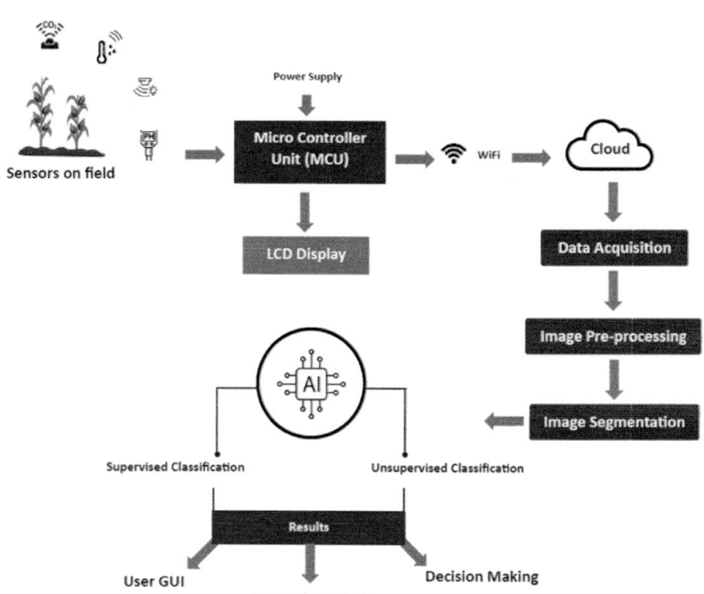

Fig. 3: AI in precision agriculture.

AI Based Crop Monitoring and Health Assessment

In recent times, Convolutional Neural Network (CNN), deep learning are gaining popularity in monitoring the crop health and detecting the target crop with health issues (Zhao et al. 2019). CNN is capable of analysing high-resolution images collected from sensors and extract the texture, variations in colour, and shapes of the objects. that indicate crop health, nutrient levels, disease presence, or pest damage. Thus, the CNN-based models represent the temporal changes in crop characteristics and performs precise prediction of crop disease. Based on the results, precise application of the resources can be ensured. However, the CNN demands powerful computational platforms like servers and high-speed internet facilities for intensive computation and transfer of data (Huang et al. 2022). With the help of deep learning algorithms, crop types and specific diseases can be identified with minimal human intervention.

Yield Prediction

Crop yield predictions driven by AI are proving revolutionary as it provides precise data-driven insights of crop growth. This facilitates in making efficient decisions for farming practices and ensuring food security (Muruganantham et al. 2022). In AI driven crop yield prediction process the high-resolution image data and the data from the field sensors are acquired as input data to the algorithms like Random Forest (RF), Support Vector Machine (SVM), Long Short-Term Memory (LSTM) or CNN etc. These models perform predictive analysis and calculates potential yield of the crops. AI also incorporates the weather data to provide precise prediction in diverse climate. Based on the results the farmers become capable of managing resources beforehand. However, one of the biggest concerns in AI-based prediction is the huge amount of data. Substantial volume of data is needed to provide for training and prediction purposes (Goel and Pandey 2024).

Irrigation and Water Management

According to Sustainable Development Goal (SDG) of United Nations, the withdrawal of freshwater for economic activities was 18.4 % of the global renewable freshwater in the year 2018 (United Nations 2024). Around 70 % of this withdrawal contributes in the agricultural activities (Food and Agriculture Organization 2023). To downscale the water use in the agricultural sectors, integration of cutting edge technologies is evident. AI integrated irrigation and water management systems provide precision, efficiency and sustainability in terms of water use (Fig. 4). AI algorithms like RF, SVM, CNN and gradient-boosted decision trees evaluate water scarcity issues by processing thermal infrared images of crops and meteorological data (Zhou et al. 2021). This approach not only ensures sustainable water use but also helps in obtaining higher crop yield by arranging precise irrigation schedules with optimal water quantity at each growth stage of the crop (Kamal et al. 2010). AI systems also provide edge to automate the irrigation system which are useful in saving time for farmers and reduce the human effort (Das et al. 2024). However, cost barrier is one of the major constrains in adopting AI in irrigation purposes by small and marginal farmers. Also, AI systems bring complexity along with, which makes it difficult to handle for a novice person (Sassenrath et al. 2008).

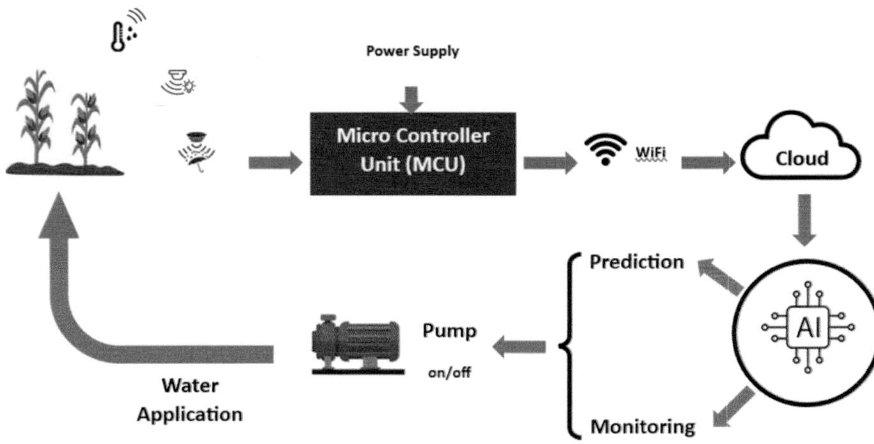

Fig. 4: AI infused irrigation system.

AI in Robotics and Automation

AI-powered robotics and automation are transforming agriculture by streamlining operations, reducing labor requirements, and increasing precision in crop management. As global demand for food rises and labor shortages persist, the application of robotics and automation offers promising solutions to meet agricultural needs efficiently and sustainably. There are also various benefits of AI in robotics and automation (Fig. 5).

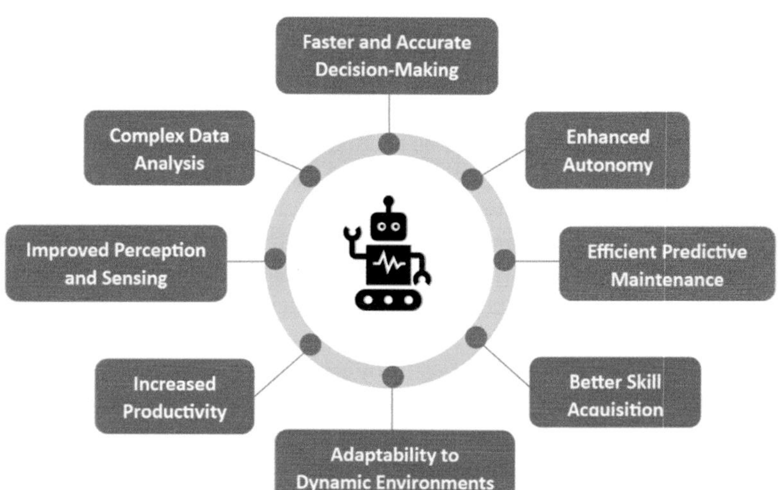

Fig. 5: Benefits of AI in robotics and automation.

Autonomous Tractors and Harvesting Robots

Autonomous tractors and robotic harvesters have become increasingly prevalent in large-scale farming operations. These machines, equipped with GPS, sensors, and machine learning algorithms, navigate fields and perform tasks with minimal human intervention (Kushwah et al. 2024a). For instance, John Deere's self-driving tractor, launched in 2022, uses six pairs of cameras and GPS-based navigation to till fields autonomously. Studies have shown that autonomous tractors can increase productivity by approximately 10–15%, as they can operate continuously without breaks and maintain precision in tasks like planting and tilling (Mail et al. 2023; Mahmud et al. 2020).

Similarly, robotic harvesters like Agrobot's E-series, designed for strawberry picking, employ computer vision to identify ripe fruits, selectively picking them to reduce waste. According to data from Agrobot, these harvesters can reduce labor costs by up to 40%, addressing the growing shortage of agricultural labor. Harvesting robots provide timely, efficient solutions, ensuring that produce is harvested at optimal ripeness, thereby improving quality and reducing post-harvest losses (Anand et al. 2024; Tituaña et al. 2024).

Precision Spraying and Weeding Robots

Precision spraying robots utilize AI algorithms and computer vision to identify specific plants, enabling targeted application of fertilizers, pesticides, and herbicides (Chowdhury et al. 2023). Blue River Technology, for example, has developed "See & Spray" robots that use cameras and AI to detect weeds and selectively apply herbicides, reducing chemical usage by up to 90% compared to traditional methods (Malkani et al. 2024). According to data from Blue River, such precision technology can save an average farm up to $12 per acre on herbicides, leading to significant cost reductions and minimized environmental impact (Liu and Bruch 2020; Gupta and Gupta 2024).

Weeding robots, like those developed by ecoRobotix, utilize machine learning and camera systems to identify and remove weeds without damaging crops. By removing weeds mechanically, these robots reduce dependency on chemical herbicides, which aligns with sustainable farming goals. A study by ecoRobotix estimates that their weeding robots can reduce herbicide usage by 95%, supporting eco-friendly practices and reducing potential soil contamination (Zhang et al. 2022).

Automated Drones for Monitoring and Analysis

Drones equipped with AI-based imaging technology play an essential role in crop health monitoring and soil analysis. These drones capture high-resolution images and use AI to detect signs of stress, disease, or nutrient deficiency in crops. For example, DJI's Agras T20 drone can cover up to 30 acres per hour, providing real-time data that helps farmers take corrective actions faster. Research has shown that using drones for crop monitoring can increase yield by up to 20% by enabling timely interventions that prevent crop losses (Drohan 2023).

In addition, drones can assess soil conditions and generate detailed maps, which allow farmers to optimize water and nutrient distribution. Studies indicate that precision irrigation enabled by drone monitoring can reduce water usage by up to 30% while maintaining optimal crop health, particularly in water-scarce regions. Automated drones also reduce labor costs associated with manual inspection and provide a scalable solution for large fields, where comprehensive, real-time monitoring would otherwise be labor-intensive and costly (Su et al. 2023).

AI-Driven Greenhouse Robotics

In controlled environments like greenhouses, AI-driven robots are used for planting, pruning, and harvesting tasks. These robots, equipped with sensors and machine learning algorithms, can adapt to the greenhouse environment, adjusting water, light, and nutrient levels based on real-time data. Iron Ox, an agricultural robotics company, uses AI-powered robotic arms that perform tasks such as transplanting seedlings and moving plants to optimal positions for growth. According to Iron Ox, their robotic greenhouse operations can achieve 30% higher productivity than traditional greenhouses by optimizing conditions for plant growth (Dietz and Drechsel 2021).

Additionally, robotic greenhouse systems can maintain consistent crop quality, and the automated monitoring of factors like humidity and temperature helps mitigate the risk of disease. This precise control over environmental factors leads to higher yields and reduced resource consumption, supporting sustainable practices within confined spaces.

AI in Supply Chain and Post-Harvest Management

The integration of Artificial Intelligence (AI) in agriculture has revolutionized various stages of food production, distribution, and management. AI plays a pivotal role in optimizing the agricultural supply chain and enhancing post-harvest management, helping address some of the industry's most pressing challenges, such as reducing waste, ensuring food safety, and improving efficiency (Fadiji et al. 2023). The potential of AI-powered tools in transforming traditional practices into data-driven, efficient systems is vast, allowing stakeholders to manage everything from planting to transportation, storage, and distribution in ways that were not possible before.

Transforming Supply Chain Efficiency with AI

The agricultural supply chain involves numerous steps, from harvesting crops to delivering them to consumers. Each of these stages presents unique challenges, including managing perishable goods, dealing with unpredictable weather, and balancing supply with demand. AI has the potential to enhance every aspect of this chain, making it more efficient and resilient.

AI can optimize supply chain operations by analyzing vast datasets, including market trends, weather forecasts, and historical crop yield information. By processing this information, machine learning algorithms can predict fluctuations in demand, allowing producers to adjust their planting and harvesting strategies accordingly. This helps to avoid overproduction, which can lead to wasted resources, or underproduction, which may result in shortages and lost revenue (Patil et al. 2024). Furthermore, these algorithms can improve the accuracy of sales forecasts, ensuring that the right amount of produce reaches the market, minimizing waste and maximizing profitability.

One of the critical areas where AI has a transformative impact is logistics. The transportation of agricultural products often involves long distances, and produce may spoil if not delivered within specific timeframes. AI can streamline logistics by analyzing real-time data on traffic conditions, fuel costs, and delivery schedules to optimize transportation routes. For example, AI-powered route optimization tools can select the most efficient paths, reducing transit times, fuel consumption, and carbon emissions. Additionally, in perishable food supply chains, AI can monitor cold storage conditions throughout transportation, providing real-time updates on temperature and humidity levels. This ensures that food quality is maintained, further reducing spoilage (Kaur and Watson 2024).

AI-driven demand forecasting also helps optimize inventory management in the supply chain. By understanding consumer behavior, AI systems predict demand spikes or declines, which allow for better planning of production and distribution cycles. Such insights enable producers and retailers to ensure that supply meets demand efficiently, minimizing waste and reducing costs associated with unsold or spoiled goods (Das et al. 2023).

Enhancing Post-Harvest Quality Control

Post-harvest management is a critical component in the food supply chain. Once crops are harvested, they must be processed, sorted, and stored under optimal conditions to maintain their quality. AI technologies are playing an increasingly important role in improving these post-harvest processes.

One of the key applications of AI in post-harvest management is in automated sorting and grading systems. Using computer vision and machine learning algorithms, AI systems can evaluate the quality of fruits, vegetables, and other crops based on parameters such as size, shape, color, and surface defects. These systems can detect imperfections and abnormalities that might not be visible to the human eye, ensuring that only the highest-quality produce reaches the market. For instance, AI can differentiate between ripe and unripe produce, sort fruits based on ripeness, or identify early signs of decay. This automated process not only saves time and labor but also ensures consistent quality, which is crucial for maintaining consumer trust and satisfaction (Chauhan et al. 2022).

AI-powered sensors and imaging technologies can also play a vital role in monitoring the quality of stored goods. Perishable goods such as fruits, vegetables, and dairy products require specific conditions to maintain freshness (Kmari and Dhingra 2024). AI-based systems can continuously monitor factors such as temperature, humidity, and gas levels in storage facilities, detecting deviations from optimal conditions that could lead to spoilage. These systems can alert managers to take corrective actions, such as adjusting the temperature or improving ventilation, to maintain the quality of stored produce.

In addition to maintaining product quality, AI can significantly reduce post-harvest losses, which account for a substantial portion of global food waste. By monitoring environmental conditions and crop quality in real time, AI tools can help farmers and distributors identify and address problems before they lead to spoilage or contamination (Anand and Barua 2022). For example, AI systems can detect early signs of mold or rot in stored crops, allowing for swift interventions that prevent further deterioration.

AI in Predictive Maintenance for Post-Harvest Machinery

Post-harvest operations rely heavily on machinery, from harvesters to sorting equipment and storage systems. The failure of any of this equipment can lead to delays, increased costs, and product loss. AI can help mitigate these risks through predictive maintenance.

By using sensors to collect data on machine performance, AI algorithms can monitor the condition of post-harvest machinery in real time. These systems can detect anomalies in vibration patterns, temperature fluctuations, and other indicators of wear and tear. Based on this data, AI can predict when equipment is likely to fail or require maintenance (Chauhan et al. 2022). This allows for timely repairs and minimizes downtime, ensuring that post-harvest processes run smoothly. Predictive maintenance also extends the lifespan of machinery by preventing unnecessary breakdowns, which can be costly and disruptive.

Reducing Post-Harvest Losses with AI

One of the most significant challenges in agriculture is post-harvest losses, particularly in developing countries, where inadequate storage and inefficient handling often lead to substantial food waste. AI offers promising solutions to reduce these losses by improving storage management and optimizing inventory control (Kollia et al. 2021).

AI-powered systems can monitor storage conditions and predict spoilage before it occurs. For example, AI models can analyze data on temperature, humidity, and pest activity to anticipate when produce is likely to degrade. This enables farmers and distributors to take proactive measures, such as adjusting environmental controls or moving products to market before spoilage sets in. AI can also help prioritize the distribution of perishable goods based on their shelf life, ensuring that older or more vulnerable products are sold or processed first (Anand and Barua 2022).

By predicting which products are at higher risk of spoilage, AI can also inform inventory management decisions. For example, AI-driven systems can suggest the order in which goods should be sold or transported based on their freshness and remaining shelf life. This reduces the likelihood of unsold produce going to waste and helps retailers maintain a steady supply of fresh products.

Supply Chain Transparency and Traceability with AI

Consumers are increasingly demanding transparency about where their food comes from, how it was produced, and whether it meets safety and ethical standards. AI, in conjunction with blockchain and other digital technologies, can provide greater traceability in the food supply chain (Fig. 6).

AI systems can track and record every step in the production, processing, and distribution of food, from farm to table. This data can be stored on blockchain platforms, ensuring that it is secure and tamper-proof. By using AI for traceability,

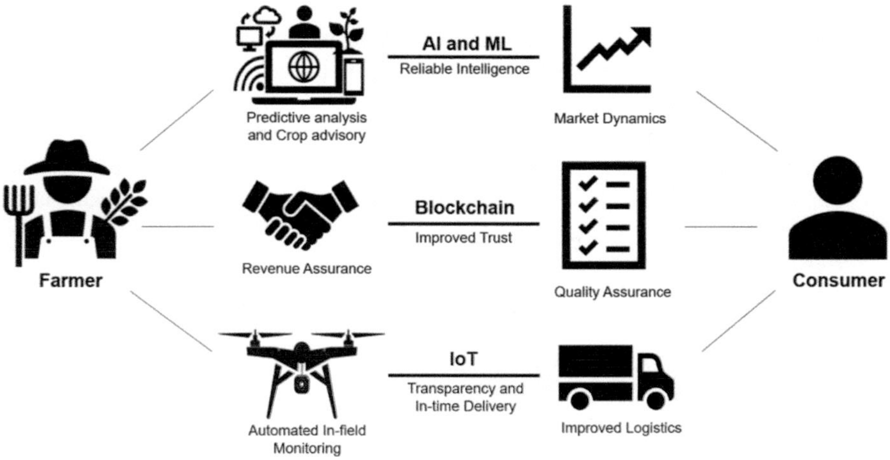

Fig. 6: Integrating digital technologies into the food supply chain.

companies can provide consumers with detailed information about the origin and journey of their food, enhancing trust and enabling informed purchasing decisions. This is particularly important in the case of food safety issues, such as contamination or recalls. In such events, AI can quickly trace the source of the problem, allowing for targeted interventions and minimizing the impact on public health (Anitha and Rai 2022).

AI is already making a significant impact on the agricultural supply chain and post-harvest management by optimizing operations, improving quality control, and reducing waste. From predictive maintenance and inventory management to real-time monitoring of storage conditions, AI is helping to transform traditional practices into efficient, data-driven systems (Ganeshkumar et al. 2023). As these technologies continue to advance, they will play an even more critical role in addressing the challenges of food production and distribution, ultimately leading to a more sustainable and resilient global food system.

AI in Sustainable Agriculture

Sustainable agriculture focuses on meeting current food needs while ensuring the ability of future generations to do the same. It involves minimizing environmental impacts, preserving resources, and promoting economic viability. Artificial Intelligence (AI) plays a crucial role in driving sustainable agricultural practices by optimizing resource use, improving yields, and reducing waste (Bhagat et al. 2022). Through technologies like machine learning, computer vision, IoT (Internet of Things), and predictive analytics, AI is transforming how we approach sustainable farming.

Optimizing Resource Use

AI-driven technologies allow farmers to use resources like water, fertilizers, and pesticides more efficiently, which is essential for sustainable agriculture (Sunil Kumar et al. 2024; Chowdhury et al. 2025).

- *Precision Farming:* AI systems analyze data from sensors, drones, and satellites to provide real-time insights on soil health, plant growth, and environmental conditions. This enables precision farming, where inputs are applied only where and when they are needed, reducing waste and resource overuse. Studies show that precision farming can reduce water usage by up to 30% and fertilizer application by 20% while maintaining or increasing crop yields (Bhatti et al. 2023).
- *Fertilizer and Pesticide Management:* AI models analyze plant data and environmental conditions to determine the precise amount of fertilizers and pesticides required. This reduces chemical runoff into water bodies and minimizes soil degradation, contributing to long-term soil health and ecosystem sustainability (Chowdhury et al. 2024b). AI solutions like Blue River Technology's "See and Spray" have reduced herbicide usage by up to 90% in trials (Liu and Bruch 2020).

Climate-Smart Agriculture

AI helps mitigate the impact of climate change on agriculture by providing farmers with real-time data and recommendations tailored to their local environmental conditions.

- *Weather Forecasting and Risk Management:* AI models can predict weather patterns and extreme weather events with increasing accuracy, helping farmers prepare for climate-related risks. Accurate weather forecasting enables better crop management, reducing the risk of crop failure due to unexpected weather changes. Research shows that AI-driven weather forecasting tools have improved farmers' decision-making and reduced crop damage by 10-30% in regions prone to climate variability (Jaseena and Kovoor 2022).
- *Carbon Footprint Reduction:* AI can help farmers adopt practices that lower carbon emissions, such as reducing fuel consumption in machinery, minimizing fertilizer use, and employing regenerative agriculture techniques. For example, AI-powered precision farming has been shown to reduce greenhouse gas emissions by up to 25% by optimizing the use of fertilizers and other inputs (Hernandez et al. 2024).

Promoting Biodiversity and Soil Health

Sustainable agriculture depends on maintaining healthy ecosystems, including promoting biodiversity and preserving soil quality. AI technologies contribute to these goals by guiding farmers in adopting practices that promote ecosystem health.

- *Soil Health Monitoring:* AI-based systems analyze data from soil sensors to monitor key indicators like nutrient levels, pH, moisture, and organic matter. These insights enable farmers to practice regenerative agriculture, which helps improve soil fertility and prevent erosion. AI-driven soil monitoring platforms like Teralytic can help reduce soil degradation by 20-30%, promoting long-term agricultural sustainability (Sharma 2021).
- *Biodiversity Conservation:* AI helps maintain biodiversity by recommending

sustainable crop rotation practices, reducing the need for monoculture. By analyzing data on local ecosystems, AI can also recommend planting cover crops or maintaining buffer zones to support pollinators and other wildlife essential for a healthy farming ecosystem (Ryo et al. 2023).

Table 1 outlines various applications of AI in agriculture, including precision farming, crop health monitoring, disease detection and yield forecasting.

Table 1: Application of AI in agriculture

AI Application Area	Technology Used	Description	Impact/Benefits	References
Precision Farming	Machine Learning, IoT, Drones	AI algorithms analyze data from sensors and drones to optimize water, fertilizer, and pesticide use.	Increased crop yield, reduced input costs, and better resource management.	Ponnusamy and Natarajan 2021
Crop Monitoring and Disease Detection	Computer Vision, Deep Learning	AI-based image recognition systems detect crop diseases, pests, and nutrient deficiencies through satellite or drone imagery.	Early identification and prevention of crop diseases, minimizing crop loss.	Roy and Bhaduri 2021
Yield Prediction	Machine Learning, Predictive Analytics, Deep Learning	AI models predict future crop yields by analyzing historical data, weather patterns, and soil conditions.	Better harvest planning and supply chain management.	Jhajharia et al. 2023
Autonomous Machinery	Robotics, Deep Learning	AI-powered machines and robots automate tasks such as planting, weeding, harvesting, and spraying.	Labor savings, increased operational efficiency, and reduced reliance on manual labor.	Saleem et al. 2021
Irrigation Management	IoT, Predictive Analytics, Machine Learning	AI systems optimize irrigation schedules based on real-time weather, soil moisture, and plant growth data.	Reduced water consumption and improved water use efficiency.	Goap et al. 2018
Soil Health Monitoring	Machine Learning, Sensors	AI-powered sensors monitor soil conditions such as pH, moisture, and nutrient levels.	Improved soil management, better crop growth, and reduced soil degradation.	Diaz-Gonzalez et al. 2022

(Contd.)

(*Contd.*)

AI Application Area	Technology Used	Description	Impact/Benefits	References
Supply Chain Optimization	Predictive Analytics, Blockchain	AI helps manage inventory, optimize transportation routes, and forecast demand in agricultural supply chains.	Reduced food waste, enhanced logistics, and improved efficiency.	Kashem et al. 2023
Post-Harvest Quality Assessment	Computer Vision, Machine Learning	AI-based systems assess the quality of harvested crops, sorting them based on ripeness, size, and defects.	Improved sorting efficiency, reduced post-harvest losses, and better product quality.	Dhiman et al. 2022
Livestock Management	AI Sensors, Machine Learning	AI systems monitor livestock health, track movement, and optimize feeding and breeding practices.	Increased productivity, reduced costs, and improved animal welfare.	García et al. 2020
Pest and Weed Control	AI-powered Robotics, Computer Vision	AI robots and drones detect and target pests and weeds, applying chemicals only where necessary.	Reduced pesticide usage and environmental impact, lower operational costs.	Adetunji et al. 2023
Weather Forecasting	Predictive Analytics, AI Models	AI analyzes historical weather patterns and real-time data to provide precise weather forecasts for farmers.	Helps farmers make better planting and harvesting decisions.	Jaseena and Kovoor 2022
Smart Greenhouses	AI Sensors, IoT, Machine Learning	AI systems automate climate control, lighting, and irrigation in greenhouses for optimal plant growth.	Increased efficiency, reduced labor, and maximized crop yield.	Rokade and Singh 2021
Crop Pricing and Market Forecasting	AI Algorithms, Big Data	Uses AI to predict future crop prices based on market trends and demand-supply dynamics.	Supports farmers in decision-making, reduces market volatility.	Fan et al. 2022

Challenges of AI Application in Agriculture

While AI offers transformative benefits to agriculture, there are several key challenges hindering its widespread adoption and effectiveness.

High Implementation Costs

Deploying AI technology, from advanced sensors to machine learning models, involves significant initial investments that can be a barrier, especially for small and medium-sized farms. Equipment, maintenance, and training costs add to the burden, limiting access to AI for resource-constrained farmers.

Data Availability and Quality

AI systems require large volumes of high-quality data to make accurate predictions, yet agriculture data can be fragmented and inconsistent, particularly in rural and under-resourced areas. Poor data quality or limited historical data can reduce the effectiveness of AI models, resulting in unreliable insights.

Lack of Technical Skills and Knowledge

Farmers and agricultural workers often need technical skills to operate AI tools effectively. However, in many regions, there is limited access to training programs, and a lack of digital literacy remains a major barrier to adopting AI-driven technologies.

Connectivity and Infrastructure Issues

Many agricultural areas lack reliable internet connectivity and the infrastructure necessary to support AI technology, such as cloud-based platforms and sensor networks. Without robust connectivity, real-time data collection, processing, and analysis become difficult, reducing the practicality of AI tools.

Data Privacy and Security Concerns

Collecting and analyzing large amounts of data on farms and crops raises privacy and security concerns. Farmers may worry about who has access to their data and how it might be used. Clear policies on data ownership and protection are essential to build trust in AI systems among farmers and stakeholders.

Climate and Environmental Variability

Agriculture is heavily impacted by climate variability, and AI models trained on historical data may struggle to adapt to new and extreme weather patterns due to climate change. This limits the accuracy and effectiveness of AI in providing reliable predictions and recommendations under rapidly changing environmental conditions.

Interoperability and Standardization Issues

Many AI applications in agriculture use proprietary software and hardware systems, making it difficult for different systems to communicate and share data effectively. Lack of standardization and interoperability between devices and platforms creates silos, limiting the integrated use of AI across the agricultural supply chain.

Ethical and Environmental Concerns

AI-driven precision farming often promotes intensive agriculture practices, which can conflict with sustainable practices. Over-reliance on certain AI technologies may lead to ecological issues like monoculture farming, affecting biodiversity and soil health.

Future Directions and Innovations

Future directions for AI in agriculture point toward enhancing efficiency, sustainability, and resilience. AI-powered autonomous systems, such as drones and robots, are expected to handle labor-intensive tasks like planting, weeding, and harvesting, helping address rural labor shortages and reducing costs. Advances in predictive analytics will improve forecasting for crop yields, weather patterns, and market demand, allowing farmers to manage risks related to climate and price fluctuations. AI-driven crop health monitoring through satellite and drone imagery will enable real-time detection of issues, allowing quicker responses to threats like pests or disease. Innovations in soil health and crop disease prediction using microbiome and genomics data will also aid in optimizing soil conditions for sustainable, long-term productivity. Overall, the future of AI in agriculture centers on creating data-driven, adaptive farming practices that address environmental challenges, reduce waste, and secure food supply for a growing population.

Conclusions

Artificial Intelligence (AI) is revolutionizing agriculture by improving productivity, efficiency, and sustainability. Technologies like machine learning and robotics address critical issues such as optimizing crop yields, managing resources, and controlling pests. Tools like precision agriculture, predictive analytics, and automation empower farmers to make informed decisions, boosting crop health while conserving resources like water, fertilizers, and pesticides. AI promotes sustainable farming practices, helping to meet growing food demands with minimal environmental impact. However, challenges like high costs, data privacy concerns, and the need for farmer education persist. With ongoing research, investment, and policy support, AI has the potential to create a more resilient and sustainable agricultural system.

References

Adetunji, C.O., T.O. Olaniyan, O.A. Anani, A. Inobeme, O.O. Osemwegie, D. Hefft et al. 2023. Artificial intelligence and automation for precision pest management. Sens. AI Solut. Food Manuf. 49-70. CRC Press.

Anand, R., S.S. Nalla, S. Nasreen, M. Chowdhury, Madhusudan B.S., A. Pandey et al. 2024. Fruit counting and analysis using artificial intelligence approaches. pp. 93-114. *In:* S.S. Chouhan, A. Saxena, U. P. Singh and S. Jain [eds.]. Artificial Intelligence Techniques in Smart Agriculture. Springer Nature, Singapore.

Anand, S. and M.K. Barua. 2022. Modeling the key factors leading to post-harvest loss and waste of fruits and vegetables in the agri-fresh produce supply chain. Comput. Electron. Agric. 198: 106936.

Anitha, R. and D. Rai. 2022. Internet of things with artificial intelligence detection and blockchains of crop availability for supply chain management. Int. J. Knowl. Based Dev. 12(3-4): 444-459.

Bhagat, P.R., F. Naz and R. Magda. 2022. Artificial intelligence solutions enabling sustainable agriculture: A bibliometric analysis. PLoS One. 17(6): e0268989.

Bhat, S.A. and N.F. Huang. 2021. Big data and AI revolution in precision agriculture: Survey and challenges. IEEE Access. 9: 110209-110222.

Bhatti, U.A., M. Masud, S.U. Bazai and H. Tang. 2023. Investigating AI-based smart precision agriculture techniques. Front. Plant Sci. 14: 1237783.

Chauhan, A., B. Brouwer and E. Westra. 2022. Robotics for a quality-driven post-harvest supply chain. Curr. Robot. Rep. 3(2): 39-48.

Chowdhury, M., E.V. Thomas, A. Jha, A. Kushwah, R. Kurmi, T.K. Khura et al. 2023. An automatic pressure control system for precise spray pattern analysis on spray patternator Comput. Electron. Agric. 214: 108287.

Chowdhury, M., R. Anand, T. Dhar, R. Kurmi, R.K. Sahni and A. Kushwah. 2024a. Digital insights into plant health: Exploring vegetation indices through computer vision. pp. 7-30. *In*: S.S. Chouhan, U.P. Singh and S. Jain [eds.]. Applications of Computer Vision and Drone Technology in Agriculture 4.0. Springer Nature, Singapore.

Chowdhury, M., T.K. Khura, R.A. Parray, H.L. Kushwaha, P.K. Upadhyay, A. Jha et al. 2024a. The use of destructive and nondestructive techniques in concrete nitrogen assessment in plants. J. Plant Nutri. 47: 2271-2294.

Chowdhury, M., T.K. Khura, P.K. Upadhyay, R.A. Parray, H.L. Kushwaha, C. Singh et al. 2024b. Assessing vegetation indices and productivity across nitrogen gradients: A comparative study under transplanted and direct-seeded rice. Front. Sustain. Food Syst. 8: 1351414.

Chowdhury, M., T.K. Khura, R.A. Parray, P.K. Upadhyay, H.L. Kushwaha, A. Lama and A. Kushwah. 2025. Multi-vegetation indices based handheld device for precise nitrogen assessment and prescription in direct seeded rice. Comput. Electron. Agric. 229: 109886.

Das, S., A. Barve, N.C. Sahu and K. Muduli. 2023. Enabling artificial intelligence for sustainable food grain supply chains: An agri 5.0 and circular economy perspective. Oper. Manag. Res. 16(4): 2104-2124.

Das, S., M. Kaur, V. Chhabra, T. Nandi, P. Mishra and S. Ghosh. 2024. A systematic review of artificial intelligence: A future guide to sustainable agriculture. Int. J. Environ. Clim. Change. 14(4): 562-573.

Dhiman, B., Y. Kumar and M. Kumar. 2022. Fruit quality evaluation using machine learning techniques: Review, motivation and future perspectives. Multimed. Tools Appl. 81(12): 16255-16277.

Diaz-Gonzalez, F.A., J. Vuelvas, C.A. Correa, V.E. Vallejo and D. Patino. 2022. Machine learning and remote sensing techniques applied to estimate soil indicators – Review. Ecol. Indic. 135: 108517.

Dietz, K. and F. Drechsel. 2021. Digital agriculture. *In:* Handbook of Critical Agrarian Studies. 568-580. Edward Elgar Publishing.

Drohan, J. 2023. Ag application drones are gaining ground in Canada, but obstacles remain. Crops & Soils 56(2): 46-54.

Fadiji, T., T. Bokaba, O.A. Fawole and H. Twinomurinzi. 2023. Artificial intelligence in postharvest agriculture: Mapping a research agenda. Front. Sustain. Food Syst. 7: 1226583.

Fan, M., M. Kang, X. Wang, J. Hua, C. He and F.Y. Wang. 2022. Parallel crop planning based on price forecast. Int. J. Intell. Syst. 37(8): 4772-4793.

Farooqui, N.A., M. Haleem, W. Khan and M. Ishrat. 2024. Precision agriculture and predictive analytics: Enhancing agricultural efficiency and yield. Intell. Tech. Predict. Data Anal. 171-188.

Food and Agriculture Organization of the United Nations. 2023. 2050: A third more mouths to feed. Retrieved from: https://www.fao.org/news/story/en/item/35571/icode/.

Ganeshkumar, C., S.K. Jena, A. Sivakumar and T. Nambirajan. 2023. Artificial intelligence in the agricultural value chain: Review and future directions. J. Agribus. Dev. Emerg. Econ. 13(3): 379-398.

García, R., J. Aguilar, M. Toro, A. Pinto and P. Rodríguez. 2020. A systematic literature review on the use of machine learning in precision livestock farming. Comput. Electron. Agric. 179: 105826.

Gardezi, M., B. Joshi, D. Rizzo, M. Ryan, E. Prutzer, S. Brugler and A. Dadkhah. 2024. Artificial intelligence in farming: Challenges and opportunities for building trust. Agron. J. 116: 1217-1228.

Gawande, V., D.R.K. Saikanth, B.S. Sumithra, S.A. Aravind, G.N. Swamy, M. Chowdhury and B.V. Singh. 2023. Potential of precision farming technologies for eco-friendly agriculture. Int. J. Plant Soil Sci. 35(19): 101-112.

Goap, A., D. Sharma, A.K. Shukla and C.R. Krishna. 2018. An IoT-based smart irrigation management system using machine learning and open-source technologies. Comput. Electron. Agric. 155: 41-49.

Goel, M. and M. Pandey. 2024. Crop yield prediction using AI: A review. *In:* 2024 2nd Int. Conf. Disrupt. Technol. (ICDT). 1547-1553. IEEE.

Gupta, N. and P.K. Gupta. 2024. Robotics and artificial intelligence (AI) in agriculture with major emphasis on food crops. Digital Agric. Solut. Sustain. Food Nutr. Secur. 577-605.

Hernandez, D., L. Pasha, D.A. Yusuf, R. Nurfaizi and D. Julianingsih. 2024. The role of artificial intelligence in sustainable agriculture and waste management: Towards a green future. Int. Trans. A. I. 2(2): 150-157.

Huang, C.H., B.W. Chen, Y.J. Lin and J.X. Zheng. 2022. Smart crop growth monitoring based on system adaptivity and edge AI. IEEE Access. 10: 64114-64125.

Jaseena, K.U. and B.C. Kovoor. 2022. Deterministic weather forecasting models based on intelligent predictors: A survey. J. King Saud Univ. Comput. Inf. Sci. 34(6): 3393-3412.

Javaid, M., A. Haleem, I.H. Khan and R. Suman. 2023. Understanding the potential applications of artificial intelligence in the agriculture sector. Adv. Agrochem. 2(1): 15-30.

Jha, A., M. Chowdhury and A.N. Satpute. 2023. Surface water quality forecasting using machine learning approach. pp. 293-315. *In*: C.B. Pande, M. Kumar and N.L. Kushwaha [eds.]. Surface and Groundwater Resources Development and Management in Semi-arid Region: Strategies and Solutions for Sustainable Water Management. Cham: Springer International Publishing.

Jhajharia, K., P. Mathur, S. Jain and S. Nijhawan. 2023. Crop yield prediction using machine learning and deep learning techniques. Procedia Comput. Sci. 218: 406-417.

Kamal, R.M. and M.S. Amin. 2010. GIS-based irrigation water management for precision farming of rice. Int. J. Agric. Biol. Eng. 3(3): 27-35.

Kashem, M.A., M. Shamsuddoha, T. Nasir and A.A. Chowdhury. 2023. Supply chain disruption versus optimization: A review on artificial intelligence and blockchain. Knowledge. 3(1): 80-96.

Kashyap, P.K., S. Kumar, A. Jaiswal, M. Prasad and A.H. Gandomi. 2021. Towards precision agriculture: IoT-enabled intelligent irrigation systems using deep learning neural network. IEEE Sensors J. 21(16): 17479-17491.

Kisliuk, B., J. Krause, H. Meemken, J. Morales, H. Müller and J. Hertzberg. 2024. AI in current and future agriculture: An introductory overview. KI – Künstl. Intell. 37: 117-132.

Kollia, I., J. Stevenson and S. Kollias. 2021. AI-enabled efficient and safe food supply chain. Electronics. 10(11): 1223.

Kumar, V., A.K. Nayak, A. Poonam, S. Aravindan and R. Khanam. 2020. Unmanned aerial vehicle and its application in Indian agriculture: A perspective. Indian Farming. 70(8).

Kumari, S. and D. Dhingra. 2024. Post-harvest management of fruits in India: A review. J. Agric. Eng. (India). 61(2).
Kushwah, A., A. Chouriya, V.K. Tewari, C. Gupta, M. Chowdhury, P. Shrivastava and P. Bhagat. 2024a. A novel embedded system for tractor implement performance mapping. Cogent Eng. 11(1): 2311093.
Kushwah, A., P.K. Sharma, H.L. Kushwaha, B.B. Sharma, A.K. Shrivastava, R.H. Nag et al. 2024b. Hybrid taguchi-grey relational analysis approach for optimizing cutter operational parameters in selective cauliflower harvesting. Sci. Rep. 14(1): 29709.
Liu, B. and R. Bruch. 2020. Weed detection for selective spraying: A review. Curr. Robot. Rep. 1: 19-26.
Mahmud, M.S., S.Z. Abidin, A.A. Emmanuel and H.S. Hasan. 2020. Robotics and automation in agriculture: Present and future applications. Appl. Model. Simul. 4: 130-140.
Mail, M.F., J.M. Maja, M. Marshall, M. Cutulle, G. Miller and E. Barnes. 2023. Agricultural harvesting robot concept design and system components: A review. Agri. Eng. 5(2): 777-800.
Malkani, P., I. Mani, P.K. Sahoo, R.A. Parray, S.S. Swain, K.R. Asha et al. 2024. Development and evaluation of a sensor enabled smart herbicide applicator for inter row crops. Smart Agric. Technol. 9: 100649.
Muruganantham, P., S. Wibowo, S. Grandhi, N.H. Samrat and N. Islam. 2022. A systematic literature review on crop yield prediction with deep learning and remote sensing. Remote Sensing. 14(9): 1990.
Niranjan, P.Y., V.S. Rajpurohit and R. Malgi. 2019. A survey on chat-bot system for agriculture domain. *In:* 2019 1st International Conference on Advances in Information Technology (ICAIT). pp. 99-103. IEEE.
OECD. 2019. Digital opportunities for better agricultural policies. OECD Publishing, Paris.
Patil, N., A. Bhise and R.K. Tiwari. 2024. Fusion deep learning with pre-postharvest quality management of grapes within the realm of supply chain management. The Sci. Temper. 15(1): 1764-1772.
Patra, K., C.M. Parihar, H.S. Nayak, B. Rana, D.R. Sena, A. Anand et al. 2023. Appraisal of complementarity of subsurface drip fertigation and conservation agriculture for physiological performance and water economy of maize. Agric. Water Manag., 283: 108308.
Ponnusamy, V. and S. Natarajan. 2021. Precision agriculture using advanced technology of IoT, unmanned aerial vehicles, augmented reality and machine learning. Smart Sensors for Industrial Internet of Things: Challenges, Solutions and Applications. pp. 207-229.
Pradhan, N.C., M.A. Naik, M. Chowdhury, A. Kushwah, K.R. Asha, T. Dhar et al. 2024. Robotic seeding or sowing system in smart agriculture. pp. 495-520. *In*: K. Pandey, N.L. Kushwaha, C.B. Pande and K.G. Singh [eds.]. Artificial Intelligence and Smart Agriculture: Technology and Applications. Springer Nature, Singapore.
Rokade, A. and M. Singh. 2021. Analysis of precise greenhouse management system using machine learning-based Internet of Things (IoT) for smart farming. *In:* 2021 2nd International Conference on Smart Electronics and Communication (ICOSEC). pp. 21-28. IEEE.
Roy, A.M. and J. Bhaduri. 2021. A deep learning-enabled multi-class plant disease detection model based on computer vision. AI. 2(3): 413-428.
Ryo, M., J. Schiller, S. Stiller, J.C. Rivera Palacio, K. Mengsuwan, A. Safonova and Y. Wei. 2023. Deep learning for sustainable agriculture needs ecology and human involvement. J. Sustain. Agric. Environ. 2(1): 40-44.
Saleem, M.H., J. Potgieter and K.M. Arif. 2021. Automation in agriculture by machine and deep learning techniques: A review of recent developments. Precision Agriculture. 22(6): 2053-2091.

Sassenrath, G.F., P. Heilman, E. Luschei, G.L. Bennett, G. Fitzgerald, P. Klesius and P.V. Zimba. 2008. Technology, complexity and change in agricultural production systems. Renew. Agric. Food Syst. 23(4): 285-295.

Satpute, A.N., K.P. Gavhane, S. Kaur, A. Jha, N.C. Pradhan and M. Chowdhury. 2024. Integration of AI and IoT in soilless cultivation to power sustainable agricultural revolution. pp. 387-411. *In*: K. Pandey, N.L. Kushwaha, C.B. Pande and K.G. Singh [eds.]. Artificial Intelligence and Smart Agriculture: Technology and Applications. Springer Nature, Singapore.

Shaikh, T.A., T. Rasool and F.R. Lone. 2022. Towards leveraging the role of machine learning and artificial intelligence in precision agriculture and smart farming. Comput. Electron. Agric. 198: 107119.

Sharma, R. 2021. Artificial intelligence in agriculture: A review. *In:* 2021 5th International Conference on Intelligent Computing and Control Systems (ICICCS). pp. 937-942.

Sharma, S., K. Verma and P. Hardaha. 2023. Implementation of artificial intelligence in agriculture. J. Comput. Cogn. Eng. 2(2): 155-162.

Sood, A., R.K. Sharma, and A.K. Bhardwaj. 2022. Artificial intelligence research in agriculture: A review. Online Information Review. 46(6): 1054-1075.

Subeesh, A. and C.R. Mehta. 2021. Automation and digitization of agriculture using artificial intelligence and the Internet of Things. Art. Int. Agric. 5: 278-291.

Sunil Kumar, T., H.M. Virdia, K.G. Patel, S. Ragi, M. Chowdhury, P. Kumar et al. 2024. Effect of summer legume residue incorporation and fertilizer regimes on rice growth, yield, and nutrient uptake. Front. Sustain. Food Syst. 8: 1467201.

Thakur, R. and B.K. Singh. 2021. Importance of artificial intelligence in agriculture. E-magazine for Agriculture & Allied Sciences. ISSN-2582-8258.

Tituaña, L., A. Gholami, Z. Xu, He, M. Karkee and R. Ehsani. 2024. A small autonomous field robot for strawberry harvesting. Smart Agric. Tech. 8: 100454.

United Nations. 2024. Integrated Monitoring Initiative for SDG 6. Retrieved from: United Nations UnWater.

Verma, M. 2018. Artificial intelligence and its scope in different areas with special reference to the field of education. Online Submission. 3(1): 5-10.

Wakchaure, M., B.K. Patle and A.K. Mahindrakar. 2023. Application of AI techniques and robotics in agriculture: A review. Art. Int. Life Sci. 3: 100057.

Zha, J. 2020. Artificial intelligence in agriculture. J. Phy.: Conf. Ser. 1693(1): 012058. IOP Publishing.

Zhang, W., Z. Miao, N. Li, C. He, and T. Sun. 2022. Review of current robotic approaches for precision weed management. Curr. Robot. Rep. 3(3): 139-151.

Zhao, Z.Q., P. Zheng, S.T. Xu, and X. Wu. 2019. Object detection with deep learning: A review. IEEE Trans. Neural Netw. Learn. Syst. 30(11): 3212-3232.

Zhou, J., Y. Han, and S.S. So. 2008. Overview of artificial neural networks. In: Livingstone, D.J. (ed.) Artificial Neural Networks. Methods Mol. Biol. 458: Humana Press.

Zhu, S.J., S.X. Li, and W.H. Chen. 2023. AI meets UAVs: A survey on AI-empowered UAV perception systems for precision agriculture. Neurocomputing. 518: 242-270.

CHAPTER
2

Big Data in Agriculture: Acquisition, Processing and Implications

Aditya Pratap Singh[1], Tanushree Biswal[2], Umakanta Maharana[3*], Abdullah Mohammad Ghazi Al Khatib[4] and Pradeep Mishra[5]

[1] Department of Genetics and Plant Breeding, Bidhan Chandra Krishi Viswavidyalaya Mohanpur - 765022, Nadia, West Bengal, India
[2] Department of Life Science, Central University of Karnataka, Aland Road - 585367 Kadaganchi, Kalaburagi, Karnataka, India
[3] School of Computer Engineering, Kalinga Institute of Industrial Technology, Bhubaneswar - 751024, Khordha, Odisha, India
[4] Department of Banking and Insurance, Faculty of Economics, Damascus University, Damascus - 00070, Syrian Arab Republic
[5] Department of Agricultural Statistics, College of Agriculture, JNKVV, Rewa - 486001 Madhya Pradesh, India

Introduction

The agricultural sector is undergoing a profound transformation, with big data playing a pivotal role in shaping its future (Mayer-Schönberger 2013). Traditionally, agriculture relied on empirical methods and historical data for decision-making, but the advent of big data has revolutionized this process. In its simplest terms, "big data" refers to vast datasets that are too large and complex to be processed using traditional methods (Yaqoob and Hashem 2016). These data set present unique challenges in terms of storage, acquisition, sharing, analytics, and visualization, as they often consist of diverse and unstructured data (Vassakis et al. 2018). In agriculture, big data encompasses the generation and analysis of massive data streams from various sources, including satellite imagery, sensor networks, farm machinery, and market trends, which collectively span the entire agricultural value chain (Maru et al. 2018).

The rise of big data in agriculture is not merely a technological shift; it is a response to pressing global challenges. With the world's population continuing to grow, ensuring food security has become an urgent priority (Beddington 2010).

*Corresponding author: umakantamaharana70@gmail.com

DOI: 10.1201/9781003536932-2

Climate change further complicates this issue by affecting weather patterns, water availability, and soil conditions, leading to uncertainties in crop production (Kang et al. 2009). Big data analytics offers a solution by enabling the agricultural sector to make more informed decisions based on real-time data. Through predictive modeling, farmers can forecast crop yields, identify pest outbreaks, and manage resources like water and fertilizers more efficiently, helping to mitigate the effects of climate change on productivity (Wolfert et al. 2017). This data-driven approach allows for a more adaptive, responsive agricultural system, capable of meeting global food demands while conserving natural resources (Paul et al. 2022).

Incorporating big data into farming practices represents a significant departure from traditional methods. Rather than relying solely on experience and intuition, farmers now have access to advanced tools that provide them with actionable insights (Evans et al. 2017). The ability to collect, analyze, and interpret vast datasets enables them to optimize their operations with unprecedented precision. For instance, by monitoring field conditions in real-time, farmers can make timely decisions about planting, irrigation, and pest control, thereby increasing crop yields and minimizing resource wastage (Koteish et al. 2022). Moreover, big data facilitates more effective supply chain management, ensuring that agricultural products move from farms to markets with greater efficiency (Himesh et al. 2018). The shift towards data-driven decision-making is not only improving productivity but also fostering sustainability by reducing inputs like water, fertilizers, and pesticides, which can have harmful environmental effects if overused (Bachmann et al. 2022; Chowdhury et al. 2025).

This ongoing digital revolution in agriculture is driven by a combination of necessity and opportunity. As global food systems face the twin challenges of feeding a growing population and combating the impacts of climate change, the role of big data will only become more critical (Shiferaw et al. 2013). By harnessing the power of big data, the agricultural sector has the potential to achieve more sustainable, efficient, and resilient farming practices, ensuring long-term food security and environmental sustainability (Delgado et al. 2019). The application of big data in agriculture marks a new era of precision farming, where every decision is informed by data, maximizing output while minimizing resource consumption (Patil Shirish and Bhalerao 2013).

Key Applications of Big Data in Agriculture

The integration of big data into agriculture has led to transformative changes across various domains (Fig. 1), allowing for smarter, more efficient, and sustainable farming practices. We will here focus on four specific applications of big data in agriculture: precision farming, predictive analytics, resource management and livestock management to understand how big data is shaping modern agriculture.

Precision Farming

Precision farming, or precision agriculture, represents a shift from uniform treatment of crops to a more tailored approach, where every part of the field receives inputs in just the right amounts based on its specific conditions (Patil Shirish and Bhalerao 2013; Gawande et al. 2023). The core idea of precision farming is to

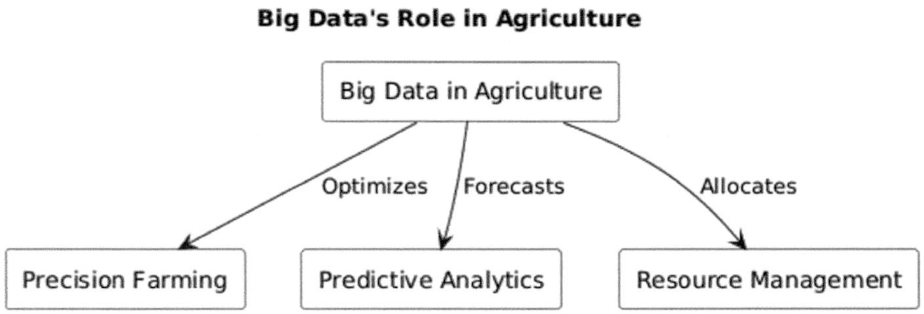

Fig. 1: Broad areas where big data has potential uses.

collect detailed data from fields, including information about soil composition, moisture levels, temperature variations, and plant health, and then use this data to guide farming decisions (Gawande et al. 2023). The goal is to enhance crop yields while minimizing resource use and environmental impact (Robert et al. 2016). A central tool in precision farming is the use of high-resolution field maps, which are generated by combining data from satellites, drones, and ground sensors. These maps provide granular information on variables like soil type, nutrient distribution, and moisture content (Pulighe and Lupia 2016). With this level of detail, farmers can precisely apply fertilizers, water, and pesticides, ensuring that each part of the field receives exactly what it needs. This not only reduces input costs but also limits the environmental risks associated with overuse of chemicals (Wu et al. 2018; Sunil Kumar et al. 2024).

The use of drones and satellite imagery for crop health monitoring is another major innovation in precision farming (Inoue 2020). Equipped with sensors that detect wavelengths of light beyond human vision, these technologies can identify early signs of plant stress or disease (Chowdhury et al. 2024a). By analyzing these data streams, farmers are alerted to potential issues long before they become visible to the naked eye (Altieri et al. 2012). This early detection allows for timely intervention, reducing crop losses and improving overall farm efficiency (Anand et al. 2024).

Predictive Analytics

Predictive analytics is one of the most powerful applications of big data in agriculture, allowing farmers to forecast outcomes and make proactive decisions. At its core, predictive analytics involves finding meaningful patterns in data using techniques like statistics, machine learning, and data mining, also called advanced analytics. It helps solve problems by applying data analysis methods (Ratner 2017). Unlike Business Intelligence (BI), where experts choose the questions and variables, predictive analytics relies on data to select models and relationships. It systematically identifies patterns between variables to optimize outcomes and minimize errors. Emerging areas like process mining analyze event logs to discover and improve processes in fields like healthcare and banking; and frameworks like Hadoop help manage large-scale data analysis (Sethy and Panda 2015).

In the context of agriculture, predictive analytics uses historical data, current environmental conditions, and machine learning models to generate forecasts for key agricultural variables such as crop yields, pest outbreaks, and weather patterns (Domingues et al. 2022; Chowdhury et al. 2023b). In the past, predicting crop yields involved a high degree of uncertainty, relying heavily on farmer intuition and experience. With increasing data science trends, forecasting started becoming backed up by statistical tools, particularly time-series modeling. However, conventional methods could consider a limited number of variables/factors at a time. Today, big data enables much more accurate predictions by analyzing a wide range of factors, including soil quality, historical crop performance, weather forecasts, and real-time sensor data (Fig. 2). These predictive models allow farmers to anticipate how much they will harvest, enabling better planning for labor, storage, and market sales (Guo et al. 2005). For instance, knowing the expected yield can help farmers negotiate better prices or optimize supply chain logistics.

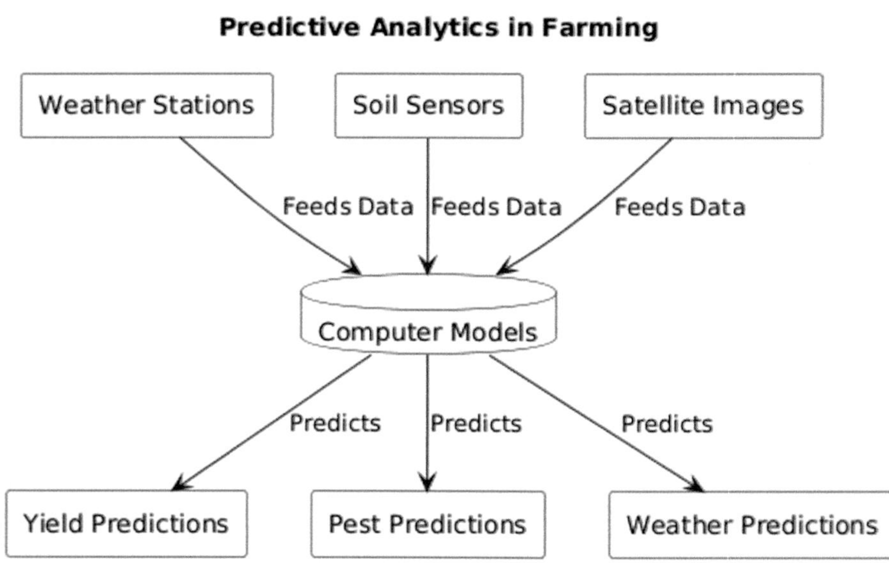

Fig. 2: Predictive analytics workflow in agriculture.

Similarly, predictive analytics has revolutionized pest and disease management. By analyzing patterns of past pest outbreaks in combination with real-time weather data, predictive models can forecast when and where pest infestations are likely to occur. Farmers can then take preventative actions, such as deploying biological controls or applying pesticides only when necessary. This targeted approach not only protects crops but also reduces the environmental footprint of pest control measures (Birch et al. 2011; Kurmi et al. 2024).

Resource Management

Efficient resource management is critical for sustainable agriculture, and big data plays a crucial role in optimizing the use of inputs like water, nutrients, and energy. In water management, big data integrates information from soil moisture sensors, weather stations, and irrigation systems to develop precision irrigation strategies. These strategies ensure that crops receive the exact amount of water they need at the right time, reducing water wastage and improving crop health (Gohil et al. 2021). Similarly, big data is used to optimize nutrient management. Data from soil tests, plant health sensors, and historical yield records help farmers determine the optimal amount and type of fertilizer to apply, improving crop productivity while reducing nutrient runoff into surrounding ecosystems (Saiz-Rubio and Rovira-Más 2020). This not only lowers input costs but also minimizes the environmental impact of farming.

Energy efficiency is another area where big data has a significant impact. Big data analytics helps to monitor energy use across farm operations, including machinery and storage facilities. Big data analytics can identify inefficiencies and suggest improvements. For example, data on machinery usage patterns can help farmers optimize fuel consumption, while insights from sensor data in storage units can prevent energy wastage, especially in temperature-sensitive environments like cold storage for perishable goods (Weersink et al. 2018).

Sources and Collection of Big Data in Agriculture

In the modern era of digital agriculture, data has become as essential as seeds and soil. The farming sector now cultivates insights from a diverse array of data sources, each contributing to a comprehensive understanding of the agricultural ecosystem. This chapter explores the primary sources of agricultural big data and the sophisticated techniques employed in its collection. Table 1 represents some use cases of IoT devices in agriculture.

Table 1: Agriculture related IoT devices and their uses

IoT Device	Uses	Reference
Smart Soil Sensors	• Measure soil moisture, temperature, and nutrient levels. • Optimize irrigation and fertilization. • Monitor soil health.	Chen and Wong 2024
Drone-based Multispectral Cameras	• Assess crop health through vegetation indices. • Detect pest infestations and diseases. • Create high-resolution field maps.	Chen and Wong 2024
Weather Stations	• Monitor local climate conditions. • Predict weather patterns. • Optimize planting and harvesting schedules.	Johnson and Patel 2023
RFID Livestock Trackers	• Monitor animal health and behavior. • Track livestock location. • Optimize feeding schedules.	Garcia and Brown 2024
Smart Irrigation Systems	• Automate water distribution. • Optimize water usage based on crop needs and weather conditions. • Reduce water waste.	Lee and Kumar 2023

Satellite Imagery

The view from above has revolutionized how we monitor and manage agricultural lands. Satellites, orbiting hundreds of kilometers above Earth, have become indispensable tools in the farmer's digital toolkit (Howitt et al. 2022).

Multispectral and hyperspectral imaging technologies (Fig. 3) aboard these satellites capture data across various electromagnetic spectrum bands. This wealth of information allows for the assessment of crop health, soil moisture content, and land use changes on a scale previously unimaginable. The Normalized Difference Vegetation Index (NDVI), derived from these spectral bands, has become a standard measure for evaluating crop vigor and predicting yields (Lu et al. 2020).

Synthetic Aperture Radar (SAR) technology has further expanded the capabilities of satellite-based agriculture monitoring. Unlike traditional optical sensors, SAR can penetrate cloud cover and operate regardless of sunlight conditions. This all-weather, day-and-night functionality makes SAR particularly valuable for monitoring crop growth stages and assessing soil moisture in deeper layers, even in regions prone to persistent cloud cover (Ranjan and Parida 2021).

Fig. 3: An example of a satellite image.
(Source: https://www.xyht.com/enviroag/satellite-imagery-precision-agriculture/)

Sensor Data and IoT Devices

The proliferation of Internet of Things (IoT) devices has transformed fields into data-rich environments (Fig. 4). These smart sensors and interconnected gadgets provide real-time, granular information about various aspects of the farming ecosystem (Ayaz et al. 2019).

Soil sensors, acting as underground sentinels, continuously monitor moisture levels, temperature, electrical conductivity (a proxy for soil salinity), and even pH levels. This detailed soil data enables precise irrigation scheduling and nutrient management, optimizing resource use and crop yields (Ahmad et al. 2022). Weather stations deployed on farms collect hyperlocal meteorological data, including

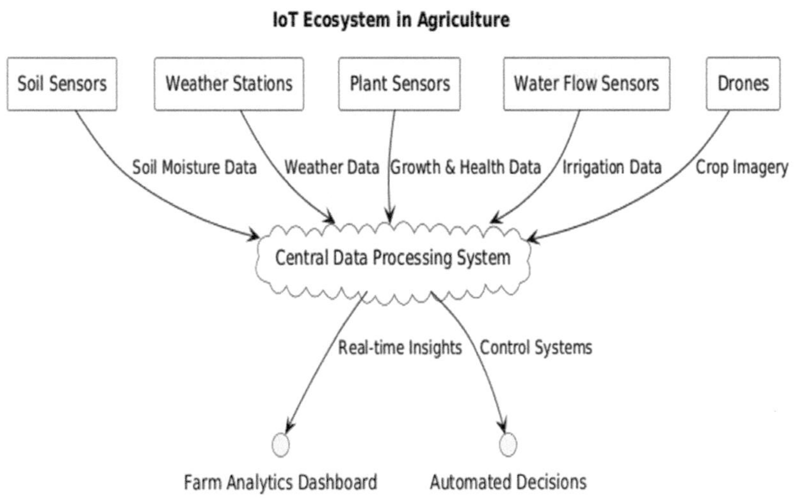

Fig. 4: IoT-based big data integration for precision agriculture.

temperature, humidity, precipitation, wind speed and direction, and solar radiation (Ghaley et al. 2014). This localized weather information improves the accuracy of farm-specific forecasts and crop modeling, allowing farmers to make more informed decisions about planting, harvesting, and crop protection.

In the realm of plant monitoring, sensors attached directly to crops or deployed within the canopy measure leaf temperature, stem water potential, and chlorophyll fluorescence. These measurements provide real-time insights into plant stress and physiological status, enabling early intervention before crop health is compromised (Zimmermann et al. 2013).

Farm Machinery and Equipment Data

The agricultural landscape has been transformed by the advent of smart machinery, turning tractors, harvesters, and other equipment into mobile data centers. These sophisticated machines are now integral components of the agricultural big data ecosystem, continuously collecting and processing information as they work the land (Xuegeng et al. 2020).

Modern combine harvesters, equipped with advanced yield monitoring systems (Fig. 5), have revolutionized our understanding of field productivity. As these machines move through the field, they gather high-resolution data on crop yield variability, providing farmers with a detailed map of their harvest. These yield monitors also measure grain moisture content in real-time, offering insights into crop maturity and optimal harvesting times (Zhao et al. 2015). Furthermore, by tracking harvesting speed and efficiency, these systems help optimize the harvesting process itself.

But the data collection doesn't stop with the harvest. Tractors and other farm equipment have become roving sensor platforms, gathering a wealth of operational data. These machines monitor fuel consumption, providing valuable information for

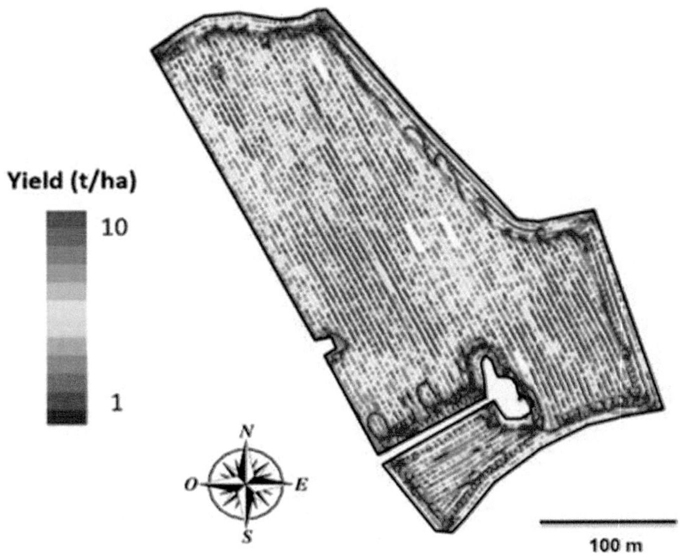

Fig. 5: Yield maps in precision agriculture.
(Source: https://www.aspexit.com/yield-maps-in-precision-agriculture)

cost management and environmental impact assessments (Kushwah et al. 2024). Engine performance data helps predict maintenance needs, reducing downtime and extending machinery lifespan. GPS-enabled tracking of operational hours and location allows for precise mapping of field operations and helps optimize route planning (Barrientos et al. 2011).

Perhaps one of the most significant advancements in farm machinery data collection comes from variable rate application systems. These smart sprayers and fertilizer applicators are at the heart of precision agriculture practices. They meticulously record the application rates of inputs like fertilizers and pesticides, creating detailed maps of where, when, and how much of each input was applied (Malkani et al. 2024; Chowdhury et al. 2024b). This spatial distribution data is crucial for fine-tuning application strategies and ensuring uniform crop treatment. Moreover, these systems keep track of tank mix compositions, providing a detailed record of the specific formulations used in each application (Rounsevell et al. 2003). The data gathered by these advanced farm machines serves multiple crucial purposes. It allows farmers to create detailed, data-driven plans for future seasons, optimizing everything from seed selection to harvesting strategies. It provides a robust framework for validating the effectiveness of precision agriculture practices, allowing for continuous improvement in resource use efficiency (Andarge and Lichtenberg 2020). Additionally, this comprehensive data trail ensures regulatory compliance, particularly important in an era of increasing environmental scrutiny in agriculture.

As we move further into the age of smart farming, the role of machinery as data collectors will only grow in importance (Shamshiri et al. 2018). The integration of artificial intelligence and machine learning algorithms with these data streams promises to unlock even greater insights, potentially leading to autonomous farming systems that can make real-time decisions based on the wealth of data they collect (Chowdhury et al. 2023a).

Data Collection and Processing Technologies

In the era of precision agriculture, the methods we use to collect and process data have become as crucial as the data itself. This section explores the cutting-edge technologies that are revolutionizing how we gather and analyze agricultural information, focusing on three key areas: aerial data collection using drones, precision soil sampling techniques, and cloud computing solutions for data storage and processing.

Drones and UAVs: Aerial Data Collection

Unmanned Aerial Vehicles (UAVs), commonly known as drones, have emerged as game-changers in agricultural data collection (Kim et al. 2019). These flying sensors offer unprecedented flexibility and detail in gathering field-level data.

Drones equipped with multispectral and thermal cameras provide high-resolution imagery (Fig. 6) that reveals crop health, water stress, nutrient deficiencies, and even early signs of pest infestations. The ability to capture this data at a plant-by-plant level allows for incredibly precise farm management (Matese and Di Gennaro 2018).

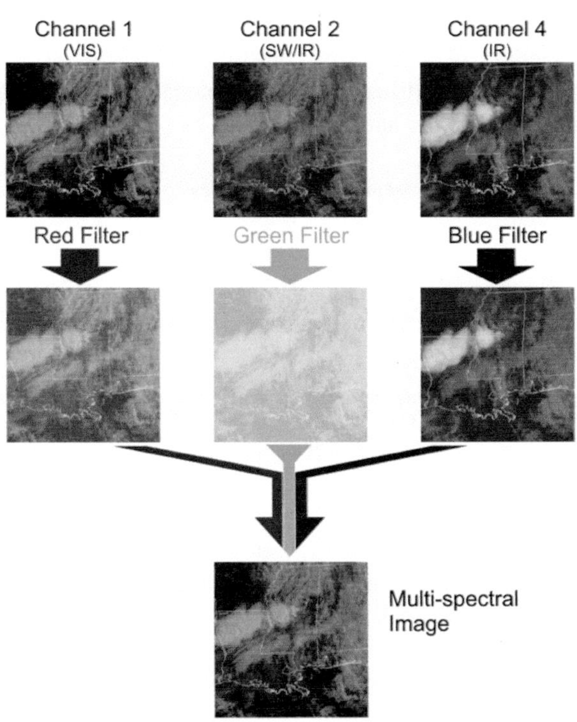

Fig. 6: Multispectral image representation. (Source: https://learningweather.psu.edu/node/70)

LiDAR (Light Detection and Ranging) technology mounted on drones takes aerial data collection a step further. By generating precise 3D models of crop canopy structure and field topography, LiDAR data aids in assessing crop biomass, planning efficient drainage systems, and optimizing orchard management (Narvaez et al. 2017).

Perhaps the most advanced sensing technology deployed on agricultural drones is hyperspectral imaging. These sensors capture data across hundreds of narrow spectral bands, enabling detailed crop phenotyping, early detection of plant diseases, and assessment of crop maturity and quality parameters. This level of detail was previously only available through laboratory analysis (Araus et al. 2022). The true power of drone technology in agriculture lies not just in the hardware, but in the sophisticated software systems that support it (Yeom et al. 2019). Automated flight planning ensures consistent data collection over time, while advanced processing algorithms quickly transform raw imagery into actionable insights through orthomosaic maps, vegetation indices, and change detection analysis.

Soil Sampling: Precision Data Gathering Techniques

While aerial data provides a comprehensive view of the field, understanding what's happening beneath the surface is equally crucial. Modern soil sampling techniques have evolved to provide highly detailed insights into soil properties and their variability across fields (Fig. 7). Grid sampling, a systematic approach where soil samples are collected at regular intervals across a field, reveals spatial patterns in soil properties. This method, typically using GPS for precise location recording, informs variable-rate application of fertilizers and amendments, allowing farmers to optimize input use efficiency (Mallarino and Wittry 2004).

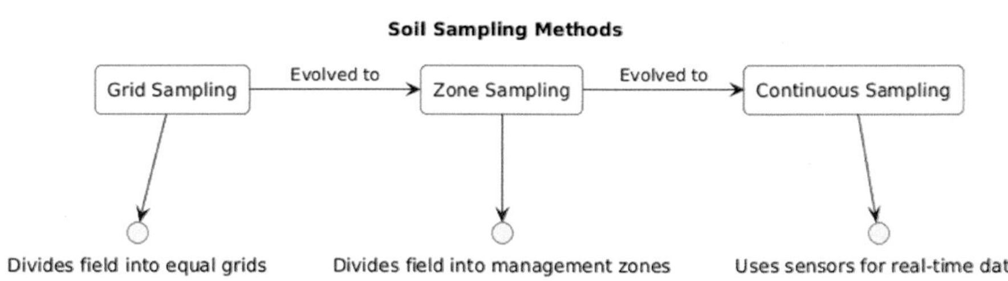

Fig. 7: Soil sampling methods.

An alternative approach, zone sampling, divides fields into management zones based on factors like topography, soil type, historical yield data, and remote sensing imagery. This method balances the need for detailed information with operational efficiency, particularly in fields with distinct areas of variability (Vitharana et al. 2008).

The frontier of soil data collection lies in on-the-go sensors and proximal soil sensing techniques. Mobile sensors mounted on vehicles allow for continuous measurement of soil properties like electrical conductivity, organic matter content, and pH as the vehicle traverses the field. Meanwhile, proximal sensing techniques such as Visible and Near-Infrared (Vis-NIR) Spectroscopy, Mid-Infrared (MIR) Spectroscopy, and Gamma-Ray Spectroscopy offer rapid, non-destructive measurements of various soil properties (Debauche et al. 2022). These advanced soil sensing techniques enable the creation of high-resolution soil property maps with reduced need for laboratory analysis, providing farmers with unprecedented insights into the variability of their soil resources.

Cloud Computing: Storage and Processing Solutions

The volume and complexity of data generated by modern agricultural technologies demand robust storage and processing solutions. Cloud computing has emerged as a critical infrastructure for agricultural big data, offering scalable, flexible, and accessible platforms for data management and analysis.

Cloud-based storage solutions provide secure, centralized repositories for the vast amounts of data generated from various sources – from drone imagery and soil sensors to weather stations and farm machinery. These platforms ensure that data is not only safely stored but also easily accessible to authorized users from any location, facilitating collaborative research and decision-making (Debauche et al. 2022). But the true power of cloud computing in agriculture lies in its processing capabilities. Scalable processing frameworks like Apache Hadoop and Apache Spark enable the analysis of large-scale agricultural datasets that would be impossible with traditional computing methods. These frameworks allow for distributed storage and processing, enabling complex analyses like crop yield predictions, disease forecasting, and optimization of farm operations (Saiz-Rubio and Rovira-Más 2020).

Cloud-based platforms also facilitate the integration and fusion of data from multiple sources – a critical step in extracting comprehensive insights from agricultural big data. Advanced data fusion algorithms, running on cloud infrastructure, can combine information from multiple sensors and sources, providing a holistic view of farm operations and enabling more informed decision-making (Liu et al. 2019). Moreover, cloud computing powers the development and deployment of sophisticated machine learning and artificial intelligence models in agriculture. These models, trained on vast datasets stored in the cloud, can provide real-time insights and predictions, from optimizing irrigation schedules to detecting early signs of crop diseases (Shaikh et al. 2022).

As we move further into the era of data-driven agriculture, the synergy between advanced data collection technologies and powerful cloud computing solutions will continue to push the boundaries of what's possible in farming. From the drones flying over our fields to the invisible algorithms processing data in the cloud, these technologies are laying the foundation for a more productive, sustainable, and resilient agricultural future.

Big Data Tools and Technologies

The agricultural sector's digital transformation is propelled by an array of sophisticated tools and technologies designed to harness the power of big data. These innovations are reshaping farming practices, enabling more informed decision-making, and driving unprecedented efficiencies across the agricultural value chain. At the heart of agricultural big data lies robust data management systems capable of handling the volume, velocity, and variety of information generated in modern farming. The Apache Hadoop ecosystem has emerged as a cornerstone for distributed storage and processing of large agricultural datasets. This framework includes the Hadoop Distributed File System (HDFS) for scalable, fault-tolerant storage, MapReduce for parallel data processing, and YARN for optimizing resource utilization across computing clusters (Fatima et al. 2023).

Complementing Hadoop are NoSQL databases, which offer flexible schemas and high performance for diverse agricultural data types. Document stores like MongoDB excel in managing semi-structured data such as sensor readings and field observations, while column-family stores like Apache Cassandra efficiently handle time-series data from weather stations and IoT devices. Graph databases, exemplified by Neo4j, prove invaluable for modeling complex relationships in agricultural supply chains or genomic data (Gilson 2019).

Cloud-based platforms have revolutionized agricultural data management by providing scalable infrastructure and specialized services. Major providers such as Amazon Web Services, Microsoft Azure, and Google Cloud Platform offer tailored solutions for IoT data ingestion, machine learning, and geospatial analysis, enabling farmers and researchers to leverage advanced analytics without significant upfront investment in IT infrastructure (Khan and Shorna 2023).

Machine Learning and AI: Data Analysis and Predictive Modeling

Machine Learning (ML) and Artificial Intelligence (AI) technologies have become indispensable tools for extracting actionable insights from agricultural big data. These advanced analytical techniques are transforming various aspects of farming, from crop management to livestock care.

Predictive modeling, powered by ML algorithms, is revolutionizing agricultural forecasting. Random Forests and Support Vector Machines are widely employed for yield prediction, crop classification, disease detection, and soil type classification. These algorithms excel in handling the complex, non-linear relationships inherent in agricultural data (Misra et al. 2020), Khaki 2022).

Computer vision, a subset of AI, has found numerous applications in agriculture. Convolutional Neural Networks (CNNs) are at the forefront of visual analysis tasks, including crop disease detection, weed identification, and fruit quality assessment (Anand et al. 2024). These models can analyze images of plants and produce with high accuracy, enabling early disease intervention, precise weed control, and automated grading of agricultural products (Fracarolli et al. 2020).

Natural Language Processing (NLP) techniques are increasingly utilized to extract valuable information from unstructured text data relevant to agriculture. Sentiment analysis algorithms gauge market sentiments from social media and news articles, providing insights into agricultural market dynamics. Information extraction methods automate the parsing of weather reports and agricultural bulletins, streamlining the gathering of critical information for farm management (Domínguez et al. 2024).

Reinforcement learning, an area of ML focused on decision-making, is being explored for optimizing complex agricultural systems. This approach shows promise in areas such as irrigation scheduling, robotic harvesting, and livestock management (Bu and Wang 2019). By learning from interactions with the environment, reinforcement learning algorithms can develop strategies that adapt to changing conditions and optimize resource utilization.

Blockchain: Data Transparency and Security

Blockchain technology is emerging as a powerful tool for enhancing transparency, traceability, and trust in agricultural supply chains. This distributed ledger technology offers a secure and immutable record of transactions and data, addressing key challenges in the agricultural sector. In supply chain management, blockchain enables transparent tracking of agricultural products from farm to table. Each step of the journey, from planting and harvesting to processing and retail, can be recorded on the blockchain, providing consumers with verifiable information about the origin and handling of their food. This transparency not only builds trust but also facilitates rapid traceability in case of food safety issues (Jahanbin et al. 2019).

Smart contracts, self-executing contracts with the terms directly written into code, are another promising application of blockchain in agriculture. These contracts can automate various processes, such as triggering payments when predefined conditions are met. For instance, crop insurance payouts could be automatically initiated based on weather data recorded on the blockchain, streamlining the claims process and reducing administrative overhead (Demestichas et al. 2020).

Blockchain also plays a crucial role in ensuring data integrity and facilitating secure data sharing in the agricultural sector. Research findings, genetic data, and sensor readings can be cryptographically secured on the blockchain, allowing for verifiable and controlled sharing of information among stakeholders (Wilson and Mutersbaugh 2015). This capability is particularly valuable for collaborative research efforts and in maintaining the integrity of certification processes for organic or fair-trade products. By providing a tamper-resistant and transparent system for recording and verifying information, blockchain technology addresses critical issues of trust and accountability in the agricultural sector. As the technology matures, it promises to revolutionize how data is managed and shared across the entire agricultural ecosystem, from individual farms to global supply chains.

Handling and Analysis of Big Data Using AI

The explosion of data in agriculture, from satellite imagery to IoT sensor data, requires advanced techniques for handling and analysis that can manage the volume, variety, and velocity of big data. Traditional statistical methods, while valuable, struggle with the complexity and scale of these datasets. Artificial Intelligence (AI) introduces a suite of algorithms that offer faster, more efficient, and more accurate methods for data handling and analysis, enhancing agricultural decision-making, forecasting, and resource management. By leveraging machine learning algorithms like Random Forests, Neural Networks, and Support Vector Machines, AI excels in predicting crop yields, diagnosing plant diseases, and optimizing soil management, especially in complex, high-dimensional datasets. Deep learning, particularly Convolutional Neural Networks (CNNs), is revolutionizing image analysis from drone and satellite sources, allowing for precise detection of issues like pest infestations or nutrient deficiencies at early stages, which is critical for timely interventions. Additionally, Natural Language Processing (NLP) enables real-time extraction of valuable information from unstructured sources, such as weather bulletins or market reports, streamlining decision-making processes in rapidly changing conditions. Predictive analytics powered by AI synthesizes data from past and current sources to forecast essential parameters like crop yields, pest risks, and water requirements, while reinforcement learning algorithms allow AI systems to adapt to field-specific conditions autonomously—optimizing irrigation, fertilization, and pest management strategies in real-time. Cloud computing platforms, integrated with AI, provide the storage and processing power to handle these enormous datasets, making it possible to run sophisticated data fusion techniques that combine information from sensors, market trends, and weather patterns, creating a holistic view of farm operations.

AI algorithms enhance big data handling in agriculture by offering real-time analysis, enabling quick responses to crop and environmental changes. They can model complex interactions among variables, facilitating a more precise understanding of field conditions than traditional techniques. Additionally, AI-driven automation in data analysis reduces labor costs, optimizes resource usage, and minimizes environmental impact, promoting sustainability and efficiency. These advantages position AI as a crucial tool in addressing the challenges of modern agriculture and improving resilience against climate variability.

Benefits of Leveraging Big Data in Agriculture

The application of big data analytics in agriculture yields a multitude of benefits, transforming traditional farming practices and enhancing overall agricultural productivity and sustainability (Table 2).

Table 2: Benefits of big data analytics in agriculture

Benefit Category	Key Areas	Specific Benefits
Increased Productivity and Efficiency	Yield Optimization	• Precision planting • Variable-rate nutrient management • Early pest & disease control
	Resource Use Efficiency	• Precision irrigation for water conservation • Energy-efficient machinery • Data-driven labor optimization
	Livestock Management	• Precision feeding • Health monitoring through IoT • Genetic improvements via data-driven breeding programs
Cost Reduction and Economic Benefits	Input Cost Reduction	• Reduced input use (fertilizers, pesticides) • Predictive equipment maintenance • Optimized inventory management

	Risk Mitigation	• Advanced weather risk management • Market risk reduction via price forecasting • Improved crop insurance through data-driven risk assessment
	Value Chain Optimization	• Enhanced supply chain efficiency • Higher quality produce through quality control • Direct-to-consumer marketing opportunities
Environmental Sustainability	Resource Conservation	• Water use optimization • Soil health improvement • Biodiversity protection through targeted pest management
	Emissions Reduction	• Reduced emissions via efficient fertilizer use • Optimized transport logistics • Carbon sequestration through soil management
	Sustainable Land Use	• Precision land management • Erosion control through data-driven practices • Habitat preservation
Food Security and Quality	Yield Stability	• Climate-resilient farming • Early pest and disease detection • Genetic improvements for stress tolerance
	Food Safety	• Improved traceability through blockchain • Contaminant detection using sensors • Adherence to food safety standards via data-driven quality control
	Nutritional Quality	• Development of nutritionally rich crop varieties • Optimal harvesting at peak value • Enhanced post-harvest management
Knowledge and Decision Support	Farmer Decision Support	• Real-time advisory systems • Scenario analysis for decision making • Continuous learning from data insights
	Policy and Planning	• Evidence-based policymaking • Data-driven resource allocation • Enhanced food security planning
	Research and Innovation	• Accelerated research with big data platforms • Precision phenotyping for crop improvement • Interdisciplinary insights from diverse datasets

Increased Productivity and Efficiency

One of the primary advantages of big data analytics in agriculture is the significant increase in productivity and efficiency. By leveraging data-driven insights, farmers can optimize their yield through precision planting, variable-rate nutrient management, and early pest and disease control. This targeted approach allows for more efficient use of resources and maximizes crop output.

Resource use efficiency is enhanced through precision irrigation for water conservation, energy-efficient machinery, and data-driven labor optimization. These improvements not only boost productivity but also contribute to cost savings and environmental sustainability (Liang and Shah 2023). In livestock management, big data enables precision feeding, health monitoring through IoT devices, and genetic improvements via data-driven breeding programs, leading to healthier animals and increased production efficiency.

Cost Reduction and Economic Benefits

Big data analytics contribute substantially to cost reduction and economic benefits in agriculture. Input costs are reduced through optimized use of fertilizers and pesticides, predictive equipment maintenance, and efficient inventory management. This not only lowers operational expenses but also minimizes waste and environmental impact (Sharma et al. 2020).

Risk mitigation strategies are improved with advanced weather risk management, market risk reduction via price forecasting, and data-driven crop insurance assessment. These tools provide farmers with better decision-making capabilities, helping them navigate uncertainties in weather patterns and market fluctuations (Jaber et al. 2022). Furthermore, value chain optimization is achieved through enhanced supply chain efficiency, improved produce quality control, and direct-to-consumer marketing opportunities. By leveraging big data insights, farmers can streamline their operations from production to distribution, ensuring higher quality products reach consumers more efficiently.

Environmental Sustainability

Big data analytics play a crucial role in promoting environmental sustainability in agriculture. Resource conservation is achieved through water use optimization, soil health improvement, and biodiversity protection through targeted pest management. These practices not only preserve natural resources but also contribute to long-term agricultural sustainability (Pretty 2008). Emissions reduction is another significant benefit, realized through efficient fertilizer use, optimized transport logistics, and carbon sequestration through improved soil management practices. By minimizing the environmental footprint of agricultural activities, big data contributes to global efforts in combating climate change. Sustainable land use is promoted through precision land management, erosion control through data-driven practices, and habitat preservation. These approaches help maintain the delicate balance between agricultural productivity and environmental conservation.

Food Security and Quality

The application of big data analytics significantly enhances food security and quality. Yield stability is improved through climate-resilient farming practices, early pest and disease detection, and genetic improvements for stress tolerance. These advancements help ensure consistent crop yields even in the face of changing environmental conditions (Carvalho 2006). Food safety is bolstered by improved traceability through blockchain technology, contaminant detection using sensors, and adherence to food safety standards via data-driven quality control. These measures not only protect consumer health but also build trust in the food supply chain.

Nutritional quality of agricultural products is enhanced through the development of nutritionally rich crop varieties, optimal harvesting at peak value, and improved post-harvest management. By leveraging data insights, farmers can produce higher quality, more nutritious food to meet the growing global demand.

Big data analytics provide invaluable knowledge and decision support to various stakeholders in the agricultural sector. Farmers benefit from real-time advisory systems, scenario analysis for decision making, and continuous learning from data insights. These tools empower farmers to make informed decisions based on up-to-date information and predictive analytics (Zhai et al. 2020). In policy and planning, big data enables evidence-based policymaking, data-driven resource allocation, and enhanced food security planning. This allows governments and organizations to develop more effective strategies to address agricultural challenges and ensure food security on a broader scale.

Research and innovation in agriculture are accelerated through big data platforms, precision phenotyping for crop improvement, and interdisciplinary insights from diverse datasets. This data-driven approach to research facilitates faster discoveries and more targeted solutions to agricultural challenges. The integration of big data analytics in agriculture offers a wide array of benefits that span productivity enhancement, economic gains, environmental sustainability, food security, and knowledge support. As technology continues to advance, the potential for big data to revolutionize agriculture and address global food challenges becomes increasingly evident. However, it is crucial to address challenges such as data privacy, accessibility, and the digital divide to ensure that these benefits are realized equitably across the agricultural sector.

Challenges in Implementing Big Data

Despite its potential, the implementation of big data in agriculture faces several challenges that need to be addressed for widespread adoption.

Data Privacy and Security Concerns

As more farm data is collected and analyzed, concerns about data ownership, privacy, and security have become prominent. Farmers may be hesitant to share data due to fears of misuse or unauthorized access (Nikander et al. 2020). Establishing clear data governance frameworks and ensuring robust cybersecurity measures are crucial to addressing these concerns (Nikander et al. 2020).

Infrastructure and Data Integration Issues

Many rural areas lack the necessary technological infrastructure to fully leverage big data solutions. Limited internet connectivity and inadequate computing resources can hinder the adoption of advanced analytics (Saidu et al. 2017). Moreover, integrating data from diverse sources and formats presents technical challenges that need to be overcome for effective big data utilization (Sagiroglu and Sinanc 2013).

The effective use of big data in agriculture requires specialized skills that many farmers and agricultural professionals may lack. This skills gap can slow down the adoption of big data technologies (Lee and Gambiza 2022). Additionally, the initial investment required for big data solutions can be a barrier for small-scale farmers.

Future Trends and Innovations

The future of big data in agriculture promises exciting advancements driven by rapid technological innovations. Advanced sensing technologies will play a key role in this transformation. Nanosensors and smart dust, for example, will enable unprecedented monitoring capabilities. These nanoscale sensors, capable of detecting plant stress at the molecular level, will help farmers make timely interventions. Distributed sensing networks, or "smart dust," will consist of tiny, wireless sensors spread across fields, creating a comprehensive, real-time picture of crop health and environmental conditions (Coble et al. 2018).

Hyperspectral imaging will take precision agriculture to new heights. Satellite-based hyperspectral sensors will allow detailed analysis of crops and soils from space, offering farmers a broader understanding of their land's needs. On the ground, portable hyperspectral devices will make it easier for farmers to perform real-time scans, providing actionable insights at their fingertips. Similarly, underground sensing technologies will offer new ways to monitor soil health. Sensors that track root zone activity will give farmers a clear view of moisture levels and nutrient uptake, while non-invasive subsoil imaging will map the physical properties of soils without disturbing them.

Artificial intelligence (AI) and machine learning (ML) will also continue to grow in importance. The development of explainable AI (XAI) will make AI models more transparent, ensuring that farmers understand the decisions these systems are making. By making AI interpretable, trust in technology will increase, fostering greater adoption. Causal AI, which can predict cause-effect relationships in agricultural systems, will further enhance decision-making. Federated learning, another AI innovation, will allow farms to collaborate on ML models without sharing sensitive data. This will help safeguard privacy while improving the accuracy of predictive models. Edge AI will process data locally on farms, enabling real-time insights without relying on cloud computing, which is crucial for regions with unreliable internet access (Chen et al. 2023).

Reinforcement learning will introduce autonomous farm management, where AI agents learn and optimize activities like irrigation and fertilization. Robotics powered by reinforcement learning will adapt to various farm conditions, making them more effective in tasks like harvesting or pest control. The integration of diverse data sources will further enrich agricultural big data systems. Multi-omics integration, combining genomics, phenomics, and environmental data, will offer a holistic view of crop performance. By incorporating microbiome data from soil and plants, farmers will manage ecosystems more sustainably. Crowdsourcing platforms will enable farmers to share observations, enriching sensor data with local knowledge, while consumer feedback integration will align farming practices with market demands.

Synthetic data generation through AI will also become prominent. Augmented datasets will help train models when real-world data is limited, and virtual field trials will allow farmers to test new methods or products in simulations before applying them to their fields. Blockchain technology will offer decentralized systems for data management and transparency. Farmer-controlled data marketplaces will give growers control over their data, allowing them to share or monetize it as they see fit. Tokenization will enable the creation of digital assets, which could revolutionize investment in agriculture. Smart contracts could automate payments for ecosystem services like carbon sequestration, and blockchain-based systems may enable peer-to-peer sharing of resources among farmers.

Quantum computing represents another frontier in agricultural big data. Its ability to model complex ecosystems in high detail will enhance our understanding of interactions between soil, crops, and climate. Quantum computing will also tackle large-scale optimization problems, such as supply chain logistics and resource allocation. As data security concerns grow, quantum-safe cryptography will be crucial in protecting sensitive agricultural data. Secure multi-party computation will allow collaborative data analysis while maintaining strict privacy controls.

Extended Reality (XR) technologies will further integrate big data into the hands-on management of farms. Augmented reality (AR) could overlay real-time data onto a farmer's view of their fields, making decision-making more intuitive. Virtual reality (VR) will provide immersive environments for data exploration, allowing users to interact with complex datasets in new ways. VR-based training programs will offer farmers a chance to practice new techniques virtually, while AR-enabled systems will allow remote experts to guide farmers in real time. Policy and regulatory frameworks will shape the future of agricultural big data. As the industry grows, there will be a push for the development of international data governance standards to ensure the responsible use of agricultural data. Ethical AI regulations will aim to protect farmers and ensure that technology benefits all stakeholders. Open data initiatives could make national agricultural data repositories accessible to all, promoting innovation while safeguarding farmers' interests.

Conclusion

Big data analytics has emerged as a paradigm-shifting force in modern agriculture, offering unprecedented opportunities for optimizing productivity, enhancing sustainability, and bolstering resilience in the face of global challenges. From precision agriculture employing geospatial data to machine learning-driven predictive analytics, big data applications are revolutionizing every facet of agronomic practices and resource management. The tangible benefits of big data integration in agriculture are multifaceted, encompassing increased crop yields, optimized resource allocation, and improved environmental stewardship. For instance, John Deere's Field Connect system utilizes IoT sensors and big data analytics to provide real-time soil moisture data, enabling farmers to make data-driven irrigation decisions. Similarly, The Climate Corporation's FieldView platform leverages machine learning algorithms to analyze historical yield data, weather patterns, and soil composition to generate field-level prescriptive planting recommendations.

However, realizing the full potential of these data-driven innovations necessitates overcoming significant hurdles, including data privacy and security concerns, inadequate rural digital infrastructure, and the requisite development of specialized data science skill sets within the agricultural sector. Looking ahead, the continued evolution of big data technologies, coupled with advancements in artificial intelligence, edge computing, and global data interoperability initiatives, promises to further revolutionize agricultural practices. As these technologies mature, it will be imperative to ensure their ethical implementation and equitable access, fostering a sustainable and resilient global food system.

References

Ahmad, U., A. Alvino and S. Marino. 2022. Solar fertigation: A sustainable and smart IoT-based irrigation and fertilization system for efficient water and nutrient management. Agronomy. 12: 1012.

Altieri, M.A., F.R. Funes-Monzote and P. Petersen. 2012. Agroecologically efficient agricultural systems for smallholder farmers: Contributions to food sovereignty. Agron. Sustain. Dev. 32: 1-13.

Anand, R., S.S. Nalla, S. Nasreen, M. Chowdhury, M.B.S., A. Pandey et al. 2024. Fruit counting and analysis using artificial intelligence approaches. pp. 93-114. *In:* S.S. Chouhan, A. Saxena, U. P. Singh and S. Jain [eds.]. Artificial Intelligence Techniques in Smart Agriculture. Springer Nature, Singapore.

Andarge, T. and E. Lichtenberg. 2020. Regulatory compliance under enforcement gaps. J. Regul. Econ. 57: 181-202.

Araus, J.L., S.C. Kefauver, O. Vergara-Díaz, A. Gracia-Romero, F.Z. Rezzouk, J. Segarra et al. 2022. Crop phenotyping in a context of global change: What to measure and how to do it. J. Integr. Plant Biol. 64: 592-618.

Ayaz, M., M. Ammad-Uddin, Z. Sharif, A. Mansour and E.H.M. Aggoune. 2019. Internet-of-Things (IoT)-based smart agriculture: Toward making the fields talk. IEEE Access. 7: 129551-129583.

Bachmann, N., S. Tripathi, M. Brunner and H. Jodlbauer. 2022. The contribution of data-driven technologies in achieving the sustainable development goals. Sustainability. 14: 2497.

Barrientos, A., J. Colorado, J.D. Cerro, A. Martinez, C. Rossi, D. Sanz and J. Valente. 2011. Aerial remote sensing in agriculture: A practical approach to area coverage and path planning for fleets of mini aerial robots. J. Field Robot. 28: 667-689.

Beddington, J. 2010. Food security: Contributions from science to a new and greener revolution. Philos. Trans. R. Soc. B: Biol. Sci. 365: 61-71.

Birch, E.A.N., G.S. Begg and G.R. Squire. 2011. How agro-ecological research helps to address food security issues under new IPM and pesticide reduction policies for global crop production systems. J. Exp. Bot. 62: 3251-3261.

Bu, F. and X. Wang. 2019. A smart agriculture IoT system based on deep reinforcement learning. Future Gener. Comput. Syst. 99: 500-507.

Carvalho, F.P. 2006. Agriculture, pesticides, food security and food safety. Environ. Sci. Policy. 9: 685-692.

Chen, H.Y., K. Sharma, C. Sharma and S. Sharma. 2023. Integrating explainable artificial intelligence and blockchain to smart agriculture: Research prospects for decision making and improved security. Smart Agric. Technol. 6: 100350.

Chowdhury, M., A. Kushwah, A.N. Satpute, S.K. Singh and A.K. Patil. 2023a. A comprehensive review on potential application of nanomaterials in the field of agricultural engineering. J. Biosyst. Eng. 48(4): 457-477.

Chowdhury, M., E.V. Thomas, A. Jha, A. Kushwah, R. Kurmi, T.K. Khura et al. 2023b. An automatic pressure control system for precise spray pattern analysis on spray patternator Comput. Electron. Agric. 214: 108287.

Chowdhury, M., R. Anand, T. Dhar, R. Kurmi, R.K. Sahni and A. Kushwah. 2024. Digital insights into plant health: Exploring vegetation indices through computer vision. pp. 7-30. *In:* S.S. Chouhan, U.P. Singh and S. Jain [eds.]. Applications of Computer Vision and Drone Technology in Agriculture 4.0. Springer Nature, Singapore.

Chowdhury, M., T.K. Khura, P.K. Upadhyay, R.A. Parray, H.L. Kushwaha, C. Singh et al. 2024b. Assessing vegetation indices and productivity across nitrogen gradients: A comparative study under transplanted and direct-seeded rice. Front. Sustain. Food Syst. 8: 1351414.

Chowdhury, M., T.K. Khura, R.A. Parray, P.K. Upadhyay, H.L. Kushwaha, A. Lama and A. Kushwah. 2025. Multi-vegetation indices based handheld device for precise nitrogen assessment and prescription in direct seeded rice. Comput. Electron. Agric. 229: 109886.

Coble, K.H., A.K. Mishra, S. Ferrell and T. Griffin. 2018. Big data in agriculture: A challenge for the future. Appl. Econ. Perspect. Policy 40: 79-96.

Debauche, O., S. Mahmoudi, P. Manneback and F. Lebeau. 2022. Cloud and distributed architectures for data management in agriculture 4.0: Review and future trends. J. King Saud Univ. Comput. Inf. Sci. 34: 7494-7514.

Delgado, J.A., N.M. Short Jr, D.P. Roberts and B. Vandenberg. 2019. Big data analysis for sustainable agriculture on a geospatial cloud framework. Front. Sustain. Food Syst. 3: 54.

Demestichas, K., N. Peppes, T. Alexakis and E. Adamopoulou. 2020. Blockchain in agriculture traceability systems: A review. Appl. Sci. 10: 4113.

Domingues, T., T. Brandão and J.C. Ferreira. 2022. Machine learning for detection and prediction of crop diseases and pests: A comprehensive survey. Agriculture. 12: 1350.

Domínguez, A.G., N. Roig-Tierno, N. Chaparro-Banegas and J.M. García-Álvarez-Coque. 2024. Natural language processing of social network data for the evaluation of agricultural and rural policies. J. Rural Stud. 109: 103341.

Evans, K.J., A. Terhorst and B.H. Kang. 2017. From data to decisions: Helping crop producers build their actionable knowledge. Crit. Rev. Plant Sci. 36: 71-88.

Fatima, S.A., S.F. Nasim and S. Ahmed. 2023. Enhancing agricultural operations: Big data analytics using distributed and parallel computing. Int. J. Emerg. Eng. Technol. 2: 1-7.

Fracarolli, J.A., F.F.A. Pavarin, W. Castro and J. Blasco. 2020. Computer vision applied to food and agricultural products. Rev. Cienc. Agron. 51: e20207749.

Gawande, V., D.R.K. Saikanth, B.S. Sumithra, S.A. Aravind, G.N. Swamy, M. Chowdhury and B.V. Singh. 2023. Potential of precision farming technologies for eco-friendly agriculture. Int. J. Plant Soil Sci. 35(19): 101-112.

Ghaley, B.B., L. Vesterdal and J.R. Porter. 2014. Quantification and valuation of ecosystem services in diverse production systems for informed decision-making. Environ. Sci. Policy. 39: 139-149.

Gilson, B.J. 2019. Examining the Relationship between Decision Lead Time and the Predictive Performance of Proactive Complex Event Processing Rules Developed with Weather Data Decision Trees. Doctoral dissertation, Colorado Technical University.

Gohil, J., J. Patel, J. Chopra, K. Chhaya, J. Taravia and M. Shah. 2021. Advent of Big Data technology in environment and water management sector. Environ. Sci. Pollut. Res. 28: 64084-64102.

Guo, H., R.W. Jolly and J. Zhu. 2005. Contract farming in China: Supply chain or ball and chain. Proc. 15th Annu. World Food & Agribus. Symp. IAMA: 6891-6892.

Himesh, S., E.P. Rao, K.C. Gouda, K.V. Ramesh, V. Rakesh, G.N. Mohapatra et al. 2018. Digital revolution and Big Data: A new revolution in agriculture. CABI Rev. 2018: 1-7.

Howitt, R., L. Karp and G. Rausser. 2022. Remote sensing technologies: Implications for agricultural and resource economics. pp. 183-217. *In:* H. de Gorter, J. McCluskey, J. Swinnen, D. Zilberman [eds]. Modern Agricultural and Resource Economics and Policy. Natural Resource Management and Policy, vol 55. Springer Int. Publ., Cham.

Inoue, Y. 2020. Satellite- and drone-based remote sensing of crops and soils for smart farming – A review. Soil Sci. Plant Nutr. 66: 798-810.

Jaber, M.M., M.H. Ali, S.K. Abd, M.M. Jassim, A. Alkhayyat, H.W. Aziz and A.R. Alkhuwaylidee. 2022. Predicting climate factors based on big data analytics based agricultural disaster management. Phys. Chem. Earth A/B/C. 128: 103243.

Jahanbin, P., S. Wingreen and R. Sharma. 2019. A blockchain traceability information system for trust improvement in agricultural supply chain. *In:* Proceedings of the 27th European Conference on Information Systems (ECIS). Stockholm & Uppsala, Sweden, June 8-14, 2019.

Kang, Y., S. Khan and X. Ma. 2009. Climate change impacts on crop yield, crop water productivity and food security – A review. Prog. Nat. Sci. 19: 1665-1674.

Khaki, S. 2022. Novel deep learning-based methods for high-throughput image-based plant phenotyping and large scale crop yield prediction. Doctoral dissertation, Iowa State Univ.

Khan, I. and S.A. Shorna. 2023. Cloud-based IoT solutions for enhanced agricultural sustainability and efficiency. AI IoT Fourth Ind. Revolut. Rev. 13: 18-26.

Kim, J., S. Kim, C. Ju and H.I. Son. 2019. Unmanned aerial vehicles in agriculture: A review of perspective of platform, control, and applications. IEEE Access. 7: 105100-105115.

Koteish, K., H. Harb, M. Dbouk, C. Zaki and C. Abou Jaoude. 2022. AGRO: A smart sensing and decision-making mechanism for real-time agriculture monitoring. J. King Saud Univ. Comput. Inf. Sci. 34: 7059-7069.

Kurmi, R., S.D. Lande, J. Kurmi, M. Chowdhury, C. Singh and P. Kumar. 2023. Comparative study on carbon footprint assessment of rice-wheat production system. Int. J. Environ. Clim. Chang. 13(11): 390-398.

Kushwah, A., A. Chouriya, V.K. Tewari, C. Gupta, M. Chowdhury, P. Shrivastava and P. Bhagat. 2024a. A novel embedded system for tractor implement performance mapping. Cogent Eng. 11(1): 2311093.

Lee, M. and J. Gambiza. 2022. The adoption of conservation agriculture by smallholder farmers in southern Africa: A scoping review of barriers and enablers. J. Rural Stud. 92: 214-225.

Liang, C. and T. Shah. 2023. IoT in agriculture: The future of precision monitoring and data-driven farming. Eigenpub Rev. Sci. Technol. 7: 85-104.

Liu, S., L. Guo, H. Webb, X. Ya and X. Chang. 2019. Internet of Things monitoring system of modern eco-agriculture based on cloud computing. IEEE Access. 7: 37050-37058.

Lu, B., P.D. Dao, J. Liu, Y. He and J. Shang. 2020. Recent advances of hyperspectral imaging technology and applications in agriculture. Remote Sens. 12: 2659.

Malkani, P., I. Mani, P.K. Sahoo, R.A. Parray, S.S. Swain, K.R. Asha et al. 2024. Development and evaluation of a sensor enabled smart herbicide applicator for inter row crops. Smart Agric. Technol. 9: 100649.

Mallarino, A.P. and D.J. Wittry. 2004. Efficacy of grid and zone soil sampling approaches for site-specific assessment of phosphorus, potassium, pH, and organic matter. Precis. Agric. 5: 131-144.

Maru, A., D. Berne, J.D. Beer, P.G. Ballantyne, V. Pesce, S. Kalyesubula et al. 2018. Digital and data-driven agriculture: Harnessing the power of data for smallholders. Glob. Forum Agric. Res. Innov. 7: 525.

Matese, A. and S.F. Di Gennaro. 2018. Practical applications of a multisensor UAV platform based on multispectral, thermal and RGB high resolution images in precision viticulture. Agriculture. 8: 116.

Mayer-Schönberger, V. and K. Cukier. 2013. Big Data: A Revolution that will Transform How We Live, Work, and Think. Houghton Mifflin Harcourt.

Misra, N.N., Y. Dixit, A. Al-Mallahi, M.S. Bhullar, R. Upadhyay and A. Martynenko. 2020. IoT, big data, and artificial intelligence in agriculture and food industry. IEEE Internet Things J. 9: 6305-6324.

Narvaez, F.Y., G. Reina, M. Torres-Torriti, G. Kantor and F.A. Cheein. 2017. A survey of ranging and imaging techniques for precision agriculture phenotyping. IEEE/ASME Trans. Mechatron. 22: 2428-2439.

Nikander, J., O. Manninen and M. Laajalahti. 2020. Requirements for cybersecurity in agricultural communication networks. Comput. Electron. Agric. 179: 105776.

Patil Shirish, S. and S.A. Bhalerao. 2013. Precision farming: The most scientific and modern approach to sustainable agriculture. Int. Res. J. Sci. Eng. 1: 21-30.

Paul, K., S.S. Chatterjee, P. Pai, A. Varshney, S. Juikar, V. Prasad et al. 2022. Viable smart sensors and their application in data driven agriculture. Comput. Electron. Agric. 198: 107096.

Pretty, J. 2008. Agricultural sustainability: Concepts, principles and evidence. Philos. Trans. R. Soc. B: Biol. Sci. 363: 447-465.

Pulighe, G. and F. Lupia. 2016. Mapping spatial patterns of urban agriculture in Rome (Italy) using Google Earth and web-mapping services. Land Use Policy. 59: 49-58.

Ranjan, A.K. and B.R. Parida. 2021. Predicting paddy yield at spatial scale using optical and Synthetic Aperture Radar (SAR) based satellite data in conjunction with field-based Crop Cutting Experiment (CCE) data. Int. J. Remote Sens. 42: 2046-2071.

Ratner, B. 2017. Statistical and Machine-learning Data Mining: Techniques for Better Predictive Modeling and Analysis of Big Data. Chapman and Hall/CRC.

Robert, M., J. Dury, A. Thomas, O. Therond, M. Sekhar, S. Badiger et al. 2016. CMFDM: A methodology to guide the design of a conceptual model of farmers' decision-making processes. Agric. Syst. 148: 86-94.

Rounsevell, M.D.A., J.E. Annetts, E. Audsley, T. Mayr and I. Reginster. 2003. Modelling the spatial distribution of agricultural land use at the regional scale. Agric. Ecosyst. Environ. 95: 465-479.

Sagiroglu, S. and D. Sinanc. 2013. Big data: A review. Proc. 2013 Int. Conf. Collaboration Technol. Syst. CTS: 42-47. IEEE.

Saidu, A., A.M. Clarkson, S.H. Adamu, M. Mohammed and I. Jibo. 2017. Application of ICT in agriculture: Opportunities and challenges in developing countries. Int. J. Comput. Sci. Math. Theory 3: 8-18.

Saiz-Rubio, V. and F. Rovira-Más. 2020. From smart farming towards agriculture 5.0: A review on crop data management. Agronomy. 10: 207.

Sethy, R. and M. Panda. 2015. Big data analysis using Hadoop: A survey. Int. J. Adv. Res. Comput. Sci. Softw. Eng. 5.

Shaikh, T.A., T. Rasool and F.R. Lone. 2022. Towards leveraging the role of machine learning and artificial intelligence in precision agriculture and smart farming. Comput. Electron. Agric. 198: 107119.

Shamshiri, R., C. Weltzien, I.A. Hameed, I. Yule, T. Grift, S.K. Balasundram et al. 2018. Research and development in agricultural robotics: A perspective of digital farming. Int. J. Agric. Biol. Eng. 11: 1-4.

Sharma, R., S.S. Kamble, A. Gunasekaran, V. Kumar and A. Kumar. 2020. A systematic literature review on machine learning applications for sustainable agriculture supply chain performance. Comput. Oper. Res. 119: 104926.

Shiferaw, B., M. Smale, H.J. Braun, E. Duveiller, M. Reynolds and G. Muricho. 2013. Crops that feed the world 10. Past successes and future challenges to the role played by wheat in global food security. Food Secur. 5: 291-317.

Sunil Kumar, T., H.M. Virdia, K.G. Patel, S. Ragi, M. Chowdhury, P. Kumar et al. 2024. Effect of summer legume residue incorporation and fertilizer regimes on rice growth, yield, and nutrient uptake. Front. Sustain. Food Syst. 8: 1467201.

Vassakis, K., E. Petrakis and I. Kopanakis. 2018. Big data analytics: Applications, prospects and challenges. Mobile Big Data: A Roadmap from Models to Technologies. 3-20.

Vitharana, U.W., M. Van Meirvenne, D. Simpson, L. Cockx and J. De Baerdemaeker. 2008. Key soil and topographic properties to delineate potential management classes for precision agriculture in the European loess area. Geoderma 143: 206-215.

Weersink, A., E. Fraser, D. Pannell, E. Duncan and S. Rotz. 2018. Opportunities and challenges for big data in agricultural and environmental analysis. Annu. Rev. Resour. Econ. 10: 19-37.

Wilson, B.R. and T. Mutersbaugh. 2015. Fair trade certification, performance and practice. pp. 281-297. *In:* Handbook of Research on Fair Trade. Edward Elgar Publ.

Wolfert, S., L. Ge, C. Verdouw and M.J. Bogaardt. 2017. Big data in smart farming – A review. Agric. Syst. 153: 69-80.

Wu, Y., X. Xi, X. Tang, D. Luo, B. Gu, S.K. Lam et al. 2018. Policy distortions, farm size, and the overuse of agricultural chemicals in China. Proc. Natl. Acad. Sci. USA. 115: 7010-7015.

Xuegeng, C.H.E.N., W. Haojun, Z. Weirong, P.A.N. Fochu and Z.H.A.O. Yan. 2020. Advances and progress of agricultural machinery and sensing technology fusion. Smart Agric. 2: 1.

Yaqoob, I., I.A.T. Hashem, A. Gani, S. Mokhtar, E. Ahmed, N.B. Anuar and A.V. Vasilakos. 2016. Big data: From beginning to future. Int. J. Inf. Manag. 36: 1231-1247.

Yeom, J., J. Jung, A. Chang, A. Ashapure, M. Maeda, A. Maeda and J. Landivar. 2019. Comparison of vegetation indices derived from UAV data for differentiation of tillage effects in agriculture. Remote Sens. 11: 1548.

Zhai, Z., J.F. Martínez, V. Beltran and N.L. Martínez. 2020. Decision support systems for agriculture 4.0: Survey and challenges. Comput. Electron. Agric. 170: 105256.

Zhao, D., T. Shen, Y. Chen and W. Jia. 2015. Fast tracking and recognition of overlapping fruit for apple harvesting robot. Trans. Chin. Soc. Agric. Eng. 31: 22-28.

Zimmermann, U., R. Bitter, P.E.R. Marchiori, S. Rüger, W. Ehrenberger, V.L. Sukhorukov et al. 2013. A non-invasive plant-based probe for continuous monitoring of water stress in real time: A new tool for irrigation scheduling and deeper insight into drought and salinity stress physiology. Theor. Exp. Plant Physiol. 25: 2-11.

CHAPTER

3

IoT Sensors and Their Applications in Precision Agriculture

Saikat Karmakar[1], Pooja Roy[2], Subrata Mandal[3*] and Susanta Das[4]

[1] Department of Agricultural and Food Engineering, Indian Institute of Technology Kharagpur, Kharagpur - 721302, India

[2] Department of Civil Engineering, Indian Institute of Technology Kanpur, Kanpur - 208016, India

[3] Division of Agricultural Engineering, Indian Agricultural Research Institute, New Delhi - 110012, India

[4] Department of Agricultural and Biological Engineering, University of Florida, Florida 32603, USA

Introduction

The origins of agriculture date back at least 22,000 years, when early humans began collecting wild grains as a food source. Archaeological evidence suggests that various crops were being cultivated as early as 9500 BC in the Levant (Raj et al. 2020; Harwood 2022). Over the millennia, continuous innovations have boosted agricultural productivity while reducing manual labor. However, the growing global population will always demand more food. It is projected that by 2050, the world's population will increase by 33% to 9.7 billion (Aryal et al. 2022), which will require global food production to rise by at least 70% to meet demand (Giller et al. 2021). Despite this need, only a small portion of the Earth's surface is suitable for farming due to limitations such as climate, topography, soil quality, and technology. Political and economic factors, including land ownership patterns, environmental laws, and population density, also influence agricultural land use (Burra et al. 2021). In fact, the amount of land used for agriculture has been declining. In 2013, around 18.6 million square miles, or 37.73% of the world's land, was used for food production, compared to 19.5 million square miles (39.47%) in 1991 (World Bank 2016).

To meet the challenge of feeding more people with less land, precision agriculture (PA) offers a promising solution. PA uses advanced techniques and technologies to

*Corresponding author: subratamandal459@gmail.com

DOI: 10.1201/9781003536932-3

improve agricultural efficiency, increase food production, reduce environmental impact, and lower costs (Gawande et al. 2023). This site-specific approach collects precise data for each field location and adjusts farming practices accordingly. Unlike traditional methods, which rely on set schedules, precision agriculture uses real-time data and predictive analytics to optimize decisions about planting, harvesting, soil quality, weather, and resource availability (Atalla et al. 2023). PA is a farm management approach that focuses on observing, measuring, and responding to the variability within agricultural fields. In PA, large fields are divided into smaller sections, each of which is managed precisely and independently to minimize wasteful use of resources like water, seeds, and nutrients, while maximizing crop yield and farm efficiency (Anand et al. 2021; Abu et al. 2022). The Internet of Things (IoT) is an emerging technology that has the potential to transform the future of agriculture. By integrating IoT into agricultural practices, traditional farming methods can be revolutionized, laying the foundation for PA.

IoT sensors can be categorized into various types based on their functions, including environmental sensors, soil moisture sensors, crop health monitors, and pest detection systems (Satpute et al. 2024). Environmental sensors measure parameters such as temperature, humidity, and light intensity, providing critical data that influences crop growth. Soil moisture sensors help determine when and how much to irrigate, conserving water and ensuring that plants receive adequate hydration. Crop health monitors, often utilizing multispectral imaging, assess plant vigor and detect nutrient deficiencies or diseases at an early stage. Pest detection systems alert farmers to pest presence, allowing for timely intervention to prevent crop loss (Chowdhury et al. 2024a).

The application of PA involves three key stages: (i) real-time data collection, (ii) data analysis and decision-making, and (iii) precise, targeted treatments. Recent advances in IoT technology have greatly enhanced all three stages (Farooq et al. 2019; Doshi and Varghese 2022). Firstly, IoT establishes the necessary network infrastructure that allows a wide range of devices, from microsensors to large agricultural machinery, to connect with each other and the Internet. This facilitates easy data collection, sharing, and transmission (Mishra and Sharma 2023). Secondly, IoT leverages internet-based solutions to enable more efficient data processing and intelligent decision-making. Services, ranging from satellite imagery to livestock health monitoring, are often cloud-based and available globally from both large corporations and small startups. This setup allows for automated, real-time, 24/7 data analysis without human intervention (Monteiro et al. 2021; Kim et al. 2020). Finally, decisions and actions derived from cloud-based analysis are transmitted back to the farm, where IoT-enabled devices, machines, and vehicles carry out precise agricultural tasks such as planting and livestock management (Kour and Arora 2020; Ayaz et al. 2019). This integration allows users to access, visualize, and interpret data more conveniently, reshaping the agricultural industry and driving substantial benefits. By incorporating IoT, PA can address critical challenges such as drought response, crop yield optimization, land management, and pest control.

This chapter aims to explore the transformative role of IoT sensors in modern farming. As precision agriculture advances, IoT technologies are increasingly vital for enhancing crop productivity, resource efficiency, and environmental sustainability. This chapter focuses on how IoT sensors collect real-time data, monitor crop health, and optimize resource use, thereby supporting informed decision-making in agriculture. The primary objective is to detail the various types of IoT sensors used in agriculture, such as soil moisture, climate, and nutrient sensors, and to demonstrate their specific applications in monitoring critical agricultural parameters. The chapter will also address challenges like data security, connectivity in remote areas, and the need for cost-effective solutions. Additionally, the chapter highlights the benefits of IoT sensor integration for farmers, including increased crop yields, reduced environmental impact, and enhanced decision-making. It will discuss how IoT sensors enable data-driven decisions that promote sustainable farming practices. Overall, this chapter serves as a concise resource for understanding how IoT sensors are shaping the future of precision agriculture, providing essential insights into the adoption and implementation of these technologies in farming.

IoT Sensors, Types and Components

The integration of IoT sensors and components in precision agriculture has revolutionized the way of data collection, analysis, and utilization. Sensors detecting temperature, humidity, soil moisture, light intensity, and real-time environmental data are incorporated with the microcontroller of IoT system. IoT sensor networks function based on these three key objectives (i) collecting the essential information of physical environment, (ii) capturing internal system signals, and (iii) obtaining valuable insights from sensor data (Deng et al. 2019; Sanyal and Zhang 2018). Wireless Sensor Networks (WSNs) establish communication within the IoT enabled applications which helps in deploying the sensors randomly and form ad hoc networks without major infrastructure upgradation (Krishnamurthi et al. 2020). There are low cost solutions available in the market to interlink the sensors with low power devices such as Zigbee, LoRaWAN (Long Range Wide Area Network), Bluetooth Low Energy (BLE) Sensors, Wi-Fi, RFID (Radio Frequency Identification), Near Frequency Communication (NFC), etc. However, chances of having missing data, outliers and noise in the acquired dataset arise with the implication of wireless connections.

In recent times, Artificial Intelligence (AI) is being considered to work with the sensors and turning them into 'smart sensors' (Ullo and Sinha 2021). AI enabled sensors are efficiently performing in weather forecasting (Vasisht et al. 2017; Shafi et al. 2021; Di Napoli et al. 2020), satellite imaging (Shafi et al. 2021; Sishodia et al. 2020), remote sensing applications (Addabbo et al. 2016; Kayad et al. 2020), hazard monitoring (Di Napoli et al. 2020) and so on. Some IoT sensors commonly used for precision agriculture are listed in Table 1.

Data Management and Analysis

In precision agriculture, IoT sensors continuously generate a large amount of raw data of soil moisture, temperature, humidity, crop health, and other environmental factors in real time. Data management as a whole includes the process of collection and analysis of raw data, and maintaining data safety. Raw data may be in structured (like temperature, humidity, etc.) or unstructured (like images) format

Table 1: Different types of IoT sensors used in precision agriculture

Sensor	Working Principle	Application	References
Soil Moisture Sensors	Measure the dielectric constant of the soil, which changes with moisture content.	Monitoring soil water levels for optimal irrigation.	Deng et al. 2019
Temperature Sensors	Use thermocouples or resistance temperature detectors (RTDs) to measure temperature.	Monitoring ambient Air and soil temperature to regulate crop growth conditions.	Shafi et al. 2021
Light Sensors	Measure the amount of light using photodiodes or phototransistors.	Ensuring crops receive optimal sunlight for photosynthesis.	Kayad et al. 2020
pH Sensors	Measure hydrogen ion concentration using electrochemical methods.	Monitoring soil acidity to optimize nutrient absorption by plants.	Di Napoli et al. 2020
Electrical Conductivity (EC) Sensors	Measure the electrical conductivity of the soil, which indicates the salt concentration.	Managing soil salinity levels and determining fertilizer needs.	Di Napoli et al. 2020
Humidity Sensors	Use capacitive or resistive elements to detect changes in air moisture content.	Controlling greenhouse environments and protecting crops from diseases related to humidity.	Krishnamurthi et al. 2020
CO_2 Sensors	Use non-dispersive infrared (NDIR) technology to measure carbon dioxide levels.	Monitoring greenhouse gas levels and optimizing crop growth in controlled environments.	Anand et al. 2021
Ultrasonic Sensors	Emit ultrasonic waves and measure the reflection time to detect distances.	Measuring plant height and growth progress.	Sishodia et al. 2020
Drones with Multispectral Sensors	Capture images using different light wavelengths.	Mapping large fields for crop health assessment, disease detection, and yield prediction.	Mishra and Sharma 2023

which needs to be cleaned and formatted before analyzing (Li et al. 2024). Once prepared, advanced methods like statistical analysis, machine learning, and data mining are applied to extract meaningful insights from the data (Parte et al. 2023). In IoT, real-time analysis is crucial for precise prediction of crop yields, optimized utilization of resources, and real-time feedback systems. This chapter elaborates on the aspects of data management and analysis of IoT systems, useful for precision agriculture.

IoT Data Collection and Storage

Data Acquisition (DAQ)

Data Acquisition refers to the measurement and collection of various parameters value through sensors like RFID, ZigBee sensors, GPS devices, etc. A DAQ system consists of sensors, microcontrollers, analog to digital converter, and application software. The system flowchart is presented in Fig. 1.

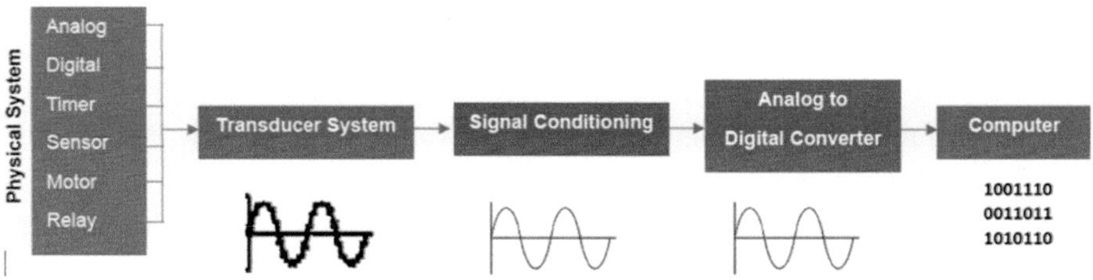

Fig. 1: Flowchart of Data Acquisition system in IoT.

Sensors observe the physical phenomena and transducer converts that into electrical signals. At this stage the signal contains noise, which is removed in the next step, signal conditioning. The noise is filtered and corrections are done to the signal. Analog sensors (thermocouple, photodiode, etc.) are also used with the IoT system which send analog output to the system. Analog to digital converter converts the analog signal to digital format for further processing. A DAQ system must ensure data acquisition at the desired speed and time (Tran et al. 2023).

Hardwares of DAQ: Hardwares are the components that are either available as modules or connect to the motherboard slots. Some useful hardware for the DAQ are USB DAQ Modules, PXI (PCI eXtensions for Instrumentation), GPIB (General Purpose Interface Bus), CAMAC (Computer Automated Measurement and Control), etc.

Software for DAQ: Software connects the DAQ hardwares with the computer. Some softwares used in DAQ are: LabVIEW, DAQmx, NI MAX (Measurement and Automation Explorer), DAQ Factory, etc.

Cloud Storage Solutions

A large amount of data is collected in the DAQ process in IoT which needs a large storage space. Cloud storage in IoT uses remote servers to store, manage, and process these vast amounts of data (Fig. 2). Sharing and isolating these data securely in cloud platform are the primary challenges in IoT data storage. Cloud storage provides scalability, remote monitoring, and real-time data access, making it as a vital element in IoT ecosystems.

Working procedure of cloud storage in IoT is as follows:

Data Transmission: Data generated by the sensors are continuously transmitted to

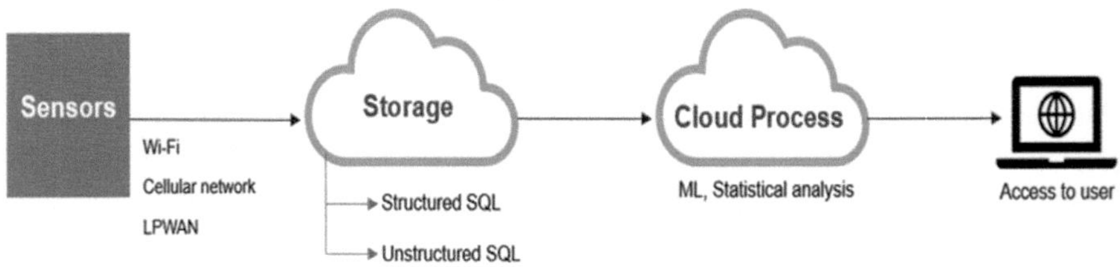

Fig. 2: Cloud storage system of IoT.

the cloud via network protocols like cellular networks, Wi-Fi or Low Power Wide Area Networks (LPWANs).

Database Type: The stored data can be structured or unstructured. Structured data are organised in relational databases (e.g., SQL) while unstructured data are stored in non-relational databases (e.g., NoSQL).

Data Processing: The cloud environment also provides the infrastructure to process raw data using machine learning or statistical analysis to extract useful information.

Data Access: After processing, the data is made accessible to users through various interfaces like dashboards, APIs, mobile apps etc., with real-time information sharing and decision-making.

Cloud Storage Module

In IoT applications, sensor-generated data demands different levels of service and security. Data storage modules are used in this case for managing and keeping this vast amount of data isolated for various applications (Cai et al. 2017). Below are some primary data storage modules used in IoT application.

Object Storage: Object storage module is typically used for storing unstructured data like images, videos, and sensor logs.

Examples: HDFS, Gluster FS, ZFS, Amazon S3, Microsoft Azure Blob Storage etc.

File Storage: File storage organizes and stores structured files data in a hierarchical folder structure, making it suitable for applications that rely on file-based storage. Used for storing structured files and documents that require file-sharing across systems or devices.

Examples: NFS, CIFS, Amazon Elastic File System (EFS), Google Filestore, Azure Files, etc.

Resource Description Framework (RDF): RDF is used for resource management of semi-structured data. RDFS schema (RDFS) is an extension of RDF, provides mechanisms to describe the relationships between resources in a semantic web (Dan 2004).

Relational Database Management System (RDBMS): RDBMS stores data in a structured format, in tables. Massive amount of data is organized through this module, according to a predefined schema. Different keys are used to maintain relationships between the data tables (Yaish et al. 2014). RDBMS is widely used for managing transactional data and Ultrawrap approach is used to encode logical representation of each RDBMS (Sequeda and Miranker 2013).

Graph DBMS: In this module data is stored and represented in a graphical format, while nodes and edges are the considered core components. Apart from traditional RDBMS, Graph DBMS creates an interconnected network of data nodes and edges which in turn helps it efficient for managing complex relationships between entities. Bazan et al. (2011) introduced a high-performance Graph DBMS consisting of large-scale nodes and edges.

NoSQL (Not Only SQL) DBMS: NoSQL DBMS is a database management system that handles unstructured data. The NoSQL DB is flexible in schema structure and thus it become suitable for distributed systems. It provides fault tolerance and high availability across distributed nodes. NoSQL is capable of handling diverse data types and real time applications. Curé et al. (2013) approached to integrate an ontology-based data access in NoSQL stores.

Accommodation of the diverse data types generated by IoT sensors is a very crucial step in managing precision agriculture. The before mentioned database modules help to integrate various data formats and ensure efficient data storage. Some key features of cloud storage in IoT are:

Scalability: Cloud storage is capable of scaling large volumes of data generated by IoT devices for efficient utilisation. With the expansion of IoT system, the cloud infrastructure needs to be expanded accordingly (Maftei et al. 2023).

Data Accessibility: Cloud storage data is accessible from anywhere, which enables users to remotely control and monitor the IoT system (Huang et al. 2023).

Real-time Processing: The cloud system can be developed for real-time data processing and analytics (Lee et al. 2024). It makes it easier and efficient in managing ideal condition for crops in precision agriculture.

Cost Efficiency: Cloud storage eliminates the need of hardware setup for storing large amount of data and thereby reduces upfront costs (Pincheira et al. 2021). However, cost of data security should also be considered in this case.

Interoperability: Cloud storage provides APIs (Application Programming Interface) which ease out the communication for IoT devices and applications. It also communicates, store, and retrieve data regardless of the underlying technology (Malathi et al. 2024).

Data Processing and Analytics

In IoT, cloud-based data processing modules such as Apache Flink, MapReduce, HDFS, etc. (Blanas et al. 2010) are used for extracting information from the collected mass data. MapReduce is commonly employed for query processing in data analysis tasks on cloud platforms such as parallelization, scalability, load balancing, and fault-tolerance (Cai et al. 2016). In this section, elaboration about various data processing and analysis technique will be discussed.

Denoising

The sensor data contains noise or irrelevant information that needs to be removed. Denoising techniques filter out these disturbances and ensures accurate analysis and better insights. Noise is an uncorrelated component of a signal that causes distortions to the original signal vectors (Krishnamurthi et al. 2020).

Parallel Processing

Parallel Processing refers to the method of simultaneous execution of multiple tasks or processes in handling large volumes of data. In IoT system, the parallel processing is useful in dealing with the semi-structural data. For an example, Spatial Partitioning Algorithm (SPA) is used for distributed processing of big Resource Description Framework (RDF) data (Lee et al. 2013). Parallel processing helps IoT systems to improve computational efficiency while real-time data collection and reduces latency.

Handling Missing Data

In IoT, data loss can occur due to sensor malfunction, network failure or any type of transmission error. This creates voids in the dataset that need to be filled prior to operations. Missing data is available in three basic conditions such as: Missing Completely At Random (MCAR), Missing At Random (MAR), and Not Missing At Random (NMAR) (Krishnamurthi et al. 2020). Various studies suggested different models for handling missing values such as: Gaussian mixture models (GMMs) (Yan et al. 2015), extended spatial and temporal correlated proximate (ESTCP) (Mary and Arockiam, 2017), etc.

Data Outlier Detection

Outliers are the points that deviate from the trend of rest of the dataset. Sensor malfunction or harsh environments are majorly responsible for the generation of outliers in the dataset. Detection of outliers is important because they can affect statistical analyses and data insights while processing. A distributed or centralized approach can be adopted to detect outliers (Samara et al. 2022). Distributed system detects outliers at sensor nodes while centralized approach detects at parent node or base station. The accuracy of distributed outlier detection is lower compared to the centralized approach due to the availability of sufficient sensor data (Zhang et al. 2010).

Complex Operations

IoT data processing system handles Complex Event Processing (CEP) which refers to the collection, correlation and analysis of huge amount of simple events to identify relevant situations in a specific domain (Luckham 2012). Complex data processing includes data ingestion, data transformation, real-time processing, parallel and distributed processing, edge computing and Complex Event Processing (CEP). CEP is software which performs data stream tracking to extract meaningful insights in real-time decision-making (Rahmani et al. 2020). However parallel processing technique is also adopted in studies for saving time and improve detection (Zhu 2021).

Machine learning and AI in data analysis

Machine Learning (ML) and Artificial Intelligence (AI) are revolutionizing farming sector by enabling data-driven decision-making, efficient operations and optimizing use of resources (Jha et al. 2023; Anand et al. 2024). Merenda et al. (2020) mentioned the efficient power utilization and extension of the average battery life due to the implication of ML algorithms. The incorporation of ML algorithms such as PCA (Shaikh et al. 2019), SVM (Shaikh et al. 2020), RNN (Shaikh et al. 2021) has been assessed in various studies. ML algorithms revolutionized the plant disease detection system based on image data (Khirade and Patil 2015; Zhang et al. 2019; Sreedhar and Kumar 2020). AI algorithms such as SVM (Guestrin 2006), deep learning (DL) (Lecun 2015, Schmidhuber 2015), Neural Networks (NNs) (Jordan and Bishop 2004; Neapolitan et al. 2018). NNs are the power saving solutions that can be used with microcontrollers to save operational cost (Merenda et al. 2020).

Data Security and Privacy

With the incorporation of IoT in smart farming sector, data security and privacy have become critical concerns. Security breach is possible in the interconnected devices and cloud platforms which may lead to data leakage or system malfunction. Cyber-attacks in a coordinated manner has the ability to destroy entire farmland or making the crops hazardous by over application of fertilizers and pesticides (Gupta et al. 2020). These phenomena can also be referred as agro-terrorism which is capable of destabilize the economy of an agrarian nation (Cupp et al. 2004). Cha et al. (2004) classified the data privacy maintaining techniques into seven categories such as; personal data protection, enforcement, pseudonymization, holistic privacy preservation, control over data, anonymous authorization, and partial data disclosure. Ferrag et al. (2020) classified IoT security surveys into four criteria which includes threat model, security and privacy countermeasures, blockchain-based solution and target IoT application. Blockchain-based solutions are being considered in recent studies (Dorri et al. 2017; Khan and Salah 2018; Panarello et al. 2018).

Applications

The Internet of Things (IoT) has become a game-changer in precision agriculture, allowing farmers to boost productivity, minimize waste, and enhance overall agricultural efficiency. By incorporating IoT technologies such as sensors, drones, and smart equipment, vast data can be gathered and analyzed, supporting better decision-making and the effective use of resources. This smart farming approach addresses many challenges faced by traditional agriculture, including land suitability, monitoring drought conditions, optimizing irrigation, controlling pests, and maximizing crop yields.

Soil and Crop Monitoring

One of the most significant applications of IoT in precision agriculture is soil and crop monitoring. Sensors can be deployed across fields to measure various parameters, including soil moisture, temperature, pH levels, and nutrient content. This real-time data allows farmers to make timely irrigation and fertilization decisions, ensuring crops receive the appropriate amount of water and nutrients. For example, soil moisture sensors can signal when irrigation is necessary, reducing water wastage and promoting healthy crop growth.

Various factors such as soil type, nutrients, irrigation, and pests influence rice yield and quality (Chowdhury et al. 2024b; Sunil Kumar et al. 2024). An IoT-based mobile application helps manage crops by providing real-time data on soil characteristics and nutrient levels. This system includes electrical conductivity (EC) and temperature sensors connected to a T-Beam microcontroller with IoT capabilities. The estimated EC value near the calibration solution is 12.88 mS/cm, while 150 mS/cm is less than 2% of the calibration solution's value. EC values, which are linearly correlated with temperature and depth, were recorded as 1.04 and 3.86 mS/cm with and without fertilizer at a 5 cm depth, and 0.656 and 420 mS/cm at a 10 cm depth (Othman et al. 2021). Plant monitoring through an IoT-based ADCON station, combined with sensors and mobile devices, allows farmers to track soil and environmental conditions like leaf wetness, temperature, and humidity, improving grape productivity and crop quality from seeding to harvest. The data transmission system further highlights soil-plant-atmosphere interactions crucial for optimizing production (Marcu et al. 2019). Additionally, in greenhouse-grown bell peppers, data from soil moisture, CO_2, light, and temperature sensors were analyzed, and day and night CO_2 levels were compared. Greenhouse ventilation (opening and closing windows and doors) was controlled based on soil moisture levels (Pallavi et al. 2017).

IoT based Smart Irrigation

Precision irrigation systems powered by IoT technology enable farmers to optimize water usage. By utilizing soil moisture sensors and weather data, these systems can automate irrigation schedules and adjust water application based on real-time conditions. This targeted approach not only conserves water but also enhances crop yields. For example, smart irrigation systems can be programmed to water only specific areas of a field that require it, preventing over-irrigation and reducing costs.

According to the UN Convention to Combat Desertification (UNCCD), by 2030, 168 countries will face severe desertification, and nearly half of the global population will live in regions with critical water shortages (Saiz-Rubio and Rovira-Más 2020). Given the growing demand for water in agriculture and other sectors, it is essential to direct water resources to areas in need. Water conservation can be achieved through more controlled irrigation systems, such as drip and sprinkler irrigation, which are more efficient. Estimating water needs for crops depends on factors like soil type, precipitation, irrigation method, crop variety, and soil moisture retention capacity. Air and soil moisture control systems using wireless sensors help optimize water use and improve crop health. In modern agriculture, significant increases in crop yields are expected with the use of IoT technologies, such as water management based on the Crop Water Stress Index (CWSI), which is determined from the crop canopy and varying air temperatures at different stages of growth (Zhang et al. 2024). Data from climate models, sensors, and satellite imagery is integrated with the CWSI model to estimate water requirements. Irrigation predictions based on these calculations can be tailored to specific fields, considering factors like slope or soil variability, thus enhancing water use efficiency.

Smart Fertilization System

With the data collected from soil sensors and other IoT devices, farmers can implement precision fertilization techniques. By applying fertilizers based on specific field conditions rather than blanket applications, farmers can significantly reduce nutrient wastage and minimize environmental impact. This targeted approach not only improves crop yield but also enhances the efficiency of fertilizer use (Chowdhury et al. 2025).

Fertilizers are substances, either natural or synthetic, that provide essential nutrients to support plant growth and enhance soil fertility. However, improper use—whether too much or too little—can harm the soil, plant health, and the environment (Köksal and Tekinerdogan 2019). In smart agriculture, site-specific nutrient application precisely estimates the amount of nutrients needed, reducing the negative effects of excessive use on both soil and the environment (Lavanya et al. 2020). Factors such as soil type, crop variety, yield goals, nutrient retention capacity, and weather conditions influence the nutrient requirements. IoT-based fertilization techniques help map the spatial distribution of nutrients, using tools like the Normalized Difference Vegetation Index (NDVI) from satellite images to assess crop health, vegetation vigor, and soil nutrient levels. Technologies like GPS, geomapping, variable rate technology (VRT), and autonomous vehicles are integral to smart fertilization practices. Additionally, fertigation and chemigation—applying water-soluble fertilizers, soil amendments, and pesticides—are effective management practices to boost fertilizer efficiency (Dhanaraju et al. 2022).

Pest and Disease Management

IoT technologies can significantly enhance pest and disease management strategies in agriculture. Smart sensors can detect pest activity and environmental conditions favorable for diseases, sending alerts to farmers. This proactive approach allows for timely interventions, such as targeted pesticide applications, thereby reducing the reliance on chemical treatments and minimizing environmental impact. Moreover, the use of drones equipped with cameras and sensors can facilitate aerial surveillance, helping farmers identify pest infestations and crop diseases early.

According to the Food and Agriculture Organization (FAO), global crop yields suffer annual losses of 20-40% due to pests and diseases, which are typically managed using pesticides and agrochemicals (Ramalingam et al. 2020). However, these chemicals pose risks to human and animal health and contribute to environmental contamination. IoT-based technologies, such as robots, drones, and wireless sensors, offer a more precise approach to pest control by providing real-time monitoring, modeling, and disease forecasting, making them more efficient than traditional methods. These IoT systems rely on detection and image processing, utilizing remote sensing and field sensors to gather data on plant health and pest presence throughout the crop cycle (Kim and Shin, 2018). Automated IoT traps are used to capture, count, and identify insect species, with the data being uploaded to the cloud for analysis.

Advances in robotic technology have led to the development of agricultural robots equipped with multispectral imaging sensors and precision spraying systems, enabling more accurate pest detection and control within the IoT framework (Venkatesan et al. 2018; Ennouri and Kallel 2019; Chowdhury et al. 2023b).

Smart Livestock Farming

In addition to crop management, IoT applications extend to livestock farming. Smart collars and wearable devices for animals can monitor health metrics such as heart rate, body temperature, and activity levels. This data enables farmers to detect health issues early, manage feeding schedules, and track the overall well-being of their livestock. For instance, IoT-enabled systems can alert farmers when an animal is sick or when it's time for vaccination, enhancing animal husbandry practices.

The use of IoT-based livestock monitoring systems presents significant opportunities to transform agricultural management practices. By equipping animals with wearable sensors, farmers can access real-time data on various health metrics such as body temperature, heart rate, and breathing patterns. This continuous, individual monitoring enables early detection of health anomalies, acting as a pre-emptive measure for potential diseases. Advanced algorithms and data analytics further process this information, providing actionable insights that help farmers make well-informed decisions regarding animal care and treatment. Additionally, IoT technology offers a detailed view of livestock behavior by tracking movements, activity levels, and social interactions within the herd. Analyzing this data helps farmers identify unusual behavioral patterns that may suggest health or welfare issues. By acting on these insights, farmers can address concerns swiftly, ensuring the animals' well-being and preventing more serious problems. Long-term data analysis also allows for trend identification, offering predictive insights that improve decision-making in livestock management (Suryavansh et al. 2021; Dineva and Atanasova 2022).

Internet of Things in Advanced Farming

The integration of Internet of Things (IoT) technology in advanced farming methods, including hydroponics and vertical farming, is revolutionizing agriculture by enhancing efficiency, sustainability, and productivity. Here are several key aspects of how IoT is applied in these innovative farming practices:

Hydroponics Farming

An automated smart hydroponics system coupled with IoT consists of three primary components: input data, cloud server and output data. The monitoring of lettuce cultivation from anywhere using the internet by evaluating parameters, such as pH level, water, nutrient-rich water-based solution, room temperature, and humidity, on a real-time basis. Plants are grown using the hydroponic system of the deep flow technique, which involves putting roots in layers of deep water and making sure that the plant nutrient solution is constantly circulated. Sensors included into Raspberry Pi collect data on plant growth elements like pH, temperature, humidity, and water level in the hydroponic reservoir. The data is processed and monitored automatically in real time to guarantee enough water circulation (Lakshmanan et al. 2020; Nurhasan et al. 2018).

Vertical Farming

Industrial agricultural practices degrade soil quality more rapidly than natural processes can restore it. The high rates of soil erosion and extensive freshwater usage in farming have contributed to the reduction of arable land and put additional strain on existing water resources. Vertical farming (VF) presents a solution by offering a controlled environment for plants, which significantly lowers resource usage while boosting production throughout the year. The method requires minimal land depending on the number of layers used. Additionally, VF leads to higher yields and conserves water more effectively than traditional methods. Monitoring carbon dioxide levels is critical, and nondispersive infrared (NDIR) CO2 sensors are crucial in regulating vertical farm conditions (Pimentel and Burgess 2013, Benke and Tomkins 2017).

Benefits of Using IoT Sensors in Agriculture

With the global rollout of WiFi 6 and 5G connectivity, IoT systems would gain improved speed in terms of connectivity (Goedde et al. 2020). The potential benefits associated with these technologies include the expansion of IoT applications, large scale deployment of IoT in precision agriculture in field crops, machinery tracking, and the monitoring of remote infrastructure (Achour et al. 2021; Jamil et al. 2022). With the growing challenges in agriculture sector related to extreme weathers, changing climatic patterns, resource scarcity, and increasing demand, the adoption of IoT sensors in this field is proving to be a game-change. Farmers are adopting these technologies to manage their operations more effectively. Thus, IoT is shaping the future of precision agriculture, making it more responsive, efficient, and environmentally sustainable.

Using IoT sensors in agriculture brings several significant benefits, revolutionizing farming by optimizing resources, increasing yields, and improving sustainability. Below are key advantages.

Enhanced Resource Efficiency

IoT sensors allow precise monitoring of resources like water, fertilizers, and energy. For example, soil moisture sensors help optimize irrigation by applying the right amount of water at the right time, preventing both over- and under-irrigation.

Improved Crop Monitoring and Health Management

IoT sensors provide real-time data on various environmental parameters, enabling farmers to monitor crop health continuously. With sensors tracking factors like soil moisture, temperature, light intensity, and humidity, farmers can identify issues early such as diseases, nutrient deficiencies, or pest infestations allowing for timely interventions.

Data-Driven Decision Making

One of the most significant benefits of IoT in agriculture is its ability to collect and analyze large amounts of data. Sensors gather data on weather conditions, soil quality, and crop performance, which is then processed using machine learning and predictive analytics.

Reduced Environmental Impact

By improving the precision of farming activities, IoT sensors contribute to sustainable agriculture. For instance, precision irrigation and nutrient management minimize water and fertilizer wastage, reducing the risk of contaminating surrounding ecosystems.

Increased Productivity and Yield

By optimizing resource use and reducing the impact of pests, diseases, and unfavorable environmental conditions, IoT sensors contribute to higher crop productivity. Yield improvements can be achieved by monitoring key growth parameters and adjusting farming practices accordingly.

Cost Reduction

The ability to monitor and control farm operations remotely can lead to significant cost reductions. Automated IoT systems reduce the need for manual labor, cutting operational costs.

Increased Scalability and Automation

IoT enables the automation of various agricultural processes, such as irrigation, fertilization, and pest control. This automation not only reduces labor costs but also allows for better scalability, enabling farms to manage larger areas with fewer resources.

Weather Monitoring and Adaptation

IoT sensors can monitor local weather conditions and provide farmers with real-time updates and forecasts. This is particularly useful for mitigating risks related to unpredictable weather patterns, such as droughts, floods, or frost. Farmers can take preventive actions like adjusting irrigation schedules, covering crops, or delaying harvests based on data collected by weather sensors.

Challenges and Considerations

When deploying IoT sensors in precision agriculture, several challenges and considerations need to be addressed for optimal performance and sustainability. Those are as follows:

Data Accuracy and Reliability

IoT sensors in agriculture must provide highly accurate and reliable data for effective decision-making. However, sensor calibration, sensitivity, and environmental factors like temperature, humidity, and soil composition can impact the precision of data collected. Malfunctioning sensors or data inconsistencies can lead to flawed interpretations, which could negatively affect crop management.

Data Management and Processing

Precision agriculture involves vast amounts of data from multiple sources (e.g., soil moisture, temperature, weather conditions, etc.). Handling this big data efficiently is a major challenge. The data must be stored, processed, and analysed in real time to enable timely decision-making. Moreover, efficient algorithms are needed to process and analyse this data to extract actionable insights, requiring robust computational infrastructure.

Interoperability and Standardization

A major challenge in IoT for agriculture is the lack of standardization across different devices and platforms. Various IoT sensors and systems are manufactured by different vendors, leading to issues of compatibility. Ensuring that data from multiple sensors can seamlessly communicate and integrate with each other is crucial for a functional and scalable IoT ecosystem in agriculture.

Power Supply and Energy Efficiency

Many IoT sensors are deployed in remote agricultural fields, making it difficult to ensure continuous power supply. Battery-powered sensors may have limited operational lifetimes, requiring frequent maintenance or replacement. Solar-powered sensors can mitigate this issue but may not be feasible in regions with limited sunlight. Developing energy-efficient sensors and alternative power solutions is essential for sustained operation.

Connectivity and Network Infrastructure

Agricultural fields, particularly those in rural or isolated regions, often lack reliable internet connectivity. Poor network coverage can hinder the real-time transmission of data from IoT sensors to cloud-based systems for analysis. This makes it crucial to develop robust communication networks, such as low-power wide-area networks (LPWANs) or satellite-based communication systems, that can support IoT in agriculture even in remote areas.

Security and Privacy Concerns

IoT sensors collect sensitive data about crop health, soil conditions, and farm management practices, which could be susceptible to cyberattacks or unauthorized access. Ensuring data security and privacy is critical, especially when using cloud-based systems for storage and analysis. This requires the implementation of strong encryption protocols, secure communication channels, and regular system audits to safeguard agricultural data.

Cost and Scalability

The initial setup cost of deploying IoT sensors, coupled with the expense of maintaining and upgrading the system, can be a financial barrier for many farmers, particularly small-scale operators. For IoT to have widespread adoption, cost-effective solutions need to be developed. Scalability is also a key issue; as the number of sensors increases, the system must be able to handle larger datasets without compromising on performance or accuracy.

Environmental Impact and Sensor Durability

IoT sensors must withstand harsh environmental conditions, such as extreme temperatures, heavy rainfall, and soil moisture. Sensor durability is a critical consideration, as frequent replacements or repairs could drive up operational costs. The materials used for sensor production should also be environmentally friendly to minimize their ecological impact, as well as sustainable to ensure long-term usability (Chowdhury et al. 2023a).

Farmers' Digital Literacy

IoT-based precision agriculture requires a certain level of technical knowledge for effective operation. Farmers need to understand how to interpret data from sensors, use digital platforms, and make informed decisions based on analytics. Lack of digital literacy or reluctance to adopt new technologies can hinder the effective implementation of IoT solutions. Adequate training and support must be provided to ensure that farmers can fully utilize these technologies.

Regulatory and Ethical Considerations

As IoT technology becomes more prevalent in agriculture, regulatory and ethical issues will arise. Data ownership, usage rights, and the potential for misuse of agricultural data are concerns that need to be addressed through proper legislation. Additionally, ethical considerations regarding the environmental impact of sensors and the long-term sustainability of IoT systems must be taken into account.

Future Trends in IoT-Based Precision Agriculture

Future trends in IoT-based precision agriculture are poised to revolutionize farming practices, driving efficiency, sustainability, and productivity. One key trend is the increasing integration of AI and machine learning with IoT sensors, enabling predictive analytics for better decision-making. This allows farmers to forecast crop diseases, pest infestations, and weather patterns with greater accuracy, leading to more proactive management of resources. Automation and robotics are becoming more prevalent, with drones and autonomous machines powered by IoT playing critical roles in monitoring crops, delivering fertilizers, and performing precise tasks like weeding and harvesting. This shift will significantly reduce labour costs and enhance the efficiency of farm operations. Another emerging trend is the development of 5G networks, which promise faster data transfer and improved connectivity for remote areas. With 5G, real-time data analysis becomes more accessible, ensuring timely responses to environmental changes and optimizing irrigation, nutrient management, and pest control. Moreover, blockchain technology is being integrated with IoT in agriculture to ensure transparency and traceability in the food supply chain. This will enable better monitoring of crop quality and provide consumers with detailed information on the origin of their food. These trends suggest that IoT will continue to drive innovation in precision agriculture, making farming smarter, more sustainable, and scalable.

Case Study

The use of different types of sensors in agricultural field is rapidly increasing to facilitate the smart farming and precision agricultural management process. This section contains a case study of IoT sensor application in soil moisture measurement system. Soil moisture i.e., the amount of water present in a soil column plays a vital role in irrigation scheduling. The objective of this case study was to develop a customized data logger with the help of PostgreSQL for continuously recording the values of soil moisture with respect to time. Towards achieving this objective, one low-cost resistance-based soil moisture sensor was used. The working principle of such sensor can be briefly found in Saleh et al. (2016).

Materials and Methods

Components

Figure 3 illustrates the pin configuration for various components utilized in this study. The Arduino UNO microcontroller was used for sensor integration and programming purposes. It receives a constant 5 V supply from one 3.7 V lithium-ion battery. A boost converter was used to step-up the voltage from the battery, and the battery was charged via TP4056 charging circuit. The soil moisture sensor was interfaced

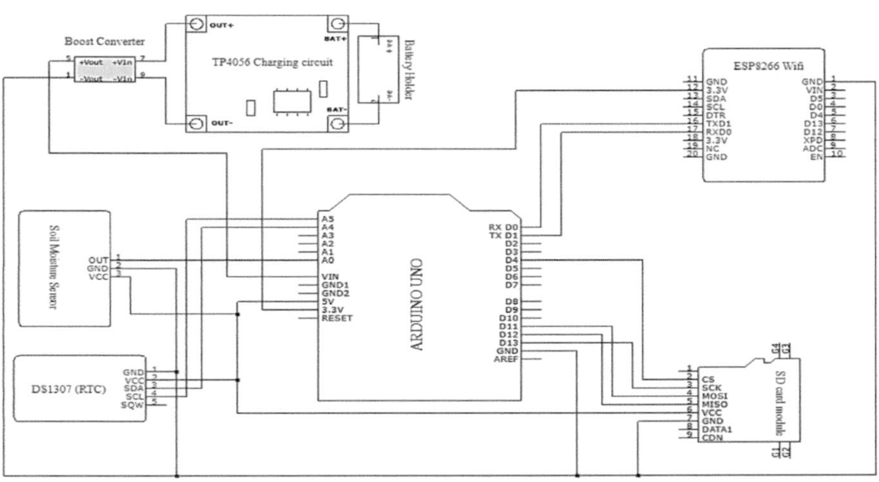

Fig. 3: Circuit connection diagram for soil moisture measuring system.

with the Arduino through its analogue pin. The ESP8266 Wi-Fi module was used for wireless communication. Precise time recording was done by a real time clock (RTC) DS1307 module. Additionally, a SD card module was used to store the soil moisture data along with timestamp. Pin configuration of various components with Arduino is shown in Table 2.

Table 2: Pin configuration of various components with Arduino

Wi-Fi Module	**Arduino**	**RTC Module**	**Arduino**
TX	RX	SDA	A4
RX	TX	SCL	A5
VCC	3.3 V	VCC	5V
GND	GND	GND	GND
SD Card Module	**Arduino**	**Soil Moisture Sensor**	**Arduino**
CS	4	Out	A0
MISO	12	VCC	5 V
MOSI	11	GND	GND
SCK	13	Boost Converter	Arduino
VCC	5 V	Vout+	Vin
GND	GND	Vout-	GND

Procedure

Figure 4 illustrates the various steps involved in the project. The initial step was the soil moisture sensor calibration, which is discussed in detail in earlier sections. Following the calibration, a temporary circuit connection was assembled on breadboard for programming and testing purposes. The RTC, Wi-Fi, and SD card modules were initialized through Arduino programming. By using the ESP8266 Wi-Fi module, the connection between the microcontroller and the cloud-based platform ThingSpeak was established, allowing to transfer the data stored in SD card to that platform. Subsequently, the soil moisture data were sent to a PostgreSQL database for analysis. Upon the successful execution of all the steps, all the components were soldered onto a zero PCB and fitted to a data logger box (Fig. 5).

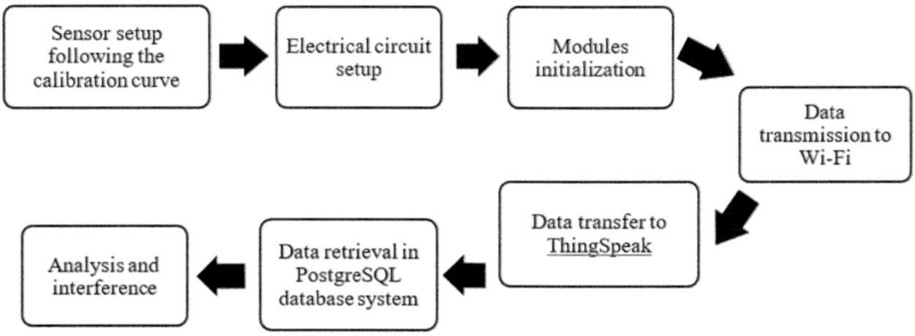

Fig. 4: Flowchart of methodology.

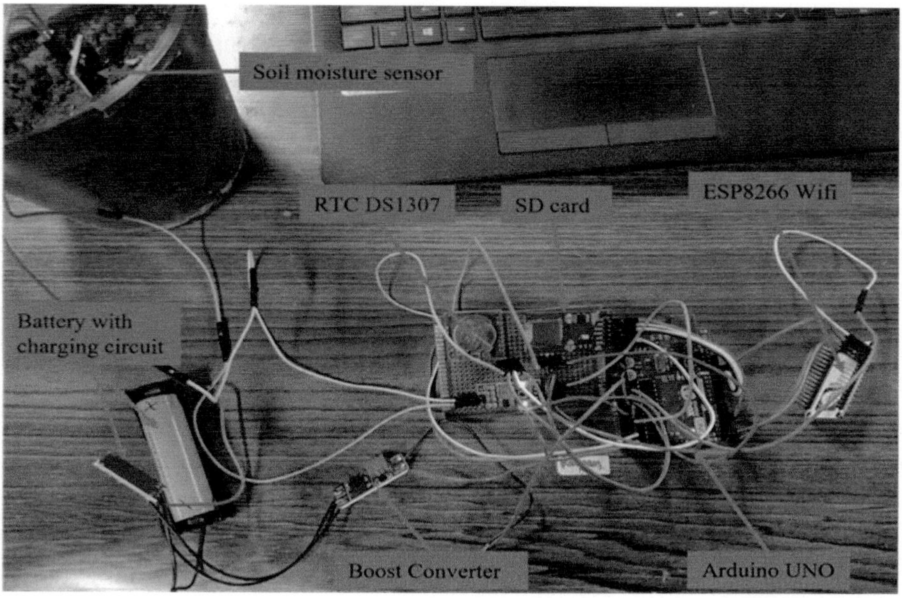

Fig. 5: Integration of the components.

Sensor Calibration and Uncertainty

Sensor calibration was done by following Saleh et al. (2016) for accurately measure the soil moisture. The working process for calibration is shown in Fig. 6 and discussed subsequently:

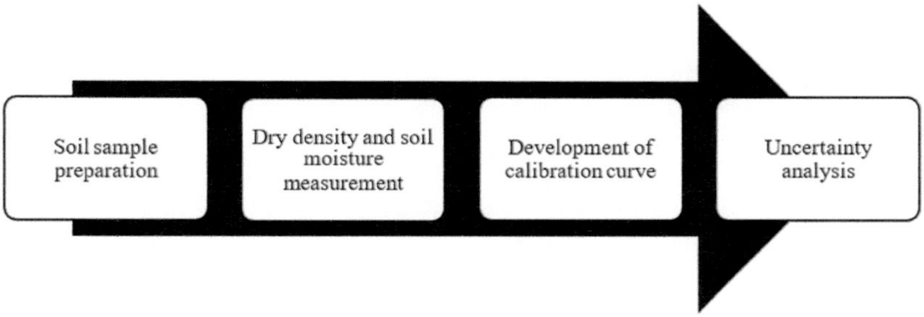

Fig. 6: Flowchart for soil moisture sensor calibration.

Soil sample preparation

Silty loam type soil sample was collected from an experimental agricultural plot at IIT Kanpur (26.5123° N, 80.2329° E). Core cutters were utilized to determine the dry density of the soil. Additionally, a portion of soil was collected separately for further laboratory analysis. The collected soil was oven-dried for 24 hours at 105°C to completely remove any moisture content present in the soil, providing a baseline for the calibration. Following drying, any debris or bigger aggregates that might have affected sensor readings were removed from the soil by crushing it and sieving it through a 2 mm sieve to produce a consistent particle size. A total of seven dried soil samples, each weighing 100 gm, were prepared for analysis. The first sample was kept dry, while a known amount of 5 gm of water was added progressively to each sample. The wet samples were mixed thoroughly to ensure uniform water distribution throughout.

Dry density and soil moisture measurement: The dry density was calculated by heating the soil, resulting in a value of 1.52 gm/cc. Ten readings of sensor value for each sample were recorded via Tera Term (https://teratermproject.github.io/index-en.html). A small amount of each wet soil type was taken for calculating the gravimetric water content (GWC). From the GWC, volumetric water content (VWC) was determined by multiplying GWC with dry density.

Calibration curve development: A calibration curve was developed between the calculated VWC and averaged sensor reading of each sample, and a linear trendline was used to derive the corresponding equation.

It can be observed from Fig. 7(a) that the calibration curve ($R^2 = 0.79$) shows non-linearity between lower sensor readings (<250) and higher sensor reading (≥250). To achieve higher accuracy, the calibration curve was adjusted to a bi-linear model, as shown in Fig. 7(b). The bi-linear model improves R^2 values for both cases. Further analysis and calculations were done based on the bi-linear model.

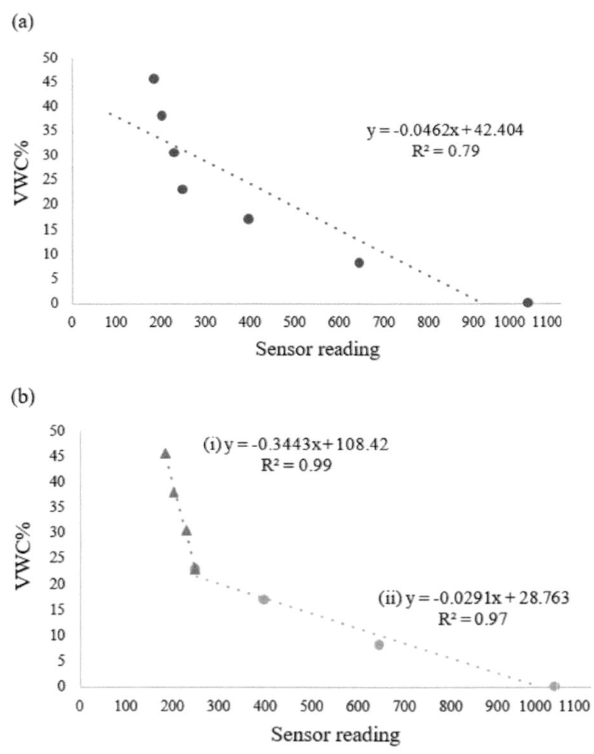

Fig. 7: Calibration curve for (a) linear model and (b) bi-linear model.

Uncertainty analysis: Type A evaluation of uncertainty associated with the calibration curve was estimated following the guidelines of JCGM (2008). Table 3 shows the result of uncertainty analysis.

Table 3: Uncertainty due to the bi-linear calibration

Factors	(i) Sensor Reading <250	(ii) Sensor Reading ≥250
Uncertainty (slope)	–0.344	–0.029
Standard Uncertainty (standard error)	0.019	0.003
Overall expanded uncertainty at 95% CI	(–0.186 ± 0.038)%	

Implementation and Limitation

After setting up the device, a testing experiment was conducted on dry soil. The time of watering was 13 November 2022, 17:21. As illustrated in Fig. 8(a), the sensor showed a late response to change in moisture content. The VWC gradually increased and became constant at 39.44% in around 5 minutes, at 17:26. The system was then implemented for continuously monitoring the plant soil moisture content as shown in Fig. 8(b) from 15 November 2022 to 20 November 2022. The measured soil moisture data was retrieved from the PostgreSQL database for further analysis and evaluation.

Fig. 8. (a) Testing experimental result and (b) Experimental setup.

A major limitation of this study is that when the sensor fully submerged into water, it showed a reading of approximately 156, corresponding to a maximum VWC of 55%. Therefore, the sensor will not be capable of measuring soil moisture levels more than this threshold.

Conclusions

The deployment of IoT sensors in precision agriculture is revolutionizing the way farming is managed. These sensors enable real-time monitoring and data-driven decision-making, allowing farmers to respond to agricultural variabilities with high precision. Key sensors, such as soil moisture, temperature, and pH sensors, provide critical insights into soil health and crop conditions, improving resource efficiency. Additionally, advanced technologies like drones and multispectral sensors offer large-scale field monitoring and yield prediction capabilities, enhancing productivity. IoT-enabled systems reduce water wastage, optimize fertilizer use, and minimize the environmental impact of pesticides, contributing to more sustainable agricultural practices. As a result, farmers can achieve higher crop yields with fewer resources, ensuring better profitability and sustainability. Despite the significant benefits, challenges such as the high cost of installation, the complexity of data integration, and the need for reliable rural connectivity persist. These issues require attention to ensure the widespread adoption of IoT technologies in agriculture.

References

Abu, N.S., W.M. Bukhari, C.H. Ong, A.M. Kassim, T.A. Izzuddin, M.N. Sukhaimie et al. 2022. Internet of things applications in precision agriculture: A review. J.R.C. 3(3): 338-347.

Achour, Y., A. Ouammi and D. Zejli. 2021. Technological progresses in modern sustainable greenhouses cultivation as the path towards precision agriculture. Renew. Sustain. Energy Rev. 147: 111-251.

Anand, R., S.S. Nalla, S. Nasreen, M. Chowdhury, Madhusudan, B.S., A. Pandey et al. 2024. Fruit counting and analysis using artificial intelligence approaches. pp. 93-114. In: S.S. Chouhan, A. Saxena, U.P. Singh and S. Jain [eds.]. Artificial Intelligence Techniques in Smart Agriculture. Springer Nature, Singapore.

Anand, T., S. Sinha, M. Mandal, V. Chamola and F.R. Yu. 2021. AgriSegNet: Deep aerial semantic segmentation framework for IoT-assisted precision agriculture. IEEE Sens. J. 21(16): 17581-17590.

Aryal, J.P., N. Manchanda and T. Sonobe. 2022. Expectations for household food security in the coming decades: A global scenario. pp. 107-131. *In:* Future foods. Academic Press.

Atalla, S., S. Tarapiah, A. Gawanmeh, M. Daradkeh, H. Mukhtar, Y. Himeur et al. 2023. IoT-enabled precision agriculture: Developing an ecosystem for optimized crop management. J. Inf. 14(4): 205.

Ayaz, M., M. Ammad-Uddin, Z. Sharif, A. Mansour and E.H.M. Aggoune. 2019. Internet-of-Things (IoT)-based smart agriculture: Toward making the fields talk. IEEE access. 7: 129551-129583.

Benke, K. and B. Tomkins. 2017. Future food-production systems: Vertical farming and controlled-environment agriculture. Sustain. Sci. Pract. Policy. 13(1): 13-26.

Blanas, S., J.M. Patel, V. Ercegovac, J. Rao, E.J. Shekita and Y. Tian. 2010. A comparison of join algorithms for log processing in MapReduce. pp. 975-986. *In:* Proc. ACM SIGMOD Int. Conf. Manag. Data, Indianapolis, IN, USA.

Brickley, D. and R.V. Guha. 2004. RDF Vocabulary Description Language 1.0: RDF Schema. W3C Recommendation, 10 February 2004. World Wide Web Consortium (W3C). https://www.w3.org/TR/rdf-schema/

Burra, D.D., L. Parker, N.T. Than, P. Phengsavanh, C.T.M. Long, R.S. Ritzema et al. 2021. Drivers of land use complexity along an agricultural transition gradient in Southeast Asia. Ecol. Indic. 124: 107402.

Cai, H., B. Xu, L. Jiang and A.V. Vasilakos. 2016. IoT-based big data storage systems in cloud computing: Perspectives and challenges. IEEE Internet Things J. 4(1): 75-87.

Cha, S.C., T.Y. Hsu, Y. Xiang and K.H. Yeh. 2019. Privacy enhancing technologies in the Internet of Things: Perspectives and challenges. IEEE Internet Things J. 6(2): 2159-2187.

Chowdhury, M., A. Kushwah, A.N. Satpute, S.K. Singh and A.K. Patil. 2023a. A comprehensive review on potential application of nanomaterials in the field of agricultural engineering. J. Biosyst. Eng. 48(4): 457-477.

Chowdhury, M., E.V. Thomas, A. Jha, A. Kushwah, R. Kurmi, T.K. Khura et al. 2023b. An automatic pressure control system for precise spray pattern analysis on spray patternator Comput. Electron. Agric. 214: 108287.

Chowdhury, M., R. Anand, T. Dhar, R. Kurmi, R.K. Sahni and A. Kushwah. 2024a. Digital insights into plant health: Exploring vegetation indices through computer vision. pp. 7-30. *In*: S.S. Chouhan, U.P. Singh and S. Jain [eds.]. Applications of Computer Vision and Drone Technology in Agriculture 4.0. Springer Nature, Singapore.

Chowdhury, M., T.K. Khura, P.K. Upadhyay, R.A. Parray, H.L. Kushwaha, C. Singh et al. 2024b. Assessing vegetation indices and productivity across nitrogen gradients: A comparative study under transplanted and direct-seeded rice. Front. Sustain. Food Syst. 8: 1351414.

Chowdhury, M., T.K. Khura, R.A. Parray, P.K. Upadhyay, H.L. Kushwaha, A. Lama and A. Kushwah. 2025. Multi-vegetation indices based handheld device for precise nitrogen assessment and prescription in direct seeded rice. Comput. Electron. Agric. 229: 109886.

Cupp, O.S., D.E. Walker and J. Hillison. 2004. Agroterrorism in the U.S.: Key security challenge for the 21st century. Biosecurity and Bioterrorism. 2(2): 97-105.

Curé, O., F. Kerdjoudj, D. Faye, C.L. Duc and M. Lamolle. 2013. On the potential integration of an ontology-based data access approach in NoSQL stores. Int. J. Distrib. Syst. Technol. 4(3): 17-30.

Dhanaraju, M., P. Chenniappan, K. Ramalingam, S. Pazhanivelan and R. Kaliaperumal. 2022. Smart farming: Internet of Things (IoT)-based sustainable agriculture. J. Agric. 12(10): 1745.

Dineva, K. and T. Atanasova. 2022. Cloud data-driven intelligent monitoring system for interactive smart farming. J. Sens. 22(17): 6566.

Dorri, A., S.S. Kanhere, R. Jurdak and P. Gauravaram. 2017. Blockchain for IoT security and privacy: The case study of a smart home. pp. 618-623. *In:* Proc. IEEE Int. Conf. Pervas. Comput. Commun. Workshops (PerCom Workshops).

Doshi, M. and A. Varghese. 2022. Smart agriculture using renewable energy and AI-powered IoT. pp. 205-225. *In:* AI, Edge and IoT-based Smart Agriculture. Academic Press.

Ennouri, K. and A. Kallel. 2019. Remote sensing: An advanced technique for crop condition assessment. Math. Probl. Eng. 2019(1): 9404565.

Farooq, M.S., S. Riaz, A. Abid, K. Abid and M.A. Naeem. 2019. A survey on the role of IoT in agriculture for the implementation of smart farming. IEEE Access. 7: 156237-156271.

Ferrag, M.A., L. Shu, X. Yang, A. Derhab and L. Maglaras. 2020. Security and privacy for green IoT-based agriculture: Review, blockchain solutions, and challenges. IEEE Access. 8: 32031-32053.

Gawande, V., D.R.K. Saikanth, B.S. Sumithra, S.A. Aravind, G.N. Swamy, M. Chowdhury and B.V. Singh. 2023. Potential of precision farming technologies for eco-friendly agriculture. Int. J. Plant Soil Sci. 35(19): 101-112.

Giller, K.E., T. Delaune, J.V. Silva, K. Descheemaeker, G. van de Ven, A.G. Schut et al. 2021. The future of farming: Who will produce our food? J. Food Secur. 13(5): 1073-1099.

Goedde, L., J. Katz, A. Ménard and J. Revellat. 2020. Agriculture's Connected Future: How Technology can Yield New Growth. McKinsey and Company. [Online] Available at: https://www.mckinsey.com/industries/agriculture/our-insights/agricultures-connected-future-how-technology-can-yield-new-growth [Accessed: 28 September, 2024].

Guestrin, C. 2006. SVMs, duality and the kernel trick. Mach. Learn. 10701: 15781.

Gupta, M., M. Abdelsalam, S. Khorsandroo and S. Mittal. 2020. Security and privacy in smart farming: Challenges and opportunities. IEEE Access. 8: 34564-34584.

Harwood, R.R. 2020. A history of sustainable agriculture. pp. 3-19. *In:* C.A. Edwards [ed.]. Sustainable Agricultural Systems. CRC Press.

Huang, X., Y. Liu, L. Huang, E. Onstein and C. Merschbrock. 2023. BIM and IoT data fusion: The data process model perspective. Autom. Constr. 149: 104792.

Jamil, F., M. Ibrahim, I. Ullah, S. Kim, H.K. Kahng and D.H. Kim. 2022. Optimal smart contract for autonomous greenhouse environment based on IoT blockchain network in agriculture. Comput. Electron. Agric. 192: 106573.

JCGM. 2008. Evaluation of measurement data—Guide to the expression of uncertainty in measurement. Int. Organ. Stand. Geneva ISBN. 50: 134.

Jha, A., M. Chowdhury and A.N. Satpute. 2023. Surface water quality forecasting using machine learning approach. pp. 293-315. *In*: C.B. Pande, M. Kumar and N.L. Kushwaha [eds.]. Surface and Groundwater Resources Development and Management in Semi-arid Region: Strategies and Solutions for Sustainable Water Management. Cham: Springer International Publishing.

Jordan, M.I. and C.M. Bishop. 2004. Neural networks. *In:* A.B. Tucker [ed.]. Computer Science Handbook, 2nd ed. CRC Press, Boca Raton, FL, USA.

Khan, M.A. and K. Salah. 2018. IoT security: Review, blockchain solutions, and open challenges. Future Gener Comput Syst. 82: 395-411.

Khirade, S.D. and A.B. Patil. 2015. Plant disease detection using image processing. pp. 768-771. *In:* International Conference on Computing Communication Control and Automation. IEEE.

Kim, S., M. Lee and C. Shin. 2018. IoT-based strawberry disease prediction system for smart farming. J. Sens. 18(11): 4051.

Kim, W.S., W.S. Lee and Y.J. Kim. 2020. A review of the applications of the internet of things (IoT) for agricultural automation. Biosyst. Eng. 45: 385-400.

Köksal, Ö. and B. Tekinerdogan. 2019. Architecture design approach for IoT-based farm management information systems. Precis. Agric. 20: 926-958.

Kour, V.P. and S. Arora. 2020. Recent developments of the internet of things in agriculture: A survey. IEEE Access. 8: 129924-129957.

Krishnamurthi, R., A. Kumar, D. Gopinathan, A. Nayyar and B. Qureshi. 2020. An overview of IoT sensor data processing, fusion, and analysis techniques. J. Sens. 20(21): 6076.

Lakshmanan, R., M. Djama, S.K. Selvaperumal and R. Abdulla. 2020. Automated smart hydroponics system using internet of things. IJECE. 10(6): 6389-6398.

Lavanya, G., C. Rani and P. GaneshKumar. 2020. An automated low-cost IoT-based fertilizer intimation system for smart agriculture. Sustain. Comput.: Inform. Syst. 28: 100300.

Lecun, Y., Y. Bengio and G. Hinton. 2015. Deep learning. Nat. 521: 436-444.

Lee, C., J. Kim, H. Ko and B. Yoo. 2024. Addressing IoT storage constraints: A hybrid architecture for decentralized data storage and centralized management. IEEE Internet Things J. 25: 101014.

Lee, K., L. Liu, Y. Tang, Q. Zhang and Y. Zhou. 2013. Efficient and customizable data partitioning framework for distributed big RDF data processing in the cloud. pp. 327-334. *In:* Proc. IEEE CLOUD, Santa Clara, CA, USA.

Li, B., G. Jiang, N. Li and C. Song. 2024. Research on large-scale structured and unstructured data processing based on large language model. pp. 111-116. *In:* Proceedings of the International Conference on Machine Learning, Pattern Recognition and Automation Engineering. Singapore.

Luckham, D. 2012. Event Processing for Business: Organizing the Real-time Enterprise. John Wiley and Sons, New Jersey, USA.

Maftei, A.A., A. Lavric, A.I. Petrariu and V. Popa. 2023. Massive data storage solution for IoT devices using blockchain technologies. J. Sens. 23(3): 1570.

Mishra, S. and S.K. Sharma. 2023. Advanced contribution of IoT in agricultural production for the development of smart livestock environments. IEEE Internet Things J. 22: 100724.

Monteiro, A., S. Santos and P. Gonçalves. 2021. Precision agriculture for crop and livestock farming—Brief review. J. Anim. 11(8): 2345.

Neapolitan, R.E. and X. Jiang. 2018. Neural networks and deep learning. 25: 15-24. *In:* Artificial Intelligence. CRC Press, Boca Raton, FL, USA.

Pimentel, D. and M. Burgess. 2013. Soil erosion threatens food production. J. Agric. 3(3): 443-463.

Samara, M.A., I. Bennis, A. Abouaissa and P. Lorenz. 2022. A survey of outlier detection techniques in IoT: Review and classification. J. Sens. Actuator Netw. 11(1): 4.

Satpute, A.N., K.P. Gavhane, S. Kaur, A. Jha, N.C. Pradhan and M. Chowdhury. 2024. Integration of AI and IoT in soilless cultivation to power sustainable agricultural revolution. pp. 387-411. *In*: K. Pandey, N.L. Kushwaha, C.B. Pande and K.G. Singh [eds.]. Artificial Intelligence and Smart Agriculture: Technology and Applications. Springer Nature, Singapore.

Schmidhuber, J. 2015. Deep learning in neural networks: An overview. Neural Netw. 61: 85-117.

Shaikh, T.A. and R. Ali. 2019. Automated atrophy assessment for Alzheimer's disease diagnosis from brain MRI images. Magn. Reson. Imaging. 62: 167-173.

Sunil Kumar, T., H.M. Virdia, K.G. Patel, S. Ragi, M. Chowdhury, P. Kumar et al. 2024. Effect of summer legume residue incorporation and fertilizer regimes on rice growth, yield, and nutrient uptake. Front. Sustain. Food Syst. 8: 1467201.

Zhang, B., J. Huang, T. Dai, S. Jing, Y. Hua, Q. Zhang et al. 2024. Assessing accuracy of crop water stress inversion of soil water content all day long. Precis. Agric. 25: 1894-1914.

CHAPTER

4

Drones in Agriculture: Aerial Intelligence for Precision Farming

Debasish Roy[1], Suman Dutta[2], Debashis Paul[3], Anshuman Das[4], Nilutpal Saikia[5] and Sumanta Das[2,6*]

[1] India Meteorological Department, Mausam Bhavan, Lodi Road, New Delhi - 110003
[2] Ramakrishna Mission Vivekananda Educational and Research Institute, West Bengal, India
[3] ICAR-Central Institute for Cotton Research, Regional Station, Sirsa, Haryana, India
[4] ICFRE-Institute of Forest Productivity, Ranchi, Jharkhand - 835303
[5] ICAR-Indian Agricultural Research Institute, New Delhi - 110012, India
[6] The University of Queensland, St Lucia, Queensland 4072, Australia

Introduction

Drones have revolutionized precision farming by providing farmers with real-time aerial intelligence to optimize crop management (Gawande et al. 2023; Dutta et al. 2024). Drones that are outfitted with cutting-edge sensors, cameras, and GPS technology boost yields, minimize resource waste, and facilitate better decision-making. Farmers can monitor crop health and identify problems such as insect infestations, water shortages, and plant stress because of the high-resolution photos they record, which include RGB, multispectral, and thermal data (Nebiker et al. 2016; Chowdhury et al. 2024a). This makes it possible to implement focused interventions like precision fertilizer administration, irrigation, and insect control (Chowdhury et al. 2025). In-depth field maps produced by drones also help with soil and topographical studies, which enable farmers to modify planting plans to optimize crop production. Drones that possess precision spraying capabilities may administer fertilizers, herbicides, or insecticides with extreme precision, minimizing chemical usage and environmental damage (Dutta et al. 2023; Malkani et al. 2024). Additionally, many modern farmers are already adopting advanced technological solutions, such as digitally-controlled farming equipment and unmanned aerial vehicles (UAVs) for monitoring and forecasting. Drones examine plant development patterns in conjunction with artificial intelligence (AI) and machine learning to

*Corresponding author: sumanta.das@uq.net.au

DOI: 10.1201/9781003536932-4

forecast yields, enhance logistics, and maximize production tactics (Oliveira et al. 2018). By determining moisture levels and directing effective irrigation programs, they also help manage soil and water. They are also used to keep an eye on animals, track movements, and guarantee resource access. Drones assist minimize agriculture's environmental impact by facilitating the effective use of water, fertilizer, and pesticides (Dutta et al. 2023). This lowers carbon emissions and prevents the abuse of chemicals, which promotes sustainability. Future farming operations might be even more optimized due to developments in drone technology and AI integration. In summary, drones offer crucial airborne intelligence that facilitates data-driven decision-making, boosting output and advancing sustainable agriculture.

Hence, objective of this chapter is to utilizing drones in agriculture is to enhance precision farming practices through aerial intelligence. Drones equipped with advanced sensors, imaging technology, and GPS capabilities enable farmers to monitor crop health, assess field conditions, and optimize resource use with greater accuracy. By providing real-time data on soil moisture, plant health, pest infestations, and crop growth patterns, drones facilitate informed decision-making, improve yields, reduce input costs, and promote sustainable agricultural practices. The goal is to revolutionize farming operations by leveraging aerial technology to maximize productivity while minimizing environmental impact. It showcases developments in deep learning, machine learning, and remote sensing methods, especially in handling high-resolution multispectral pictures for feature extraction and object categorization. The chapter delves deeper into the various applications of Unmanned Aerial Vehicles (UAVs) in agriculture, exploring how these advanced technologies are transforming farming practices. Highlighting their potential to increase crop yield, lower input costs, and support sustainable farming through duties like pesticide spraying, irrigation control, and crop health monitoring.

Agricultural Drones: Types

UAVs, generally weighing 25 kg or less, are controlled remotely and do not need a pilot onboard. They are effective for conducting surveys over large areas and are used for various purposes, including image analysis, ground monitoring, and detailed crop assessments. UAVs can be categorized according to various factors, including the number of rotors, speed, intended use, and design features. UAVs weighing 25 kg or more are subject to specific regulations. The categories of UAVs by weight include: Heavy UAVs weighing around 2 tons or more, these are mainly employed for military applications because of their capability to carry significant fuel loads. Medium UAVs Ranging from 200 to 2000 kg and 50 to 200 kg, these UAVs are employed in various applications and can sustain longer flight durations due to their fuel capacity. Lightweight UAVs weighing between 5 and 50 kg, are commonly used in agriculture. Micro-UAVs, weighing under 5 kg, are simpler to deploy and typically more affordable compared to their larger counterparts. UAVs can also be categorized by their design and flight capabilities, such as fixed-wing, single-rotor, multi-rotor, and hybrid vertical take-off and landing (VTOL) configurations. Fixed-wing and multi-rotor UAVs differ significantly in structure, flight time, endurance, and energy requirements. Single-rotor UAVs feature one large rotor on top and a smaller rotor at the tail, while multi-rotor UAVs come in various forms, including Tricopters, Quadcopters, Hexacopters, and Octocopters, depending on the number of rotors and their applications. Different types of UAVs are discussed in Fig. 1 and Table 1.

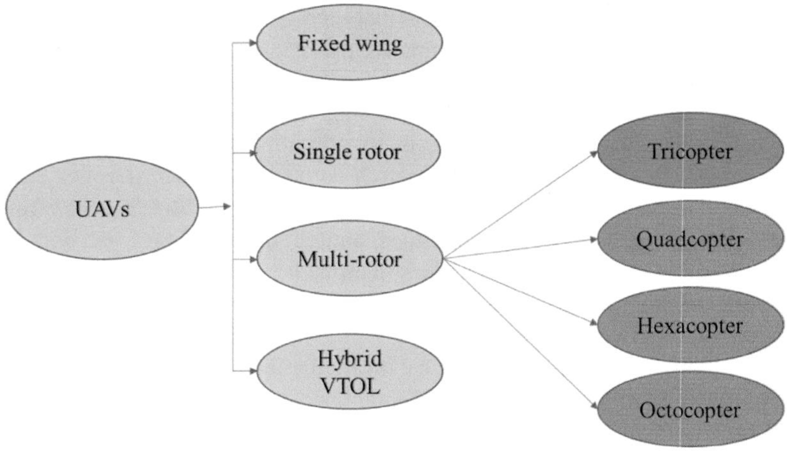

Fig. 1: Various UAV configurations.

Fixed-Wing Drones

A fixed-wing UAV designed for data collection operates via remote control. This UAV has a 195 cm wingspan, a carbon-fiber frame, and is powered by a single propeller engine. These design choices enhance aerodynamics and contribute to extended flight times, especially at higher speeds over surveyed areas. Typically, these UAVs are equipped with high-resolution cameras to provide enhanced mapping and surveillance functions from higher altitudes. Additionally, the flight system is uncomplicated, and the UAV's design and upkeep are relatively straightforward (Ziliani et al. 2018).

Single-Rotor Drones

A single-rotor system comprises two distinct components: the helicopter itself and the ground-based control system. The helicopter is equipped with several integral parts, including a flight controller, gyroscope, GPS receiver, image and telemetry transmitter, a heading sensor, and components for spraying. On the ground, the control system features a telemetry receiver and a remote-control transmitter. Some systems are also equipped with forced-air engine cooling to manage engine temperature at high altitudes or during slow flight. To measure the aircraft's roll and pitch angles and to detect its 3D velocity, high-precision vertical gyroscopes are employed. A magnetic heading sensor helps make fine adjustments for changes in flight direction. The UAV's altitude and position are monitored using a pressure altimeter. Various control parameters are calculated using Kalman filters and Proportional Integral Derivative (PID) algorithms (Xue et al. 2016).

Table 1: Types of drones and their specifications

Drone	Type	Flight Time (min)	Flow Rate (l/min)	Max Distance (km)	Tank Capacity (l)	Camera
DJI Agras	Hexacopter	15	6	3	20	
DJI MG-1P	Octocopter	20	0.525	5	10	
MS10 Max	Hexacopter	15	1.5	3	10	
Oryctes Dual	Hexacopter	15	10	2	16	
Hylio AG-122	Octocopter	15	4.3	2	22	
DJI Phantom 4 RTK	Quadcopter	30		7		20 MP CMOS sensor (with a ground sampling distance as low as 5 cm).
Delair UX11	Fixed-Wing	52		47		RGB and multispectral camera (with a ground sampling distance as small as 5 cm).
SenseFly eBee Ag	Fixed-Wing	55		36		A dual-purpose RGB and 4-band multispectral camera (with a ground sampling distance as low as 2.8 cm).
DJI Matrice 200 V2	Quadcopter	38		8		Camera comes separately

Multi-Rotor Drones

Multi-rotors can be categorized based on the number of rotors and their arrangement. Common types of multi-copters include the tricopter, quadcopter, hexacopter, and octocopter.

Tri Copter

As illustrated in Fig. 2, a tricopter features a general design with three rotors that help maintain balance during flight. The rotors are configured so that the right rotor spins clockwise, while the other two rotors rotate in the opposite direction. To counteract the imbalance caused by the clockwise torque, a servo mechanism tilts the rear rotor. This setup enables effective pitch control through the coordinated movement of the three rotors, facilitating forward motion. By adjusting the thrust of the left and right rotors, the tricopter can achieve rolling. Additionally, the vehicle is capable of moving sideways as well (McArthur et al. 2018).

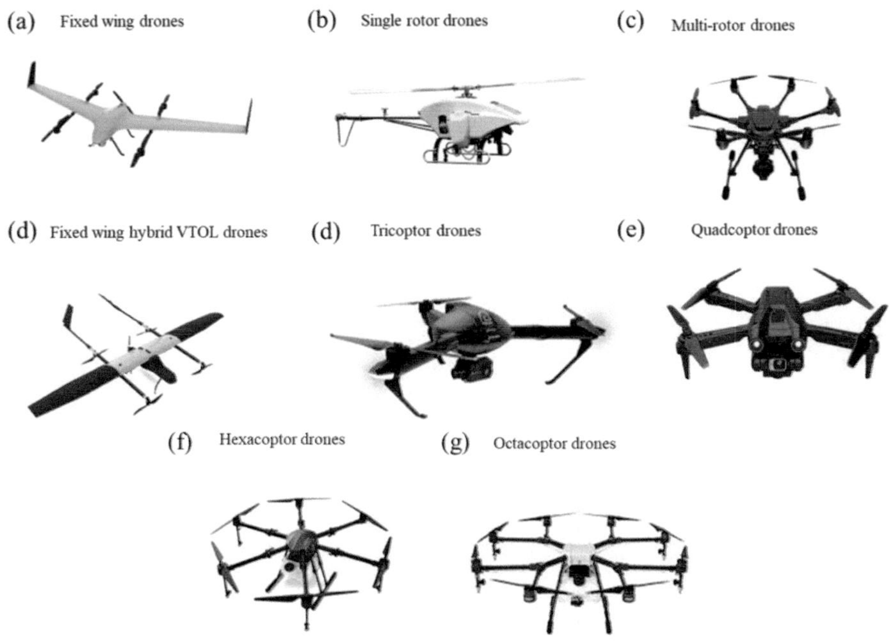

Fig. 2: Pictorial representation of different types of drones.

Quad Copter

The quadcopter represents an advanced UAV design featuring four rotors that generate lift for the aircraft. As depicted in Fig. 2, among these four rotors, two that are positioned opposite each other spin in a One rotates in a clockwise direction, while the other two spin counter clockwise. This configuration allows the quadcopter to maneuver around its axes: it can tilt forward or backward, known as 'pitch'; move laterally to the left or right, referred to as 'roll'; and rotate clockwise or counter clockwise, termed 'yaw'. Quadcopters come in two main configurations based on their shape: the plus and cross models. The cross model is more commonly used than the plus model due to its superior stability (Sharma 2021).

Hexa Copter

The Greek term "Hexa" means six, and a hexacopter is a type of drone that has six arms, with each arm housing a high-speed BLDC motor. As illustrated in Fig. 2, the drone's frame is constructed from fiberglass. Aluminum tubes are attached to the outer edge of the frame at the arm mounts. The six motors are affixed to the ends of these tubes. The frame plate serves as the primary support structure for other components of the drone, including the batteries, motor, flight control GPS antenna, and high-speed tube. Additionally, it contains FPV cameras, ESCs, circuit boards, and other essential components. This particular model is designed for agricultural use, specifically for pesticide spraying. It features a fluid tank with a maximum capacity of 5 liters mounted on the underside of the frame, with the tank's outlet connected to the inlet of the spray motor. The spray motor's outlet is connected to the spray nozzles. A U-shaped aluminum pipe, with dimensions of 14 × 1.5 mm, is used to secure components like the fluid tank, spray motor, and spray lance. The spray lance features four nozzles, each positioned 45 cm apart, covering a total span of 1.3 meters. The landing gear is positioned beneath the spray unit to ensure safe take-off and landing operations.

Octocopter

An octocopter, which features eight rotors, is utilized for agricultural spraying like the Hexa UAV. As depicted in Fig. 2, this model has a diagonal wheelbase measuring 1630 mm in diameter and is capable of operating for 15 minutes while carrying a 10 kg payload. It is equipped with six nozzles that provide a spray width of 5 to 8 meters. The movement and deposition of the sprayed droplets were analyzed using the Time-Resolved Particle Image Velocimetry (TRPIV). This method revealed that two key factors—rotor speed and nozzle position—affect the movement and deposition of the spray (Tang Qing et al. 2017).

Hybrid Drones

Hybrid Vertical Take-Off and Landing (VTOL)

Hybrid VTOL UAVs integrate the benefits of both fixed-wing and multi-rotor systems, offering enhanced versatility and performance. As illustrated in Fig. 2, these vehicles excel at take-offs, akin to multi-rotor systems with the flight efficiency of fixed-wing systems. However, the integration of these features makes the development and maintenance of hybrid systems, as well as their control mechanisms, quite complex. Specifically, three distinct controllers such as horizontal, vertical, and transition are employed to manage their operations (Delavarpour et al. 2021).

Applications of Drones in Farming

Drones offer the capability to monitor various crops across different locations. By incorporating drone technology, crop yields can be improved, time efficiency can be increased, land management can be made more sustainable, and long-term performance can be optimized. Drones are essential in a variety of Internet of Things (IoT) applications, offering real-time data collection and analysis to support smarter, more efficient decision-making across different sectors. Drones play a crucial role in numerous Internet of Things (IoT) applications. In agriculture and smart farming, drones are particularly significant, impacting the daily lives of farmers and people around the globe. This chapter focuses on several key applications of drone technology in these areas.

Crop Monitoring and Health Assessment

UAV-based crop monitoring represents a sophisticated technology that enhances crop vegetation indices through high-resolution imagery, providing insights into the condition of various crops (Oliveira et al. 2018, Chowdhury et al. 2024b; Barman et al. 2024). Effective crop monitoring is essential for predicting and managing potential disruptions in food supplies and fluctuations in global crop market prices. UAV imagery plays a pivotal role in precise crop monitoring, with wireless communication systems enabling seamless data transfer. For example, drones like the DJI Inspire 2 capture high-resolution images that help detect early signs of issues, enabling prompt interventions before significant damage can occur. Geo-tagged aerial images provide valuable insights that not only reduce operational costs but also significantly boost crop yields. The monitoring approach relies on capturing a broad spectrum of wavelengths, with UAV cameras proficiently capturing five main wavelength bands: Blue (0.44–0.51 μm), Green (0.52–0.59 μm), Red (0.63–0.68 μm), Red-edge (0.69–0.73 μm), and Near-infrared (0.76–0.85 μm) are specific spectral bands used in remote sensing for capturing detailed information about vegetation, soil, and other environmental factors. These bands are crucial for various agricultural applications, such as monitoring crop health, detecting stress, and assessing overall field conditions. The process of identifying stressed crop areas using multispectral indices involves several detailed steps, followed by field validation. The NDVI (Normalized Difference Vegetation Index) is one of the most commonly used indices (Purevdorj et al. 1998; Hashimoto et al. 2019). NDVI makes it easy to locate areas of healthy vegetation, thereby streamlining the targeted spraying process. The NDVI formula is detailed in Equation (1), and its application is extensively covered in various research studies (Chowdhury et al. 2024c, 2025).

$$\mathrm{NDVI} = \frac{\mathrm{NIR} - \mathrm{R}}{\mathrm{NIR} + \mathrm{R}} \tag{1}$$

In this equation, R represents the reflectance of the red band. The NDVI values always range between −1 and +1. If the result is close to 0, it suggests that the vegetation in the area is sparse. Typically, areas with low vegetation have NDVI values between 0.2 and 0.5. On the other hand, if the result approaches +1, it indicates high vegetation density. This makes it easy for farmers to quickly assess the health and condition of their fields. There are some well-known vegetation indices related to crop health monitoring Table 2. Early detection of crop damage is crucial for ensuring high yields. Drone pilots can gather high-resolution data to assess and document damage caused by floods, fires, pests, and weather events. Remote sensing and photogrammetry data from drones serve as evidence for crop insurance claims or estimating damages. Advanced sensors provide 2D or 3D orthomosaics that help farmers explore new strategies to enhance yields and reduce crop damage. Thermal or infrared cameras can further enhance agricultural management by assessing the thermal properties of plants and detecting harmful wildlife. Thermal imaging is also useful for monitoring plant diseases, water deficiencies, and other physiological conditions.

Soil Analysis and Mapping

Drones are advanced, reliable tools that can assist farmers in assessing farm conditions at the start of each crop season (Fig. 3). They create 3-D maps for soil analysis, which are crucial for planning seed planting. By analyzing soil and field conditions, drones provide valuable data for irrigation and nitrogen management,

Table 2: Vegetation indices related to Crop health Monitoring

Index	Abbreviation	Formula	Reference
Difference Vegetation Index	DVI	$s \times NIR - R$	Richardson and Wiegand 1977
Enhanced Vegetation Index	EVI	$G \times \frac{NIR - R}{NIR + C1 \times R - C2 \times B + L}$	Huete et al. 1999
Green Normalized Difference Vegetation Index	GNDVI	$\frac{NIR - G}{NIR + G}$	Gitelson et al. 1996
Modified Soil Adjusted Vegetation Index	MSAVI	$NIR + 0.5 - (0.5\sqrt{(2NIR+1)^2 - 8(NIR-2R)}$	Qi et al. 1994
Normalized Difference Vegetation Index	NDVI	$\frac{NIR - R}{NIR + R}$	Purevdorj et al. 1998
Normalized Difference Water Index	NDWI	$\frac{G - NIR}{G + NIR}$	McFeeters 1996
Ratio Vegetation Index	RVI	$\frac{R}{NIR}$	Richardson and Wiegand 1977
Soil Adjusted Vegetation Index	SAVI	$\frac{NIR - R}{NIR + R} \times (1 + L)$	Huete 1988
Transformed Vegetation Index	TVI	$\sqrt{\frac{NIR - R}{NIR + R}} + 0.5$	DW 1975

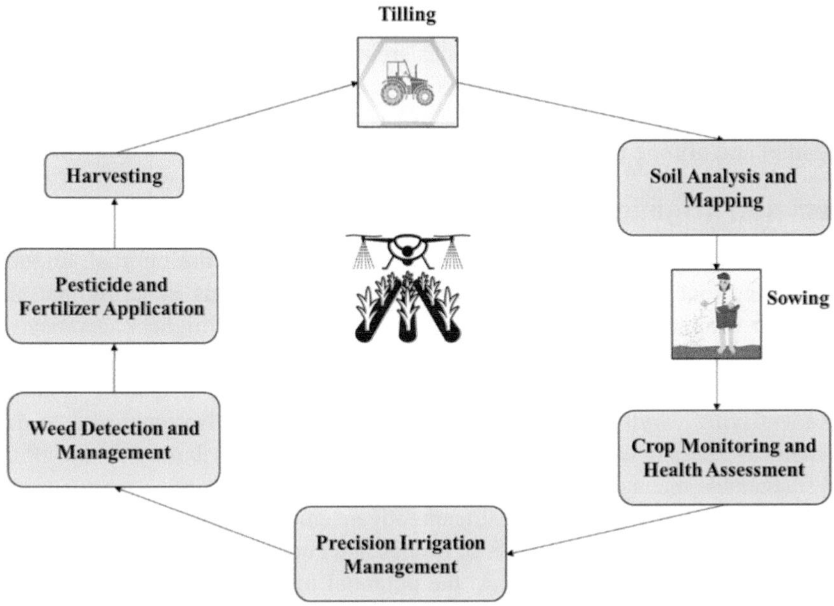

Fig. 3: Drone application for farming.

which enhances crop growth (Mondal et al. 2025). Currently, most fieldwork in broadacre farming is performed using human-operated machinery. However, in the context of modern, mechanized agriculture, farmers require proactive involvement to monitor field conditions effectively. The quality of the soil can significantly impact a farmer's productivity (Kushwah et al. 2024).

Soil analysis is a vital step in the crop cycle. DJI Phantom 4 Pro, offers real-time and precise assessments of soil health. Precision agriculture enables the identification of issues related to soil quality, nutrient management, or areas with poor soil (Kumari et al. 2023). This data helps farmers determine optimal planting patterns, manage crops, and maintain soil health. Additionally, field soil analysis improves worker safety and health. Similar to IoT (Internet of Things), the concept of IOS (Internet of the Soil) involves monitoring soil conditions like moisture, temperature, and electrical conductivity (Satpute et al. 2024). Remote sensing techniques support precision agriculture by collecting data that allows for seasonal growth monitoring. Satellite imaging, such as that from Sentinel-2 satellites provided by the European Space Agency (ESA), is also available. While satellites offer valuable data, drones can capture images in cloudy conditions where satellite imaging might be restricted. However, operating drones involves more effort in terms of pre-flight preparation, flying, and post-processing of images compared to satellite imaging. Drone imaging provides higher resolution, down to the centimeter level, whereas satellite images have a resolution measured in meters.

Drone-based remote sensing systems operate under various farming conditions, typically flying at an altitude of 100 meters and a speed of 4 meters per second. Radiometric and geometric calibrations, as well as orthomosaic processing, are conducted semi-automatically using GPS and calibration data. Soil mapping, which categorizes regions with similar soil properties, plays a crucial role in agriculture. Factors such as soil pH, salinity, texture, slope, water availability, and erosion risk can be assessed from these maps. Drones are particularly useful at the beginning of the crop cycle, providing accurate 3-D maps for soil analysis that aid in planning planting strategies. Post-planting, drone-based soil analysis offers insights into irrigation and nitrogen management.

Precision Irrigation Management

Precision irrigation water management involves providing the optimal amount of water to the root zone of crops at the most effective intervals, ensuring that plants meet their evapotranspiration and metabolic needs with the highest accuracy. This approach aims to maximize crop productivity while using minimal water (Chitra 2021). There is considerable potential for using drones in irrigation monitoring and identifying varying levels of plant stress. Drones are also employed to detect mechanical issues in canal operations and maintenance, contributing for the efficient and sustainable management of water resources.

Researchers have made significant advancements by creating a structured framework that incorporates a fleet of drones and coordinating vehicles for efficient canal evaluation. This approach has the potential to cover larger areas and reduce inspection times, thereby improving canal repair and maintenance. This model could be implemented in India to enhance precision irrigation management in canal systems. Although the use of drones in precision irrigation is currently limited, there is significant potential for growth, particularly with new regulations from the Government of India supporting this technology. Encouraging public-private partnerships is also crucial, as companies like Tata Kisan Kendra are already advancing digital farming solutions for precision agriculture and irrigation.

Robots fitted with hyperspectral, multispectral, or thermal sensors can detect dry areas or regions needing improvement within a field (Jha et al. 2023, 2024). Additionally, as crops develop, drones can measure the vegetation index, which indicates overall crop density and health, and detect thermal signatures, reflecting the energy or heat emitted by the crops. These drones help identify which areas need more water and which need less, leading to more efficient water use. This not only saves costs, especially in areas facing rising water prices or droughts but also reduces the risk of fertilizer runoff into streams and rivers can be mitigated, preventing harm to aquatic life and reducing the risk of harmful algal blooms (Patra et al. 2023).

Pesticide and Fertilizer Application

UAVs are revolutionizing agricultural practices by enhancing the precision and efficiency of pesticide and fertilizer application. Equipped with advanced spraying systems, UAVs can apply chemicals with pinpoint accuracy, it helps reduce waste and minimize environmental impact (Table 3). Their ability to cover vast areas quickly and access challenging terrains offers significant time and labour savings. Additionally, many UAVs are integrated with sensors and cameras that provide valuable data on crop health and field conditions, enabling more informed decision-making. While the technology offers numerous benefits, including improved application efficiency and reduced chemical usage, it requires adherence to regulatory standards and ongoing maintenance to ensure optimal performance. For pesticide and fertilizer applications, UAVs typically fall into two categories: multi-rotor and fixed-wing drones. Multi-rotor UAVs, such as the DJI Agras series, are favoured for their maneuverability and ability to hover, making them ideal for precise, localized applications and small to medium-sized fields. They are equipped with large tanks and sophisticated spraying systems that ensure even distribution of chemicals. In contrast, fixed-wing UAVs, like the sense FlyeBee X, excel in covering extensive areas efficiently due to their longer flight times and higher speeds. While they are less adept at precise hovering, their ability to map and survey large fields complements their application capabilities. Each type offers distinct advantages depending on the scale and specific needs of the agricultural operation.

The rapid adoption of high-tech UAVs for agricultural use is evident, particularly due to their precision and quick response in spraying systems. According to (Wen et al. 2018), five critical components of a spraying UAV are emphasized, including:

a. Prescription map system and spraying controller for precise application
b. Chemical tank
c. Flow sensor
d. Compact diaphragm pump
e. Nozzle with pressure control

Table 3: Aerial (manned aircraft) based remote sensing

Crop Name	UAV Type	Camera Type	Number of Rotors	Pest Name	Observations	References
Pine	Multi-Rotor	Multispectral	6	Bursaphelenchus	Visual Images	Wu et al. 2021
Maize	Multi-Rotor	Multispectral	6	Lepidoptera	Visual Images	Ishengoma et al. 2021
Lemon	Multi-Rotor	Hyperspectral	4	Phytophthora Gummosis	Visual inspection of images	Moriya et al. 2021
Rice	Multi-Rotor	Hyperspectral	4	Ustilaginoidea virens	Damage assessments	An et al. 2021
Grapevine	Multi-Rotor	Hyperspectral	4	Grapevine vein-clearing virus	Visual Images	Nguyen et al. 2021
Wheat	Multi-Rotor	Hyperspectral	4	Fusarium head blight	Visual inspection of images	Ma et al. 2019
Potato	Multi-Rotor	Hyperspectral	4	Phytophthora infestans	Visual Images	Yu et al. 2021
Soya	Multi-Rotor Quad copter	RGB	4	Defoliant pests such as insects and mollusks	Pest Segmentation and Classification	Tetila et al. 2020
Banana	Multi-Rotor Quad copter	RGB	4	Yellow sigatoka	Visual Images, Degree of Sever	Calou et al. 2020
Grape	Multi-Rotor	RGB, Multispectral, Hyperspectral	4	Grapephylloxera	Ground trap sand root digging, visual vigour assessments	Vanegas et al. 2018
Cotton	Multi-Rotor	Multispectral	4	Two-spotted spidermite	Damage assessments	Huang et al. 2018
Sugar Beet	Multi-Rotor	Hyperspectral	4	Beet Cyst Nematode	Visual Images	Joalland et al. 2018
Sorghum	Fixed Wing	Multispectral	1	Sugarcane aphid	Arthropod counts	Stanton et al. 2017
Canola	Multi-Rotor	Multispectral	8	Green peachaphid	Arthropod counts, soil and plant tissue nutrient analyses	Severtson et al. 2016
Onion	Fixed Wing	Multispectral	1	Thrips	-	Nebiker et al. 2016

edge technologies such as artificial intelligence (AI), nano-fertilizers, Internet of Things (IoT) sensors, and high-quality image capabilities. They increase crop productivity, lessen their impact on the environment, make better use of resources, and eradicate pests (Das 2024, Jayathi et al. 2024). Real-time data, accurate decision-making, and insights for improved crop management are all made possible by these technologies, which also contribute to cost savings, environmental sustainability, and higher productivity. Also, by reducing environmental impact and enhancing pest control effectiveness, the integration of IoT technologies with drones for accurate pesticide administration has transformed sustainable farming. The strategic use of drones in precision farming holds enormous promise to alter conventional agricultural techniques, providing food security for the expanding global population while reducing environmental degradation, despite obstacles such as high initial costs and regulatory constraints (Balaji et al. 2023). Key enabling technologies, such as IoT and UAV, can successfully and efficiently support smart farming techniques in this difficult environment by addressing the following constraints: profit improvement, user-friendliness, system simplicity and scalability, and ease of installation. Because these technologies can meet human requirements in both urban and rural settings, they will eventually turn traditional farming methods into an evolutionary agricultural ecosystem. Instead for them to be implemented successfully, their effects on the economy and environment must be fully recognized.

Effects on the Economy

Efficiency has been enhanced by using drones for data collection, spraying, and agricultural monitoring can drastically save labour expenses (Singh et al. 2024, J. Gao et al. 2024). Drones help increase crop yields and enhance product quality by facilitating the accurate application of inputs and the early detection of pests and illnesses (Tahir et al. 2023). Farmers may be able to make more money by using drone data for services like environmental monitoring, land appraisal, and crop insurance. This will possibly offer new revenue streams (Benami et al. 2021).

Effects on the Environment

By precisely applying fertilizer and pesticides, drones can minimize waste and lower pollution levels in the environment (Hu et al. 2022). Drones help maintain healthy soil and avoid erosion by maximizing irrigation and fertilizer management (Singh 2022). Drone-assisted precision agriculture can help biodiversity by lowering the usage of toxic pesticides and encouraging environmentally friendly farming methods (Toscano et al. 2024). Government officials, farmers, and academics may maximize the benefits of drone adoption while minimizing any negatives by undertaking thorough economic and environmental impact evaluations.

Future Trends and Innovations

In this section, we present various potential research avenues for progressing UAV technology.

AI and Machine Learning Integration

Although, Artificial Intelligence methods, such as machine learning (ML) and neural networks, have been utilized in UAVs, the full implementation of deep learning and reinforcement learning techniques is still limited due to constraints in power and processing capabilities. Researchers should focus on creating innovative deep learning approaches for UAVs, particularly for Synthetic Aperture Radar (SAR) applications. These strategies could enhance contextual decision-making and learning based on trajectory data, supporting more efficient autonomous piloting. Given the constraints of storage and energy, there is a need for lightweight and portable machine learning (ML) solutions overcome these challenges.

Energy Harvesting and Battery Innovation

Research should concentrate on identifying innovative energy harvesting techniques for UAVs and advancing materials for UAV batteries to prolong mission durations. The development of new, lightweight, and efficient batteries is essential for enhancing UAV flight times. Additionally, future studies should investigate optimal power control strategies and energy consumption management mechanisms.

Efficient Data Processing Algorithms

Power-efficient algorithms are needed for real-time processing of UAV data, including imagery and sensor data. Key areas include swarm coordination, optimization, collision avoidance, and trajectory planning for enhanced UAV performance.

Standardization and Integration

As UAV technology progresses, improving standardization, privacy regulations, image processing algorithms, low-cost sensors, extended flight times, and higher payload capacities is crucial. Integration of UAVs into applications like field crop phenotyping should also be prioritized.

Imaging Processing Challenges

Future research should address challenges in UAV imaging, such as varying orientations, overlaps, and different scales, to improve accuracy.

Transmission Lines Inspection

Efficient methods for inspecting transmission lines using UAVs need to be developed. This encompasses data analysis tools, collaborative platforms, reliable tracking and detection methods, vision-based inspection, and low-altitude photogrammetry techniques.

Advanced Sensing and Mapping

Researchers should explore new sensing devices, multispectral imagery, multi-decision-making algorithms, and edge computing to enhance agricultural monitoring and detection tasks.

Guidance, Navigation, and Localization

Research should focus on developing more effective strategies for sensing, guidance, navigation, and localization. Issues in these areas can cause inaccuracies and delays in parcel delivery. Therefore, low-cost, efficient sensing devices and localization systems are essential.

Public Safety Networks

Future research should explore the deployment of UAVs in robust public safety networks, focusing on advanced communication systems, disaster response, and integrating blockchain technology for health monitoring solutions. Recent studies have explored UAVs in public safety networks, highlighting the need for continued development.

Cooperative Multi-UAV Systems

Adaptive control and collaborative algorithms are crucial for multi-UAV systems, particularly as large numbers of UAVs are deployed for activities like light shows, QR code creation, and aerial promotions. Strong coordination mechanisms are essential to ensure the success of these applications.

Integration with Artificial Intelligence and Machine Learning

The incorporation of machine learning and artificial intelligence (AI) into drone technology is transforming precision agriculture. With advanced sensors and cameras, these cutting-edge drones are becoming vital assets for precision farming, environmental monitoring, and emergency response. This chapter examines the key progressions in machine learning and AI, emphasizing their crucial contribution to developing intelligent systems for smart agriculture projects. The synergy between drones, AI, and smart agriculture presents significant potential to enhance sustainability and operational effectiveness. Key areas affected by this transformation include smart agriculture, energy efficiency, noise reduction, environmental monitoring, and autonomous navigation. Machine learning and deep learning models play a vital role in addressing these challenges and exploring new possibilities (Taha and Shoufan 2019). AI and machine learning are revolutionizing agricultural development and management by optimizing resource usage, predicting growth patterns, and enhancing drone sustainability. The collaboration between AI-powered drones and agricultural innovation offers a future that is smarter, more sustainable, safer, and more efficient, leading to the creation of more precise and intelligent environments. Smart agriculture strives to establish sustainable, efficient, and technologically advanced environments by utilizing technology and data to improve quality of life and the performance of agricultural systems.

Potential Challenges and Opportunities

Unmanned Aerial Vehicles (UAVs) represent a rapidly advancing technology that has given rise to a range of innovative applications across various fields. The current trend in this technology is shifting from the use of large, remotely-controlled drones to networks of smaller, autonomous drones working together to perform complex tasks more efficiently and cost-effectively. A key challenge in this evolution is the development of effective sensing, communication, and control algorithms that can handle the dynamic nature of UAV networks, which feature varying levels of mobility. Recently, the integration of Artificial Intelligence (AI) into learning-based networking has gained traction. This approach leverages the cognitive capabilities of nodes within UAV networks to make more informed networking decisions by incorporating computational intelligence. One notable example of this trend is the creation of AI-powered routing protocols. These protocols use machine learning techniques to model and predict changes in network topology, channel conditions, and environmental factors, leading to improved routing and overall network performance.

Conclusion

Drones have revolutionized precision agriculture by efficiently collecting data, monitoring crops, and managing resources. Their ability to deliver accurate and timely insights, powered by advanced technologies like AI, machine learning, IoT, and high-resolution imagery, has significantly boosted farming productivity and sustainability. Drones are particularly valuable in applications such as crop health assessment, irrigation management, and pesticide application, showcasing their versatility across various agricultural practices.

Looking ahead, advancements in energy harvesting, data processing, and multispectral imaging hold the potential to further enhance drone performance, signaling a promising future for this technology. However, challenges such as regulatory restrictions, financial barriers, and environmental concerns remain. To maximize the benefits of UAVs while minimizing risks, innovative solutions and ongoing research are essential. The future direction of drone technology in agriculture lies in overcoming these obstacles, enabling the transformation of traditional farming methods into smart, sustainable systems. This progress could be key to ensuring food security for a growing global population.

References

Aasen, H., E. Honkavaara, A. Lucieer and P.J. Zarco-Tejada. 2018. Quantitative remote sensing at ultra-high resolution with UAV spectroscopy: A review of sensor technology, measurement procedures, and data correction workflows. Remote. Sens. 10(7): 1091.

Al-Thani, N., A. Albuainain, F. Alnaimi and N. Zorba. 2020, June. Drones for sheep livestock monitoring. 2020 IEEE 20th Mediterranean Electrotechnical Conference (MELECON): 672-676.

Alfatni, M.S.M., A.R.M. Shariff, H.Z.M. Shafri, O.M.B. Saaed and O.M. Eshanta. 2008. Oil palm fruit bunch grading system using red, green and blue digital number. J. Appl. Sci. 8(8): 1444-1452.

An, G., M. Xing, B. He, H. Kang, J. Shang, C. Liao et al. 2021. Extraction of areas of rice false smut infection using UAV hyperspectral data. Remote Sens. 13(16): 3185.

Anand, R., S.S. Nalla, S. Nasreen, M. Chowdhury, Madhusudan, B.S, A. Pandey et al. 2024. Fruit counting and analysis using artificial intelligence approaches. pp. 93-114. *In:* S.S. Chouhan, A. Saxena, U.P. Singh and S. Jain [eds.]. Artificial Intelligence Techniques in Smart Agriculture. Springer Nature, Singapore.

Andrew, W., C. Greatwood and T. Burghardt. 2017. Visual localisation and individual identification of holstein friesian cattle via deep learning. Proceedings of the IEEE International Conference on Computer Vision Workshop: 2850-2859.

Andrew, W., C. Greatwood and T. Burghardt. 2019. Aerial animal biometrics: Individual friesian cattle recovery and visual identification via an autonomous UAV with onboard deep inference. 2019 IEEE/RSJ International Conference on Intelligent Robots and Systems (IROS): 237-243.

Anthony, D., S. Elbaum, A. Lorenz and C. Detweiler. 2014. On crop height estimation with UAVs. 2014 IEEE/RSJ International Conference on Intelligent Robots and Systems, 4805-4812.

Singh, A., B. Ganapathysubramanian, A.K. Singh and S. Sarkar. 2016. Machine learning for high-throughput stress phenotyping in plants. Trends Plant Sci. 21(2): 110-124.

Balaji, P., D. Devasena, S. Kayalvizhi and N. Aarthikannan. 2023. IoT-empowered precision agricultural multi-rotor drones: A revolutionary approach for sustainable farming. 2023 International Conference on Recent Advances in Science and Engineering Technology (ICRASET): 1-5.

Barman, A., S. Dutta, A. Bera, P. Saha, M. Roy Choudhury, M. Bera et al. 2024. Synergizing sustainability: A critical review on harnessing agroforestry for biomass, carbon sequestration, and water-food-energy nexus. Energy Ecol. Environ. 1-35.

Benami, Elinor, et al. 2021. Uniting remote sensing, crop modelling and economics for agricultural risk management. Nature Reviews Earth & Environ. 2(2): 140-159.

Bendig, J., A. Bolten and G. Bareth. 2012. Introducing a low-cost mini-UAV for thermal- and multispectral-imaging. ISPRS Archives. 39: 345-349.

Blaschke, T. 2010. Object based image analysis for remote sensing. ISPRS J. Photogramm. Remote Sens. 65(1): 2-16.

Cai, G., B.M. Chen and T.H. Lee. 2010. An overview on development of miniature unmanned rotorcraft systems. Front. Electr. Electron. Eng. China, 5: 1-14.

Calou, V.B.C., A. dos Santos Teixeira, L.C.J. Moreira, C.S. Lima, J.B. de Oliveira and M.R.R. de Oliveira. 2020. The use of UAVs in monitoring yellow sigatoka in banana. Biosyst. Eng. 193: 115-125.

Chen, C., B. Zhang, H. Su, W. Li and L. Wang. 2016. Land-use scene classification using multi-scale completed local binary patterns. Signal Image Video Process. 10: 745-752.

Chitra, S. 2021. Automated irrigation system using IOT. IJIRCCE, 9(4).

Chollet, F. and F. Chollet. 2021. Deep Learning with Python. Simon and Schuster.

Chowdhury, M., R. Anand, T. Dhar, R. Kurmi, R.K. Sahni and A. Kushwah. 2024a. Digital insights into plant health: Exploring vegetation indices through computer vision. pp. 7-30. *In*: S.S. Chouhan, U.P. Singh and S. Jain [eds.]. Applications of Computer Vision and Drone Technology in Agriculture 4.0. Springer Nature, Singapore.

Chowdhury, M., T.K. Khura, P.K. Upadhyay, R.A. Parray, H.L. Kushwaha, C. Singh et al. 2024b. Assessing vegetation indices and productivity across nitrogen gradients: A comparative study under transplanted and direct-seeded rice. Front. Sustain. Food Syst. 8: 1351414.

Chowdhury, M., T.K. Khura, R.A. Parray, H.L. Kushwaha, P.K. Upadhyay, A. Jha et al. 2024c. The use of destructive and nondestructive techniques in concrete nitrogen assessment in plants. J. Plant Nutri. 47: 2271-2294.

Chowdhury, M., T.-K. Khura, R.A. Parray, P.K. Upadhyay, H.L. Kushwaha, A. Lama and A. Kushwah. 2025. Multi-vegetation indices based handheld device for precise nitrogen assessment and prescription in direct seeded rice. Comput. Electron. Agric. 229: 109886.

Das, S. 2024. Transforming agriculture: Harnessing robotics and drones for sustainable farming solution. J. Exp. Agric. Int. 46(7): 219-231.

de Oliveira Turci, L. 2017. Real-time operating system FreeRTOS application for fire alarm project in reduced scale. IJCDS (International Journal of Computing and Digital Systems) 6(4).

Delavarpour, N., C. Koparan, J. Nowatzki, S. Bajwa and X. Sun. 2021. A technical study on UAV characteristics for precision agriculture applications and associated practical challenges. Remote Sens. 13(6): 1204.

Dong, X., M.C. Vuran and S. Irmak. 2013. Autonomous precision agriculture through integration of wireless underground sensor networks with center pivot irrigation systems. Ad Hoc Netw. 11(7): 1975-1987.

Dutta, H., S. Dutta, H. Yadav, I.G. Prabhu, A. Roy, N.B. Chowdary and S. Das. 2024. A promising and forward-looking advancement using drones: Perspectives from Indian Sericulture. Reg. Rural Dev. 2(3): 10014.

Dutta, S., A.K. Singh, B.P. Mondal, D. Paul and K. Patra. 2023. Perspective chapter: Digital inclusion of the farming sector using drone technology. *In:* Human-Robot Interaction-Perspectives and Applications. IntechOpen.

DW, D. 1975. Measuring forage production of grazing units from Landsat MSS data. Proceedings of 10th International Symposium on Remote Sensing of Environment, 1975: 1169-1178.

Gao, J., C.K. Bambrah, N. Parihar, S. Kshirsagar, S. Mallarapu, H. Yu et al. 2024. Analysis of various machine learning algorithms for using drone images in livestock farms. Agric. 14(4): 522.

Gao, P., Y. Zhang, L. Zhang, R. Noguchi and T. Ahamed. 2019. Development of a recognition system for spraying areas from unmanned aerial vehicles using a machine learning approach. Sens. 19(2): 313.

Gašparović, M., M. Zrinjski, Đ. Barković and D. Radočaj. 2020. An automatic method for weed mapping in oat fields based on UAV imagery. Comput. Electron. Agric. 173: 105385.

Gawande, V., D.R.K. Saikanth, B.S. Sumithra, S.A. Aravind, G.N. Swamy, M. Chowdhury and B.V. Singh. 2023. Potential of precision farming technologies for eco-friendly agriculture. Int. J. Plant Soil Sci. 35(19): 101-112.

Gitelson, A.A., Y.J. Kaufman and M.N. Merzlyak. 1996. Use of a green channel in remote sensing of global vegetation from EOS-MODIS. Remote Sens. Environ. 58(3): 289-298.

Goolsby, J.A., J. Jung, J. Landivar, W. McCutcheon, R. Lacewell, R. Duhaime and A. Schwartz. 2016. Evaluation of unmanned aerial vehicles (UAVs) for detection of cattle in the cattle fever tick permanent quarantine zone. Sub. Agri. Environ. 67: 24-27.

Hashimoto, N., Y. Saito, M. Maki and K. Homma. 2019. Simulation of reflectance and vegetation indices for unmanned aerial vehicle (UAV) monitoring of paddy fields. Remote Sens. 11(18): 2119.

Hinas, A., J.M. Roberts and F. Gonzalez. 2017. Vision-based target finding and inspection of a ground target using a multirotor UAV system. Sens. 17(12): 2929.

Hu, H., Y. Kaizu, J. Huang, K. Furuhashi, H. Zhang, M. Li and K. Imou. 2022. Research on methods decreasing pesticide waste based on plant protection unmanned aerial vehicles: A review. Front. Plant Sci. 13: 811256.

Huang, H., J. Deng, Y. Lan, A. Yang, X. Deng, L. Zhang et al. 2018. A two-stage classification approach for the detection of spider mite-infested cotton using UAV multispectral imagery. Remote Sens Lett. 9(10): 933-941.

Huang, Y., W.C. Hoffmann, Y. Lan, W. Wu and B.K. Fritz. 2009. Development of a spray system for an unmanned aerial vehicle platform. Appl. Eng. Agric. 25(6): 803-809.

Huang, Y., S.J. Thomson, W.C. Hoffmann, Y. Lan and B.K. Fritz. 2013. Development and prospect of unmanned aerial vehicle technologies for agricultural production management. Int. J. Agric. Biol. Eng. 6(3): 1-10.

Huete, A., C. Justice and W. Van Leeuwen. 1999. MODIS vegetation index (MOD13). Algorithm Theoretical Basis Document. 3(213): 295-309.

Huete, A.R. 1988. A soil-adjusted vegetation index (SAVI). Remote Sens. Environ. 25(3): 295-309.

Ishengoma, F.S., I.A. Rai and R.N. Said. 2021. Identification of maize leaves infected by fall armyworms using UAV-based imagery and convolutional neural networks. Comput. Electron. Agric. 184: 106124.

Jayanthi, G., W. Nancy, B. Umamaheswari, R. Chithrakkannan, R. Sujith and S.S. Prasanna. 2024. Intelligent agricultural drones utilizing nano-fertilizer dispensation for precision farming. 2024 International Conference on Communication, Computing and Internet of Things (IC3IoT), 1-6.

Jensen, J.R. 1996. Introductory Digital Image Processing: A Remote Sensing Perspective. 2nd Edition, Prentice Hall, Inc., Upper Saddle River, NJ.

Jha, A., M. Chowdhury and A.N. Satpute. 2023. Surface water quality forecasting using machine learning approach. pp. 293-315. *In*: C.B. Pande, M. Kumar and N.L. Kushwaha [eds.]. Surface and Groundwater Resources Development and Management in Semi-arid Region: Strategies and Solutions for Sustainable Water Management. Cham: Springer International Publishing.

Jha, A., S. Garg, S. Satpute, J.P. Singh, D. Singh, U.G. Phutela et al. 2024. Isotherm and kinetic modelling for assessing the effectiveness of economical adsorbents in sewage water treatment. Indian J. Chem. Technol. 31: 438-450.

Joalland, S., C. Screpanti, H.V. Varella, M. Reuther, M. Schwind, C. Lang et al. 2018. Aerial and ground based sensing of tolerance to beet cyst nematode in sugar beet. Remote Sens. 10(5): 787.

Kamusoko, C. 2017. Importance of remote sensing and land change modeling for urbanization studies. Urban Development in Asia and Africa: Geospatial Analysis of Metropolises. 3-10.

Kang, Y., W. Joo, S. Lee and D. Shin. 2017. Priority-driven spatial resource sharing scheduling for embedded graphics processing units. J. Syst. Archit. 76: 17-27.

Kerkech, M., A. Hafiane and R. Canals. 2018. Deep leaning approach with colorimetric spaces and vegetation indices for vine diseases detection in UAV images. Comput. Electron. Agric. 155, 237-243.

Kumari, A., P. Ashoka, P. Tiwari, P. Sachan, A.K. Malla, A. Tripathy and M. Chowdhury. 2023. Zero tillage lead to enhanced productivity and soil health. Int. J. Environ. Clim. Chang. 13(10): 3707-3715.

Kushwah, A., A. Chouriya, V.K. Tewari, C. Gupta, M. Chowdhury, P. Shrivastava and P. Bhagat. 2024. A novel embedded system for tractor implement performance mapping. Cogent Eng. 11(1): 2311093.

LeCun, Y., Y. Bengio and G. Hinton. 2015. Deep learning. Nat. 521(7553): 436-444.

Longmore, S.N., R.P. Collins, S. Pfeifer et al. 2017. Adapting astronomical source detection software to help detect animals in thermal images obtained by unmanned aerial systems. Int. J. Remote Sens. 38(8-10): 2623-2638.

López-Granados, F. 2011. Weed detection for site-specific weed management: Mapping and real-time approaches. Weed Res. 51(1): 1-11.

Ma, H., W. Huang, Y. Jing, C. Yang, L. Han, Y. Dong et al. 2019. Integrating growth and environmental parameters to discriminate powdery mildew and aphid of winter wheat using bi-temporal Landsat-8 imagery. Remote Sens. 11(7): 846.

Madhusudan, B.S., H.L. Kushwaha, A. Kumar, R.A. Parray, S.S. Swain, M. Chowdhury et al. 2024. Critical assessment of furrow openers and operational parameters for optimum performance under conservation tillage. Sci. Rep. 14(1): 20928.

Malkani, P., I. Mani, P.K. Sahoo, R.A. Parray, S.S. Swain, K.R. Asha et al. 2024. Development and evaluation of a sensor enabled smart herbicide applicator for inter row crops. Smart Agric. Technol. 9: 100649.

Marinello, F., A. Pezzuolo, A. Chiumenti and L. Sartori. 2016. Technical analysis of unmanned aerial vehicles (drones) for agricultural applications. Eng. Rural Dev. 15(2): 870-875.

Maurya, P. 2015. Hardware implementation of a flight control system for an unmanned aerial vehicle. Retrieved, 6(01): 2015.

McArthur, D.R., A.B. Chowdhury and D.J. Cappelleri. 2018. Design of the interacting-boomcopter unmanned aerial vehicle for remote sensor mounting. J. Mech. Robot. 10(2): 25001.

McFeeters, S.K. 1996. The use of the Normalized Difference Water Index (NDWI) in the delineation of open water features. Int. J. Remote Sens. 17(7): 1425-1432.

Mink, R., A. Dutta, G.G. Peteinatos, M. Sökefeld, J.J. Engels, M. Hahn and R. Gerhards. 2018. Multi-temporal site-specific weed control of Cirsium arvense (L.) Scop. and Rumex crispus L. in maize and sugar beet using unmanned aerial vehicle based mapping. Agric. 8(5): 65.

Molaei, F., E. Rahimi, H. Siavoshi, S.G. Afrouz and V. Tenorio. 2020. A comprehensive review on internet of things (IoT) and its implications in the mining industry. Am. J. Eng. Appl. Sci. 13(3): 499-515.

Mondal, B.P., G.B. Raj, M. Dixit, T. Sadhu, V. Pandit, L. Prasanna et al. 2025. Role of soils for satisfying global demands for food, water, and bioenergy. *In:* Environmental Nexus for Resource Management (pp. 302-316). CRC Press.

Moriya, É.A.S., N.N. Imai, A.M.G. Tommaselli, A. Berveglieri, G.H. Santos, M.A. Soares et al. 2021. Detection and mapping of trees infected with citrus gummosis using UAV hyperspectral data. Comput. Electron. Agric. 188: 106298.

Nair, V. and G.E. Hinton. 2010. Rectified linear units improve restricted boltzmann machines. Proceedings of the 27th International Conference on Machine Learning (ICML-10): 807-814.

Nebiker, S., N. Lack, M. Abächerli and S. Läderach. 2016. Light-weight multispectral UAV sensors and their capabilities for predicting grain yield and detecting plant diseases. Int. Arch. Photogramm. 41: 963-970.

Nguyen, C., V. Sagan, M. Maimaitiyiming, M. Maimaitijiang, S. Bhadra and M.T. Kwasniewski. 2021. Early detection of plant viral disease using hyperspectral imaging and deep learning. Sens. 21(3): 742.

Nissimagoudar, P., V. Mane, K. Shamshuddin, H. Gireesha, R. Joshi, M. Kiran and B. Desai. 2016. Integrated course projects in automotive electronics and RTOS. J. Eng. Educ. Transform. 30(2): 71-78.

Oerke, E.C. 2006. Crop losses to pests. J. Agric. Sci. 144(1): 31-43.

Oliveira, H.C., V.C. Guizilini, I.P. Nunes and J.R. Souza. 2018. Failure detection in row crops from UAV images using morphological operators. IEEE Geosci. Remote. Sens. Lett. 15(7): 991-995.

Pajares, G. 2015. Overview and current status of remote sensing applications based on unmanned aerial vehicles (UAVs). Photogramm. Eng. Remote Sen. 81(4): 281-330.

Pastor, E., J. Lopez and P. Roy. 2007. UAV payload and mission control hardware/software architecture. IEEE Trans. Aerosp. Electron. Syst. 22(6): 3-8.

Patra, K., C.M. Parihar, H.S. Nayak, B. Rana, D.R. Sena, A. Anand et al. 2023. Appraisal of complementarity of subsurface drip fertigation and conservation agriculture for physiological performance and water economy of maize. Agric. Water Manag., 283: 108308.

Pradhan, N.C., M.A. Naik, M. Chowdhury, A. Kushwah, K.R. Asha, T. Dhar et al. 2024. Robotic seeding or sowing system in smart agriculture. pp. 495-520. *In*: K. Pandey, N.L. Kushwaha, C.B. Pande and K.G. Singh [eds.]. Artificial Intelligence and Smart Agriculture: Technology and Applications. Springer Nature, Singapore.

Purevdorj, T.S., R. Tateishi, T. Ishiyama and Y. Honda. 1998. Relationships between percent vegetation cover and vegetation indices. Int. J. Remote Sens. 19(18): 3519-3535.

Qi, J., A. Chehbouni, A.R. Huete, Y.H. Kerr and S. Sorooshian. 1994. A modified soil adjusted vegetation index. Remote Sens. Environ. 48(2): 119-126.

Qing, T., Z. Ruirui, C. Liping, X. Min, Y. Tongchuan and Z. Bin. 2017. Droplets movement and deposition of an eight-rotor agricultural UAV in downwash flow field. J. Agric. & Biol. Eng. 10(3): 47-56.

Richardson, A.J. and C.L. Wiegand. 1977. Distinguishing vegetation from soil background information. Photogramm. Eng. Remote Sens. 43(12): 1541-1552.

Rivas, A., P. Chamoso, A. González-Briones and J.M. Corchado. 2018. Detection of cattle using drones and convolutional neural networks. Sens. 18(7): 2048.

Satpute, A.N., K.P. Gavhane, S. Kaur, A. Jha, N.C. Pradhan and M. Chowdhury. 2024. Integration of AI and IoT in soilless cultivation to power sustainable agricultural revolution. pp. 387-411. *In*: K. Pandey, N.L. Kushwaha, C.B. Pande and K.G. Singh [eds.]. Artificial Intelligence and Smart Agriculture: Technology and Applications. Springer Nature, Singapore.

Saxena, L. and L. Armstrong. 2014. A survey of image processing techniques for agriculture. *In:* Proceedings of Asian Federation for Information Technology in Agriculture (pp. 401-413). Perth, W.A. Australian Society of Information and Communication Technologies in Agriculture.

Severtson, D., N. Callow, K. Flower, A. Neuhaus, M. Olejnik and C. Nansen. 2016. Unmanned aerial vehicle canopy reflectance data detects potassium deficiency and green peach aphid susceptibility in canola. Precis. Agric. 17: 659-677.

Sharma, R. 2021. Review on application of drone systems in precision agriculture. J. Adv. Res. Elec. Eng. Technol. 7(2): 5-7.

Singh, A.K. 2022. Precision agriculture in india – Opportunities and challenges. Ind. J. Fert. 18(4): 308-331.

Singh, E., A. Pratap, U. Mehta and S.I. Azid. 2024. Smart agriculture drone for crop spraying using image-processing and machine learning techniques: Experimental validation. IoT, 5(2): 250-270.

Smith, A.M. and R.E. Blackshaw. 2003. Weed–crop discrimination using remote sensing: A detached leaf experiment. Weed Technol. 17(4): 811-820.

Snow, C. 2016. The truth about drones in precision agriculture. Skylogic Res. 1: 1-6.

Srivastava, K., P.C. Pandey and J.K. Sharma. 2020. An approach for route optimization in applications of precision agriculture using UAVs. Drones. 4(3): 58.

Stanton, C., M.J. Starek, N. Elliott, M. Brewer, M.M. Maeda and T. Chu. 2017. Unmanned aircraft system-derived crop height and normalized difference vegetation index metrics for sorghum yield and aphid stress assessment. J. Appl. Remote Sens. 11(2): 26035.

Taha, B. and A. Shoufan. 2019. Machine learning-based drone detection and classification: State-of-the-art in research. IEEE Access. 7: 138669-138682.

Tahir, M.N., Y. Lan, Y. Zhang, H. Wenjiang, Y. Wang and S.M.Z.A. Naqvi. 2023. Application of unmanned aerial vehicles in precision agriculture. *In:* Precision Agriculture (pp. 55-70). Academic Press.

Tetila, E.C., B.B. Machado, G. Astolfi, N.A. de Souza Belete, W.P. Amorim, A.R. Roel and H. Pistori. 2020. Detection and classification of soybean pests using deep learning with UAV images. Comput. Electron. Agric. 179: 105836.

Toscano, F., C. Fiorentino, N. Capece, U. Erra, D. Travascia, A. Scopa et al. 2024. Unmanned Aerial Vehicle for Precision Agriculture: A Review. IEEE Access 12: 69188-69205.

Tyagi, R. and P.C. Pandey. 2024. Applications of drones in precision agriculture: Future of smart and sustainable farming. *In:* Remote Sensing in Precision Agriculture (pp. 429-453). Academic Press.

Vanegas, F., D. Bratanov, K. Powell, J. Weiss and F. Gonzalez. 2018. A novel methodology for improving plant pest surveillance in vineyards and crops using UAV-based hyperspectral and spatial data. Sens. 18(1): 260.

Vardhan, P.H., S. Dheepak, P.T. Aditya and S. Arul. 2014. Development of automated aerial pesticide sprayer. Int. J. Eng. Sci. Res. Technol. 3(4): 458-462.

Verhoeven, G. 2011. Taking computer vision aloft–archaeological three-dimensional reconstructions from aerial photographs with photoscan. Archaeol. Prospect. 18(1): 67-73.

Wang, W., Y. Wang, J. Dai and Z. Cao. 2019. Dynamic soft real-time scheduling with preemption threshold for streaming media. Int. J. Digit. Multimed. Broadcast. 2019(1): 5284968.

Weiss, M., F. Jacob and G. Duveiller. 2020. Remote sensing for agricultural applications: A meta-review. Remote Sens. Environ. 236: 111402.

Wen, S., Q. Zhang, J. Deng, Y. Lan, X. Yin and J. Shan. 2018. Design and experiment of a variable spray system for unmanned aerial vehicles based on PID and PWM control. Appl. Sci. 8(12): 2482.

Wu, B., A. Liang, H. Zhang, T. Zhu, Z. Zou, D. Yang et al. 2021. Application of conventional UAV-based high-throughput object detection to the early diagnosis of pine wilt disease by deep learning. For. Ecol. Manag. 486: 118986.

Xue, X., Y. Lan, Z. Sun, C. Chang and W.C. Hoffmann. 2016. Develop an unmanned aerial vehicle based automatic aerial spraying system. Comput. Electron. Agric. 128(2016): 58-66.

Yanliang, Z., L. Qi and Z. Wei. 2017. Design and test of a six-rotor unmanned aerial vehicle (UAV) electrostatic spraying system for crop protection. J. Agric. & Biol. Eng. 10(6): 68-76.

Yu, R., Y. Luo, Q. Zhou, X. Zhang, D. Wu and L. Ren. 2021. Early detection of pine wilt disease using deep learning algorithms and UAV-based multispectral imagery. For. Ecol. Manag. 497: 119493.

Zhao, B., X. Chen, X. Zhao, J. Jiang and J. Wei. 2018. Real-time UAV autonomous localization based on smartphone sensors. Sens. 18(12): 4161.

Ziliani, M.G., S.D. Parkes, I. Hoteit and M.F. McCabe. 2018. Intra-season crop height variability at commercial farm scales using a fixed-wing UAV. Remote Sens. 10(12): 2007.

CHAPTER

5

Robotics on the Farm: Automating Tasks for Increased Productivity

Mude Arjun Naik[1,2*], Achugatla Kesav Kumar[1], Sajja Poojith[1], Rachamalla Naveen[1], V. Shoban Naik[1], Sunil Kumar Rathod[1], A. Subba Rami Reddy[2] and Y. Pavan Kumar Reddy[2]

[1] ICAR-Indian Agricultural Research Institute, New Delhi - 110012, India
[2] Agricultural College, Mahanandi, ANGRAU, Andhra Pradesh - 518502, India

Introduction

The world is facing a significant challenge in feeding its rapidly growing population. According to the United Nations' World Population Prospects report (2022), the global population is expected to reach 8.5 billion by 2030, 9.7 billion by 2050, and 10.4 billion by 2100. To meet the food demands of this growing population, annual cereal production needs to increase by about 3 billion tonnes, and annual meat production must rise by over 200 million tonnes (FAO 2017, Mishra and Sahni 2024).

However, this poses a considerable challenge as the availability of resources such as water, land, and labour continue to dwindle (Singh and Sahni 2019; Jha et al. 2023). To address this issue, it is essential to adopt innovative technological interventions, develop high-yielding hybrid varieties and integrating robotics and automation in agricultural practices. Recent decades have seen rapid advancements in technology, transforming various industries, including agriculture (Satpute et al. 2024; Malkhani et al. 2024). One of the most significant innovations is the introduction of robotics, which has revolutionized agricultural practices by automating critical tasks that are tedious, labour-intensive, and time-consuming (Jat et al. 2023; Bali et al. 2023).

To address the above-mentioned issues, this chapter explores the role of robotics in agriculture, tracing its historical evolution, distinguishing agricultural robots from their industrial counterparts, and analyzing the components and applications of robots used in farming today. Additionally, it discusses the current status, challenges, and future prospects of this technology, concluding with an assessment of its potential to reform the future of agriculture.

*Corresponding author: arjunnaik133@gmail.com

DOI: 10.1201/9781003536932-5

History of Robotics from Industry to Agriculture

The concept of robotics has its roots in ancient times, with the term "robot" deriving from the Czech word "robota", meaning forced labour (Anonymous 2024a). However, the development of functional robots is a relatively recent phenomenon, driven primarily by advancements in computer science, electrical engineering and mechanical engineering. The introduction of modern robotics is often traced back to the mid-20th century, with the invention of the first programmable robotic arm in 1961 by George Devol and Joseph Engel Berger (Anonymous 2024a). This innovation marked the beginning of robotics in the manufacturing industry and over the years, robotics has evolved to become an integral part of various industries, including automobile manufacturing, electronics assembly and medical surgery. According to the International Federation of Robotics (IFR), the operational stock of industrial robots reached approximately 3,903,633 units by the end of 2022 (Anonymous 2024b). In 2021 the industrial sector accounted for 42% of the world's robotics revenue, while agricultural robots had an 8% share (Fig. 1).

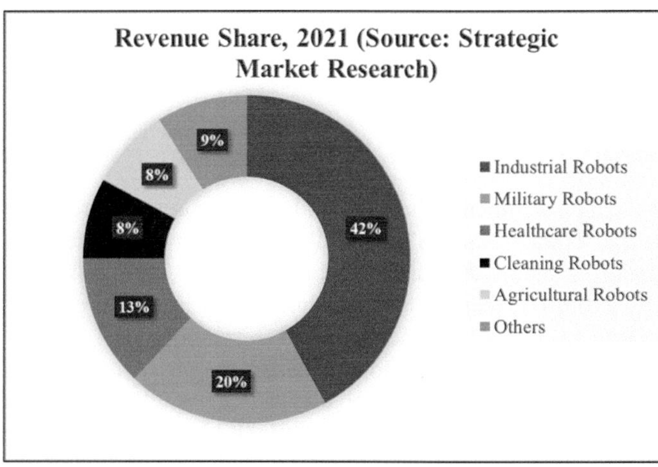

Fig. 1: Revenue share of various robotic applications in the world.

The agricultural robotics market has been expanding rapidly, with a compound annual growth rate (CAGR) of almost 19% (Anonymous 2024b). The market was valued at USD 4.9 billion in 2021 and is expected to reach USD 10.5 billion by the end of the forecast period. Additionally, more than 200,000 agricultural robots were shipped in 2020, and the agricultural drone industry is projected to reach a valuation of USD 6.2 billion by 2024 (Anonymous 2024c).

Although robotics was initially adapted in the industrial sector, its application in agriculture has been gaining momentum (Billingsley et al. 2007). The use of robotics in agriculture began to take shape in the late 20th century and accelerated in the 21st century with advancements in artificial intelligence (AI), machine learning, sensor technology and automation (Pradhan et al. 2024). Agricultural robots, also known as agribots have evolved significantly from their early prototypes, which were primarily used for tasks like milking cows or plowing fields (Benos et al. 2023). Today, robots are being designed for various agricultural activities, including seedbed preparation, intercultural operations, crop health monitoring, harvesting, poultry and livestock management. The application of robots in different farming tasks are shown in Fig. 2.

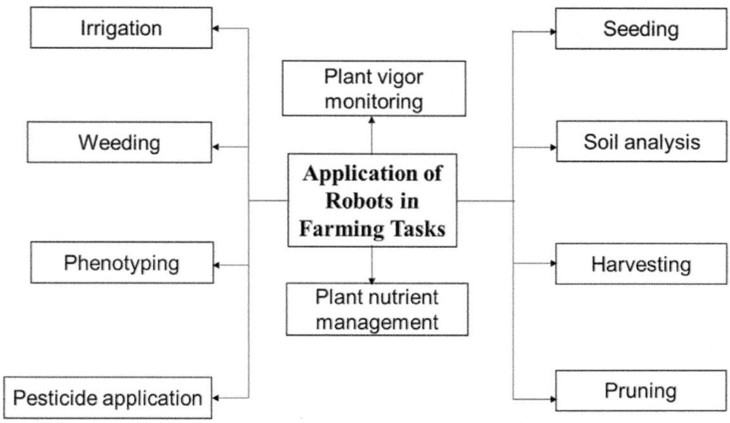

Fig. 2: Application of robots in different farming tasks.

Table 1: Difference between industrial and agricultural robots

Feature	Industrial Robots	Agricultural Robots
Primary function	Designed for precision tasks in manufacturing such as assembly, handling and packing.	Designed for farming tasks such as seeding, intercultural, crop monitoring and harvesting.
Environment	Works in a controlled environment.	Works in a dynamic environment.
Mobility	Often stationary or moves along fixed paths (e.g., robotic arms).	Typically mobile, utilizing autonomous navigation to traverse varied terrains.
Task complexity	Performs repetitive, high-speed tasks requiring precision and consistency.	Performs complex tasks that may vary based on environmental conditions (e.g., fruit ripeness).
Sensors used	Equipped with high-precision sensors for exact measurements (e.g., laser, vision).	Uses a range of sensors for environmental monitoring (e.g., soil moisture and crop health).
Human interaction	Generally operated with limited human interaction; often require safety measures.	Can be designed for collaborative use alongside humans, enhancing efficiency in farming.
Adaptability	Less adaptable, programmed for specific tasks and environments.	Highly adaptable to different crops and farming conditions through advanced AI and machine learning.
Cost considerations	High initial investment but can reduce labour costs significantly over time.	Investments vary widely, focus on reducing labour costs and increasing yield efficiency.

Agricultural Robots vs. Industrial Robots: Key Differences

Agricultural robots differ significantly from industrial robots due to the distinct environments and tasks they are designed for as shown in Table 1. The industrial robots operate in structured environments, performing repetitive and predictable tasks, whereas agricultural robots navigate dynamic, uncontrolled and unstructured environments (Billingsley et al. 2007). Agricultural robots must be adaptable, capable of handling uneven terrains, varying weather conditions and plant matter of different sizes, shapes and maturity (Benos et al. 2023). In contrast, industrial robots prioritize speed, accuracy and precision in controlled environments. The complexity of agricultural tasks and environmental variability make agricultural robots more reliant on advanced artificial intelligence (AI) and sensor technology than industrial robots (Jat et al. 2023; Anand et al. 2024).

Classification and Components of Agricultural Robots

An agricultural robot is a complex machine made of several key components that work together to perform a specific task. The classification of agricultural robots based on different purposes are shown in Fig. 3. The flowchart of different components and applications of agricultural robots are shown in Fig. 4. The components include:

Sensors

Sensors are a crucial component of agricultural robots, enabling them to collect data on crop conditions and environmental factors such as soil moisture levels, temperature,

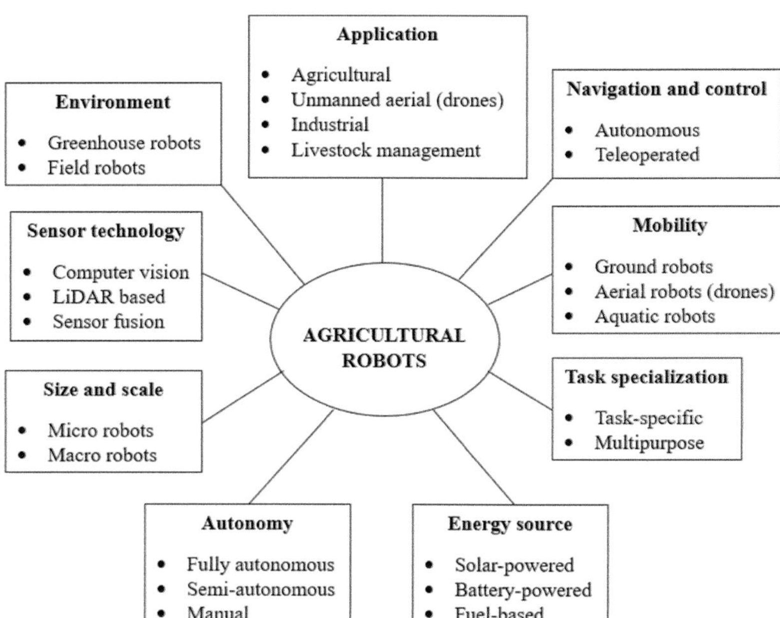

Fig. 3: Classifications of agricultural robots.

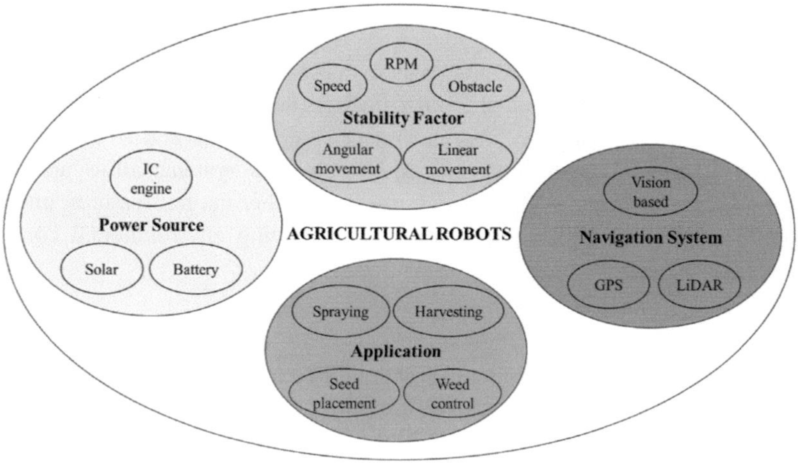

Fig. 4: Flowchart of different components and applications of agricultural robots.

humidity, and light intensity. Advanced sensors, including multispectral and hyper spectral imaging sensors, allow robots to detect crop diseases, nutrient deficiencies, and water stress. This enables the robots to apply targeted inputs, such as fertilizers or pesticides, on a site-specific basis, reducing waste and improving crop yields. Additionally, sensors like LiDAR (Light Detection and Ranging), ultrasonic sensors, and vision-based cameras enable robots to perceive their surroundings, navigate through fields, and avoid obstacles (Xie et al. 2022).

Actuators

Actuators play a vital role in agricultural robots, enabling them to perform physical tasks and interact with their environment. These actuators are responsible for the movement and mechanical actions of the robot, allowing it to execute specific functions such as picking fruits and vegetables using robotic arms and tilling soil with plows or cultivators (Xie et al. 2022).

Artificial Intelligence (AI) and Machine Learning (ML)

These technologies are integral components of agricultural robots, enabling them to make informed decisions based on data collected by sensors. These technologies empower robots to identify patterns and recognize targets, such as crops, weeds, pests and diseases to distinguish between targets and non-targets, ensuring precise actions (Talaviya et al. 2020).

Navigation Systems

Accurate navigation is essential for agricultural robots to operate efficiently and effectively in the field. To achieve this, sophisticated navigation systems (Vougioukas et al. 2005), featuring, RTK-GPS based (Bakker et al. 2011), computer vision-based guidance (Åstrand and Baerveldt 2005), sensor fusion based (Bechar and Vigneault

2017), obstacle detection and avoidance systems (Singh et al. 2020) navigation systems are extensively used that enable them to precisely position themselves within the field, navigate through rows of crops without causing damage, and perform tasks such as planting, spraying, and harvesting in straight lines and over large areas. Advanced navigation algorithms integrated with GPS systems allow agricultural robots to avoid obstacles, such as rocks, trees, and other natural features, and move efficiently across uneven terrain, adapting to changing environmental conditions (Bell et al. 2000).

Power Supply

Agricultural robots require a reliable power source to operate autonomously in the field for extended periods. Many robots are powered by batteries, while some larger robots may use hybrid systems that combine batteries with solar panels or traditional fuel sources to extend their operating time.

Software and Control Systems

The software used to control agricultural robots is equally important as the hardware, playing a vital role in optimizing robot performance and farm management. This software enables farmers to monitor the robot's performance in real-time, schedule tasks, assign duties and analyse data collected from the field, providing valuable insights into crop health, soil conditions and other factors. Many modern agricultural robots are now integrated with cloud-based platforms, allowing for remote control and monitoring, real-time updates and notifications and enhanced data analysis and decision-making capabilities.

Need for Agricultural Robots in Agriculture

The integration of robotics in farming offers a promising solution to support the modern agricultural industry, providing a robust platform to address pressing challenges (Balaska et al. 2023). Agricultural robots, or "agribots," address issues such as labour shortages, efficiency, precision, sustainability, and climate change, ultimately enhancing food security and reducing environmental impact. The key benefits of agricultural robots are:

Addressing Labour Shortage

Robots help fill the labor gap by automating tasks, ensuring timely completion of farming operations, particularly during peak seasons when labor shortages are most acute (Christiaensen et al. 2020).

Increasing Efficiency and Precision

Advanced technologies like sensors, AI, and GPS enable robots to perform tasks with high accuracy, reducing input waste and maximizing crop output (Pradhan et al. 2024).

Sustainability and Environmental Benefits

Robots help mitigate negative environmental impacts by optimizing water and fertilizer usage, reducing pesticide application, and promoting soil health (Zeddies et al. 2024).

Adapting to Climate Change

Robots provide real-time data, enabling precise interventions and informed decision-making to optimize crop yields despite challenging climate conditions (Inoue 2020; Kumari et al. 2023).

Data-Driven Decision-Making

Robots collect and analyse data, offering valuable insights into soil health, crop growth, weather patterns, and pest activity, supporting precision agriculture practices and improving farm management (Pope and Sonka 2020).

Cost Savings and Economic Viability

While initial investment costs are high, agricultural robots offer long-term economic viability through reduced labour costs, optimized input usage, and increased yields, making them a viable option for farmers of all sizes (Das 2023; Kushwah et al. 2024). By leveraging agricultural robots, farmers can streamline operations, increase efficiency, and make data-driven decisions to optimize crop yields, reduce costs, and promote sustainability, ultimately contributing to a more productive and environmentally friendly agricultural industry.

Applications of Robots in Agriculture: Transforming Farming Practices

The applications of robotics in agriculture are diverse and rapidly expanding. Some of the most common uses include:

Planting and Seeding

Robots can plant seeds with high precision, ensuring optimal spacing and depth which leads to better crop yield and reduced seed waste (Pradhan et al. 2024).

Robotic Irrigation Systems

Irrigation is a critical component of agriculture, and effective irrigation and water management techniques are crucial for optimizing crop yields. This is especially important since water is a scarce resource in agriculture. Conventional irrigation techniques, however, frequently lead to inefficient use of water and unequal dispersal to crop growth.

Robotic irrigation systems have emerged as a promising solution to address these challenges. These systems utilize a combination of sensors, actuators, and control algorithms to automatically modify irrigation parameters based on real-time data on soil moisture, weather, and crop requirements. Robotic irrigation systems can range from basic sensor-based controllers to fully autonomous robots that can precisely navigate fields and deliver water, providing farmers with exact control and optimization over their water use (Mahapatra et al. 2024).

Robotic Weeding System

Traditional methods of weed control, such as manual labour and chemical herbicides, have long posed significant challenges, including high labour costs, environmental concerns and increasing herbicide resistance (Rai et al. 2023). However, the emergence of robotic weeding systems has revolutionized the approach to these challenges, offering precise, sustainable and efficient weed control solutions. These systems combine advanced technologies in machine vision, artificial intelligence, and mechanical engineering to deliver targeted weed management while minimizing environmental impact and reducing labour requirements (Upadhyay et al. 2024; Wang et al. 2022).

Modern robotic weeding systems represent a convergence of multiple technological innovations. Various weed detection methods are employed, as illustrated in Table 2. These systems typically incorporate sophisticated sensing technologies, advanced processing capabilities, and precise control mechanisms. The integration of these components enables accurate weed detection, identification, and targeted control actions, which can be achieved through mechanical, chemical, or alternative methods such as laser or electrical treatment (Sa et al. 2018). The incorporation of artificial intelligence in robotic weeding has significantly advanced agricultural automation. Prior to 2015, weed detection primarily relied on conventional methods such as morphological feature extraction, textural analysis, and basic machine learning classifiers (Hasan et al. 2021).

Traditional approaches focused on analysing specific plant features, such as leaf shape, colour, and texture patterns, to differentiate between crops and weeds. Early machine learning implementations utilized various classical algorithms, including Support Vector Machines (SVM), Random Forests, Decision Trees, and K-Nearest Neighbors (KNN). However, these methods had limitations in handling environmental variations and complex field conditions (Rai et al. 2023). The introduction of deep learning has revolutionized weed detection capabilities. Contemporary systems extensively utilize Convolutional Neural Networks (CNNs) and transfer learning techniques, enabling more robust and accurate weed identification. Studies have demonstrated significant improvements in detection accuracy through deep learning approaches, particularly in challenging field conditions.

Some notable advancements include the use of advanced CNN architectures, YOLO (You Only Look Once) implementations for real-time detection (Zhu et al. 2022), region-based CNNs for precise weed localization, semantic segmentation networks for detailed plant mapping, and transfer learning applications. Pre-trained models are adapted for specific crop-weed scenarios, and domain adaptation techniques are employed to handle varying environmental conditions. Fine-tuning strategies are also used to improve accuracy (Li et al. 2022).

Table 2: Different advanced weed detection methods

Weed Detection	Different Detectors	Working Principle	References
Camera-based	RGB Cameras	Basic visual detection with controlled lighting	Hemming and Rath 2001
	Multispectral Imaging	Enhanced plant discrimination capabilities	Wendel and Underwood 2016
	Hyper spectral Imaging	Detailed spectral analysis for species identification	Bioucas-Dias et al. 2013
	Thermal Imaging	Temperature-based detection methods	Liu et al. 2020
	RGB-D Cameras	Depth perception and 3D mapping	McCool et al. 2018
Sensor-based	LIDAR sensor	Three-dimensional plant mapping and navigation	Sa et al. 2018
	Ultrasonic Sensor	Distance measurement and obstacle detection	Alatise and Hancke 2020
	Time-of-Flight Sensors	Precise depth detection	Vázquez-Arellano et al. 2016
	Sensor Fusion	Combination of multiple sensor types for enhanced accuracy	Bechar and Vigneault 2017
Machine learning based	Convolutional Neural Networks (CNNs)	-	Pound et al. 2017
	YOLO architectures for real-time detection	-	Upadhyay et al. 2024
	Transfer Learning techniques	-	Rai et al. 2023
	Domain adaptation approaches	-	Das et al. 2021

Crop Health Monitoring

Crop health monitoring plays a pivotal role for food security, enhancing agricultural productivity, managing pests, diseases, and nutrient deficiencies, and promoting sustainable resource use (Chowdhury et al. 2025). Crop health monitoring mainly covers several critical aspects namely management of pest and disease, nutrient deficiency management, crop vigour monitoring and phenotyping. The traditional approach to crop health monitoring has served farmers for centuries, relying heavily on manual labour, experience and intuition. Traditional crop health monitoring includes visual inspection, pest traps, basic tools (hand lenses or magnifying glasses) experienced-based decision-making, Callender-based approaches. While it can be effective for small-scale farms or regions with well-understood environmental conditions, it has several limitations, including being time-consuming, subjective, and lacking precision (Gawande et al. 2023). As large-scale farming grows, traditional methods fall short, causing delayed issue detection, reduced crop yields, and increased input costs for farmers.

Effective crop health monitoring is essential for modern agriculture, which is increasingly challenged by unpredictable climate conditions, soil degradation, and pest infestations (Chowdhury et al. 2023). Timely and accurate identification and quantification of crop stresses are crucial for effective farm management and optimization of crop variety selection (Barbedo 2023). Modern approaches to crop health monitoring use cutting-edge technology to provide precise, real-time insights into crop conditions. By leveraging tools such as remote sensing, drones, IoT sensors, AI, and robotics, farmers can make data-driven decisions that optimize resource use, increase productivity, and reduce environmental impact. They can precisely target affected areas, apply treatments, or perform targeted interventions, reducing the need for widespread application of chemicals and minimizing environmental impact (Oberti et al. 2016).

This section explores several types of robots have been developed and deployed by integrating various sensors, machine learning algorithms, and autonomous systems for crop health monitoring.

Crop Disease and Insect Management

Disease detection is an important part of the production cycle, as diseases can cause significant economic damage if not detected during the early stage. Insect detection has not gained a lot of attention, as many insects are located underneath the leaf, inside the buds or even underground, which makes it extremely difficult to accurately detect them and, as a result, monitor them (Chowdhury et al. 2024).

Disease detection is mainly a visual task, and therefore all robots incorporate a vision-based system. Detection accuracy is defined as the ratio of the identified diseased plants to the total number of diseased plants. Crop disease and insect detection robotics equipped with cameras, sensors, artificial intelligence, and autonomous systems for identification, decision making and control measures. Robots mainly use RGB cameras and multispectral and/or hyperspectral sensors for visual tasks. RGB cameras are basic colour cameras that help to detect visible signs of disease, such as leaf spots or discoloration, while multispectral and hyperspectral sensors capture wavelengths of light beyond the visible spectrum (e.g., infrared), enabling the detection of subtle changes in the plants' physiology. For the detection of powdery mildew and tomato spotted wilt virus (TSWV) in greenhouse pepper plants, using a colour camera showed high accuracy, while the use of a multispectral camera achieved an accuracy of 80% and 61% for these two diseases, respectively (Schor et al. 2016; Schor 2017). Similarly, RGB cameras have been used for Pyralidae insect detection on tomato and rice plants with an accuracy of about 94% (Liu et al. 2019). A Spectral sensor-based device has been used for real-time detection and severity estimation of groundnut bud necrosis virus in tomatoes with an accuracy of about 93.65% (Kumari, 2024). Moreover, Mustafa et al. (2024) developed wheat fusarium spectral and texture indices (WFSI, WFTI) using hyperspectral imaging and data, achieving high classification accuracy for Fusarium head blight (FHB) detection in wheat. Classifiers like Xgboost and neural nets showed strong performance, highlighting hyperspectral imaging's potential for early FHB monitoring in precision agriculture. A RobHortic field robot has been developed for inspecting the presence of pests and diseases in horticultural crops using proximal sensing, presented about 66.4% and 59.8% detection rates for laboratory and field conditions respectively (Cubero at al. 2020).

Plant Vigour Monitoring

Plant vigour monitoring is the process of monitoring plant stress that is visible symptoms, such as colour changes, withering, spots, or any other anomaly on the leaves or overall plant. This can be monitored with remote sensing data of various spatial, temporal and spectral resolutions, acquired at different platforms. A large number of studies are employed to crop growth indicators, such as leaf area index (LAI), leaf nitrogen concentration (LNC) and their relationship with the crop spectral properties. The availability of free remote sensing data as well as commercial satellite data acquisition and the use of ground-based sensors, provides the time series data for crop growth monitoring. The above ground biomass is an important index for the crop growth and can be the basic factor in detecting and estimating the yield of crop. The traditional methods in the crop biomass estimation need a large resource. Nowadays, canopy temperature is determined using infrared thermometers and thermal cameras (Bayati et al. 2018). Many studies, utilizes, satellite-based vegetation indices (VIs) are the most commonly used models for the estimation of biomass (Zheng et al. 2004; Schlerf et al. 2005), LAI (Turner et al. 1999; Patra et al. 2023) and plant N content (Xue and Su 2017; Chowdhury et al. 2025). VI is the mathematical transformation of the original spectral reflectance, which are used for interpreting vegetation biomass and canopy (Rahman et al. 2003; He et al. 2006; Kokhan 2011). The vegetation indices are widely used to remove the variations caused in spectral reflectance while measuring biophysical properties caused due to the soil background, sun view angles, and atmospheric conditions (Niu et al. 2000). Additionally, plant monitoring systems are beginning to employ costly hyperspectral sensors (Wang et al. 2018).

Currently, crop state is effectively monitored using a variety of VIs based on remote sensing (RS) data. They give regular feedback on the condition and output of agricultural fields. The primary issue with the majority of research is a requirement to employ a comparative evaluation of the values of various VIs produced from optical satellite systems that are required to undergo atmospheric correction. A valuable source of objective information for managing agricultural production and crop yield estimations is the information on crop growth and dynamics that the RS data provides (Huete et al. 2002). When vegetation is remotely sensed through passive sensors, data on several plant parameters (Liu et al. 2016), which are influenced by chemical and morphological traits as well as leaf surface characteristics are obtained (Zhang and Kovacs 2012).

Most agricultural monitoring robots are designed especially for vineyards or orchard/tree crops. Using a variety of sensors, a crop monitoring robot has been experimentally tested in semi-structured environments (Vidoni et al. 2017) and laboratory circumstances (Bietresato et al. 2016), and it has shown to be quite reliable for estimating orchard canopy volume and vigor. Vineyard crop monitoring robots have been developed extensively, despite the lack of well-defined performance indicators. In every instance, a range of vision-related sensors are integrated, including LiDAR sensors on the robots and several kinds of cameras (RGB, hyperspectral, multispectral, etc.), to monitor volume, vegetation, and health of the plants. More generic crop monitoring robots handle agricultural operations, such as soil sampling (Vaeljaots et al. 2018) CO_2 concentration levels may have a significant effect on crop growth. Under this scope, a hexapod robot has been developed for the collection of agronomical information, such as soil nutrients and plant growth. The monitored CO_2 concentration levels were up to 2500 ppm (Iida et al. 2008).

Phenotyping

Phenotyping robots have been developed in tandem with plant monitoring robot development. Plant phenotyping and plant monitoring are closely related because they both use vision systems to collect as much data as possible about the plant or fruit being inspected. The distinction is that, whereas plant monitoring robots concentrate on the condition, health and chemical makeup of the plant, phenotyping robots pay greater attention to morphology, plant growth and features that affect crop production. Several sensors namely, LiDAR, RGB camera, thermal infra-red camera and imaging spectroradiometer have been utilized in robot phenotyping (Deery et al. 2014).

Fully autonomous phenotyping robots have been developed for sorghum field trials, achieving up to 96% stalk detection (Mueller-Sim et al. 2017). In addition, these robots monitor GRVI (Green-Red Vegetation Index), light penetration, and leaf erectness and necrosis. Comparable autonomous phenotyping robots (Mobile and Bettybot) in the sugar beet crop analyze the geometric and colorimetric evolution of the plants in great detail (Benet et al. 2018). The autonomous phenotyping robot, Bonirob measures many plant parameters such as height, stem thickness, plant distribution, and population density by fusing data from multiple sensors (Ruckelshausen et al. 2009). Apart from the autonomous robots, a stationary, completely automated phenotyping platform has been utilized in wheat, maize and rice plants (Virlet et al. 2016). This platform may offer extremely precise measurements of canopy growth throughout crop life cycles, with plant height root mean square (RMS) error as low as around 1.9 cm. Plant height, width, convex hull volume, and plant surface area are among the very repeatable measures made by an autonomous robot mounted on a tractor in sorghum crop that are fairly similar to hand measurements made in the field (Bao et al. 2019).

All phenotyping robots are not fully automated. Collaboratively, the two phenotyping robots can operate as an autonomous ground vehicle to gather data on individual plants and as a mobile observation tower to survey the entire field (Shafiekhani et al. 2017). The robots can also perform joint tasks such as RMS (Root mean square) errors for plant heights of less than 0.5 cm, RGB to IR image calibrations of 2.5 pixels, and temperature readings of less than 1°C. Finally, an energy sorghum crop has been used to test a second semi-autonomous phenotyping robot (Young et al. 2019). The robot's algorithms yielded an average absolute inaccuracy of 13% when estimating stem width and 15% when estimating plant height.

Harvesting and Picking

Harvesting robots, especially those designed for delicate crops like fruits and vegetables, are becoming increasingly sophisticated. These robots can identify ripe produce, pick it without causing damage, and operate efficiently even in challenging conditions.

Livestock Management

Robots are also being used to manage livestock, from automated milking systems to drones that monitor the health and location of animals in large pastures.

Fruit Sorting Robot

The fruit sorting robot is based on size and colour in a packaging system. Image processing where colour is identified by HSV analysis and the diameter is known in the greyscale image and thresholding is set allows for the sorting. Red and green grapes as well as red and green tomatoes need to be sorted. For red tomatoes, 11.76 seconds are needed for green tomatoes, 12.56 seconds for red grapes and 12.92 seconds for green grapes for the robot to complete the task. The location of the boxes holding the sorted fruit is what causes the time difference.

Grading Robot

A grading robot system that automatically pulls fruits out of containers and examines each fruit from every angle. Two fruit-providing robots and a grading robot make up the robot system. Twelve suction pads served as the robots' end-effectors for moving fruits, and both had three degrees of freedom, Cartesian coordinate manipulators. As a machine vision system, the grading robot was equipped with 12 colour TV cameras and 28 lighting devices. Twelve fruits were suctioned up at a time by the grading robot, and twelve bottom views of every fruit were taken while the manipulator was moving from a halfway stage to carriers on a conveyor line. The suction pads were rotated for 270° rotations to obtain 4 side images of each fruit before the fruits were released onto the carriers. The manipulator's stroke was about 1.2 m and it took about 4.3 seconds to move back to the initial position which meant that this robot could grade three fruits per second.

Status of Agricultural Robotics: A Growing Field

As of 2024, agricultural robotics is increasing rapidly, with several companies and research institutions developing new technologies to address the challenges of modern farming. Major agricultural machinery manufacturers such as John Deere, Kubota, and AGCO are investing heavily in robotic solutions, while startups and research institutions are pushing the boundaries of innovation. Despite the progress, agricultural robotics is still in its early stages, particularly when compared to industrial robotics. However, the field is expected to grow exponentially in the coming years as technology becomes more accessible, reliable, and affordable.

Challenges of Agricultural Robots

Despite the numerous advantages of agricultural robots, there are still several limitations illustrated in Fig. 5. Initially, they are costly, and robots must operate in highly variable environments, which require advanced AI and sensor systems to function effectively. Additionally, the use of robotics in agriculture raises ethical questions about the future of human labour in farming, particularly in regions where agriculture is a major source of employment. Further, operating and maintaining robotic systems requires specialized knowledge in robotics, AI, and data analysis and also they have data privacy and security problems. There are also challenges including lighting, backdrop separation and data modification present difficulties for robots that monitor plant vitality. Lack of image databases for each disease to train the detection/classification models.

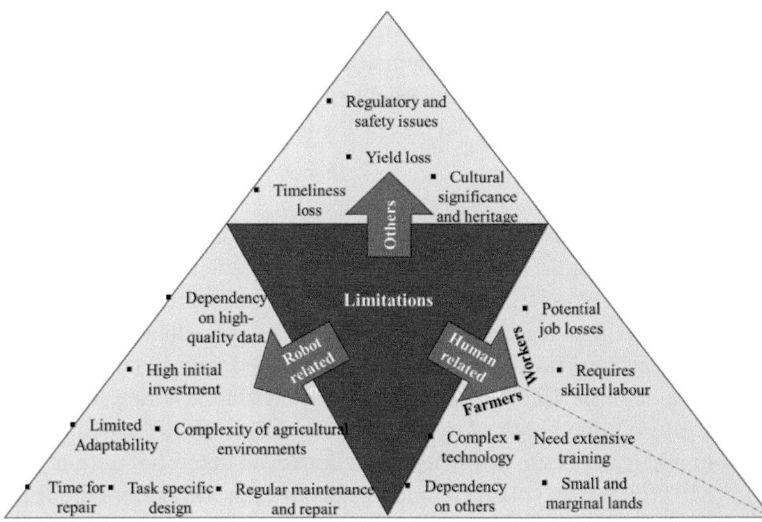

Fig. 5: Limitations of agricultural robots.

Future Prospects: Robotics in the Era of Smart Farming

The future of agricultural robotics is closely tied to the broader trend of "smart farming" or precision agriculture, which involves the use of data, automation and advanced technologies to optimize farming practices. In the coming years, robots are expected to become more autonomous, capable of working alongside other technologies such as drones, IoT sensors and big data analytics to create fully integrated farming systems. As robots become more affordable and user-friendly, their adoption is likely to increase, transforming agriculture into a more efficient, sustainable, and technologically advanced industry.

Conclusion

The demand for agricultural robots is rising due to global concerns like population growth, climate change, and labour shortages. Agricultural robots are poised to play a vital role in the future of farming, addressing labour shortages, increasing food demand, climate change, and the need for sustainable practices. The adoption of robotics in farming can significantly enhance the efficiency and precision of agricultural operations. This leads to timely task execution, targeted input application, increased productivity, reduced cultivation costs, and minimized environmental impact. To address the limitation of single-purpose robots, developing multipurpose robots capable of performing various tasks can reduce costs for farmers. Moreover, swarm robotics is gaining attention in agriculture, enabling multiple operations to be performed simultaneously, further improving efficiency and productivity.

References

Alatise, M.B. and G.P. Hancke. 2020. A review on challenges of autonomous mobile robot and sensor fusion methods. IEEE Access. 8: 39830-39846.

Anand, R., S.S. Nalla, S. Nasreen, M. Chowdhury, Madhusudan, B.S., A. Pandey et al. 2024. Fruit counting and analysis using artificial intelligence approaches. pp. 93-114. *In:* S.S. Chouhan, A. Saxena, U.P. Singh and S. Jain [eds.]. Artificial Intelligence Techniques in Smart Agriculture. Springer Nature, Singapore.

Åstrand, B. and A.J. Baerveldt. 2005. A vision-based row-following system for agricultural field machinery. J. Mechatron. 15(2): 251-269.

Bali, A., S.S. Chouhan and G. Kumar. 2023. Adaptive fault-tolerant control for pure-feedback stochastic nonlinear systems with sensor and actuator faults. Circuits Syst. Signal Process. 42: 5838–5867.

Bao, Y., L. Tang, M.W. Breitzman, M.G. Salas Fernandez and P.S. Schnable. 2019. Field-based robotic phenotyping of sorghum plant architecture using stereo vision. J. Field Robotic. 36(2): 397-415.

Barbedo, J.G.A. 2023. Crop health sensing: Disease, pest, nutrient, and water stresses. *In:* Encyclopedia of Digital Agricultural Technologies (pp. 198-207).

Bayati, M. and R. Fotouhi. 2018. A mobile robotic platform for crop monitoring. Advances in Robotics & Automation, 7(1): 1000186.

Bechar, A. and C. Vigneault. 2017. Agricultural robots for field operations. Part 2: Operations and systems. Bio. Engg. 153: 110-128.

Bell, T. 2000. Automatic tractor guidance using carrier-phase differential GPS. Comput. Electron. Agric. 25(1-2): 53-66.

Benet, B., C. Dubos, F. Maupas, G. Malatesta and R. Lenain. 2018. Development of autonomous robotic platforms for sugar beet crop phenotyping using artificial vision. Available online: https://hal.archives-ouvertes.fr/hal 02155159/document (accessed on 6 May 2020).

Bietresato, M., G. Carabin, R. Vidoni, A. Gasparetto and F. Mazzetto. 2016. Evaluation of a LiDAR-based 3D-stereoscopic vision system for crop-monitoring applications. Comput. Electron. Agric. 124: 1-13.

Bioucas-Dias, J.M., A. Plaza, G. Camps-Valls, P. Scheunders, N.M. Nasrabadi and J. Chanussot. 2013. Hyperspectral remote sensing data analysis and future challenges. IEEE Geoscience and Remote Sensing Magazine 1(2): 6-36.

Chowdhury, M., E.V. Thomas, A. Jha, A. Kushwah, R. Kurmi, T.K. Khura et al. 2023. An automatic pressure control system for precise spray pattern analysis on spray patternator. Comput. Electron. Agric. 214: 108287.

Chowdhury, M., T.K. Khura, P.K. Upadhyay, R.A. Parray, H.L. Kushwaha, C. Singh et al. 2024. Assessing vegetation indices and productivity across nitrogen gradients: A comparative study under transplanted and direct-seeded rice. Front. Sustain. Food Syst. 8: 1351414.

Chowdhury, M., T. K. Khura, R.A. Parray, P.K. Upadhyay, H.L. Kushwaha, A. Lama and A. Kushwah. 2025. Multi-vegetation indices based handheld device for precise nitrogen assessment and prescription in direct seeded rice. Comput. Electron. Agric. 229: 109886.

Cubero, S., E. Marco-Noales, N. Aleixos, S. Barbe and J. Blasco. 2020. Robhortic: A field robot to detect pests and diseases in horticultural crops by proximal sensing. Agriculture, 10(7): 276.

Das, T.K., S. Paul and N.T. Yaduraju. 2021. Integrated weed management in crops and cropping systems: concept, needs and challenges. Ind. J. of Agro. 66(1): 1-13.

Deery, D., J. Jimenez-Berni, H. Jones, X. Sirault and R. Furbank. 2014. Proximal remote sensing buggies and potential applications for field-based phenotyping. Agronomy, 4(3): 349-379.

Gawande, V., D.R.K. Saikanth, B.S. Sumithra, S.A. Aravind, G.N. Swamy, M. Chowdhury and B.V. Singh. 2023. Potential of precision farming technologies for eco-friendly agriculture. Int. J. Plant Soil Sci. 35(19): 101-112.

Global Burden of Crop Loss. 2020. Available online: https://www.cabi.org/projects/global-burden-of-crop-loss/.

Hasan, A.M., F. Sohel, D. Diepeveen, H. Laga and M.G. Jones. 2021. A survey of deep learning techniques for weed detection from images. Comput. Electron. Agric. 184: 106067.

He, Y., X. Guo and J. Wilmshurst. 2006. Studying mixed grassland ecosystems I: Suitable hyperspectral vegetation indices. Canadian J. Remote Sen. 32(2): 98-107.

Hemming, J. and T. Rath 2001. Computer-vision-based weed identification under field conditions using controlled lighting. J. Agril. Engg. Res. 78(3): 233-243.

Hirayama, T. and K. Shinozaki. 2010. Research on plant abiotic stress responses in the post-genome era: Past, present and future. The Plant Journal 61(6): 1041-1052.

Huete, A., K. Didan, T. Miura, E.P. Rodriguez, X. Gao and L.G. Ferreira. 2002. Overview of the radiometric and biophysical performance of the MODIS vegetation indices. Remote Sens. Environ. 83(1-2): 195-213.

Iida, M., D. Kang, M. Taniwaki, M. Tanaka and M. Umeda. 2008. Localization of CO_2 source by a hexapod robot equipped with an anemoscope and a gas sensor. Comput. Electron. Agric. 63(1): 73-80.

Jat, D., K. Dubey, R.R. Potdar, S.K. Chakraborty, S.P. Kumar, N.S. Chandel et al. 2023. Development of an automated mobile robotic sprayer to prevent workers' exposure of agro-chemicals inside polyhouse. J. Field Robot. 40(6). Available from: https://doi.org/10.1002/rob.22181.

Jha, A., M. Chowdhury and A.N. Satpute. 2023. Surface water quality forecasting using machine learning approach. pp. 293-315. *In:* C.B. Pande, M. Kumar and N.L. Kushwaha [eds.]. Surface and Groundwater Resources Development and Management in Semi-arid Region: Strategies and Solutions for Sustainable Water Management. Cham: Springer International Publishing.

Kokhan, S.S. 2011. Application of vegetation indexes derived from satellite images IRS-1D LISS-III for determination of crop status. Space Science and Technology – Kosmicna Nauka Tecnologia, 17(5): 58-63.

Kumari, A., P. Ashoka, P. Tiwari, P. Sachan, A.K. Malla, A. Tripathy and M. Chowdhury. 2023. Zero tillage lead to enhanced productivity and soil health. Int. J. Environ. Clim. Chang. 13(10): 3707-3715.

Kumari, K., R. Parray, Y.B. Basavaraj, S. Godara, I. Mani, R. Kumar et al. 2024. Spectral sensor-based device for real-time detection and severity estimation of groundnut bud necrosis virus in tomato. J. Field Robot. 42(1): 5-19.

Kushwah, A., P.K. Sharma, H.L. Kushwaha, R.H. Nag, G. Carpenter, M.K. Choudhary et al. 2024. Economic evaluation of precise intelligent cauliflower harvester: A comparative study with manual harvesting. J. Sci. Res. Rep. 30(1): 33-42.

Li, Y., Z. Guo, F. Shuang, M. Zhang. and X. Li. 2022. Key technologies of machine vision for weeding robots: A review and benchmark. Comput. Electron. Agric. 196: 106880.

Liu, B., Z. Hu, Y. Zhao, Y, Bai and Y. Wang. 2019. Recognition of pyralidae insects using intelligent monitoring autonomous robot vehicle in natural farm scene. arXiv preprint arXiv:1903.10827.

Liu, C., P.S. Sun and S.R. Liu. 2016. A review of plant spectral reflectance response to water physiological changes. Chin. J. Plant Ecol. 40(1): 80.

Liu, Y., C. Xie and M. Li. 2020. Research progress of spectral imaging technology in agricultural product quality and safety detection. Spectrosc. Spect. Anal. 40(1): 1-7.

Mahapatra, J., R.K. Sahni, P.K. Sahoo and D.J. Shrinivasa. 2024. IoT-based real-time farm management system for smart agriculture. *In:* Artificial Intelligence Techniques in Smart Agriculture (pp. 267-287). Springer Nature, Singapore.

Malkani, P., I. Mani, P.K. Sahoo, R.A. Parray, S.S. Swain, K.R. Asha et al. 2024. Development and evaluation of a sensor enabled smart herbicide applicator for inter row crops. Smart Agric. Technol. 9: 100649.

McCool, C., T. Perez and B. Upcroft. 2018. Mixtures of lightweight deep convolutional neural networks: Applied to agricultural robotics. IEEE Robotics and Automation Letters 3(2): 1344-1351.

Mishra, N. and Sahni, R.K. 2024. Introduction to artificial intelligence techniques in agricultural applications and their future aspects. *In:* Artificial Intelligence Techniques in Smart Agriculture (pp. 35-54). Springer Nature, Singapore.

Mueller-Sim, T., M. Jenkins, J. Abel and G. Kantor. 2017, May. The Robotanist: A ground-based agricultural robot for high-throughput crop phenotyping. *In:* 2017 IEEE International Conference on Robotics and Automation (ICRA) (pp. 3634-3639). IEEE.

Mustafa, G., H. Zheng, I.H. Khan, J. Zhu, T. Yang, A. Wang et al. 2024. Enhancing fusarium head blight detection in wheat crops using hyperspectral indices and machine learning classifiers. Comput. Electron. Agric. 218, 108663.

Niu, Z., Y.H. Chen, H.Z. Sui, Q.Y. Zhang and C.J. Zhao. 2000. Mechanism analysis of leaf biochemical concentration by high spectral remote sensing. J. Remote Sens, 4(2): 125-129.

Oberti, R., M. Marchi, P. Tirelli, A. Calcante, M. Iriti, E. Tona et al. 2016. Selective spraying of grapevines for disease control using a modular agricultural robot. Biosyst. Eng. 146: 203-215.

Patra, K., C.M. Parihar, H.S. Nayak, B. Rana, D.R. Sena, A. Anand et al. 2023. Appraisal of complementarity of subsurface drip fertigation and conservation agriculture for physiological performance and water economy of maize. Agric. Water Manag. 283: 108308.

Pound, M.P., J.A. Atkinson, D.M. Wells, T.P. Pridmore and A.P. French. 2017. Deep learning for multi-task plant phenotyping. IEEE/RSJ International Conference on Intelligent Robots and Systems (IROS), 2055-2060.

Pradhan, N.C., M.A. Naik, M. Chowdhury, A. Kushwah, K.R. Asha, T. Dhar et al. 2024. Robotic seeding or sowing system in smart agriculture. pp. 495-520. *In:* K. Pandey, N.L.

Kushwaha, C.B. Pande and K.G. Singh [eds.]. Artificial Intelligence and Smart Agriculture: Technology and Applications. Springer Nature, Singapore.

Rahman, A.F., J.A. Gamon, D.A. Sims and M. Schmidts. 2003. Optimum pixel size for hyperspectral studies of ecosystem function in southern California chaparral and grassland. Remote Sens. Environ. 84(2): 192-207.

Rai, N., R.R. Nidamanuri and A. Upadhyay. 2023. Deep learning for site-specific weed management: A systematic review. Comput. Electron. Agric. 206: 107698.

Rajbhoj, A., S. Deshpande, J. Gubbi, V. Kulkarni and P. Balamuralidhar. 2019. A system for semi-automatic construction of image processing pipeline for complex problems. *In:* Enterprise, Business-Process and Information Systems Modeling: 20th International Conference, BPMDS 2019, 24th International Conference, EMMSAD 2019, Held at CAiSE 2019, Rome, Italy, June 3-4, 2019, Proceedings 20 (pp. 295-310). Springer International Publishing.

Ruckelshausen, A., P. Biber, M. Dorna, H. Gremmes, R. Klose, A. Linz et al. 2009. BoniRob: An autonomous field robot platform for individual plant phenotyping. *In:* Precision agriculture '09 (pp. 841-847). Wageningen Academic.

Sa, I., Z. Chen, M. Popović, R. Khanna, F. Liebisch, J. Nieto and R. Siegwart. 2018. WeedNet: Dense semantic weed classification using multispectral images and MAV for smart farming. IEEE Robotics and Automation Letters. 3(1): 588-595.

Satpute, A.N., K.P. Gavhane, S. Kaur, A. Jha, N.C. Pradhan and M. Chowdhury. 2024. Integration of AI and IoT in soilless cultivation to power sustainable agricultural revolution. pp. 387-411. *In*: K. Pandey, N.L. Kushwaha, C.B. Pande and K.G. Singh [eds.]. Artificial Intelligence and Smart Agriculture: Technology and Applications. Springer Nature, Singapore.

Savary, S., L. Willocquet, S.J. Pethybridge, P. Esker, N. McRoberts and A. Nelson. 2019. The global burden of pathogens and pests on major food crops. Nat. Ecol. Evol. 3(3): 430-439.

Schlerf, M., C. Atzberger and J. Hill. 2005. Remote sensing of forest biophysical variables using HyMap imaging spectrometer data. Remote Sens. Environ. 95(2): 177-194.

Schor, N., A. Bechar, T. Ignat, A. Dombrovsky, Y. Elad, and S. Berman. 2016. Robotic disease detection in greenhouses: Combined detection of powdery mildew and tomato spotted wilt virus. IEEE Robotics and Automation Letters, 1(1): 354-360.

Schor, N., S. Berman, A. Dombrovsky, Y. Elad, T. Ignat and A. Bechar. 2017. Development of a robotic detection system for greenhouse pepper plant diseases. Precis. Agric. 18: 394-409.

Shafiekhani, A., S. Kadam, F.B. Fritschi and G.N. DeSouza. 2017. Vinobot and vinoculer: Two robotic platforms for high-throughput field phenotyping. Sensors, 17(1): 214.

Singh, R.S. and R.K. Sahni. 2019. Transformation of Indian agriculture through mechanization. Econ. Affairs. 64(2): 297-303.

Singh, V., A. Rana, M. Bishop, A.M. Filippi, D. Cope, N. Rajan and M. Bagavathiannan. 2020. Unmanned aircraft systems for precision weed detection and management: Prospects and challenges. Advan. Agro. 159: 93-134.

Talaviya, T., D. Shah, N. Patel, H. Yagnik and M. Shah. 2020. Implementation of artificial intelligence in agriculture for optimisation of irrigation and application of pesticides and herbicides. Artif. Intell. Agric. 4: 58-73.

Turner, D.P., W.B. Cohen, R.E. Kennedy, K.S. Fassnacht and J.M. Briggs. 1999. Relationships between leaf area index and Landsat TM spectral vegetation indices across three temperate zone sites. Remote Sens. Environ. 70(1): 52-68.

Upadhyay, A., A. Kumar and K.K. Jain. 2024. Recent advances in agricultural robots for automated weeding: A comprehensive review. Sensors. 24(2): 6743.

Vacavant, A., M.A. Lebre, H. Rositi, M. Grand-Brochier and R. Strand. 2019. New definition of quality-scale robustness for image processing algorithms, with generalized uncertainty modelling, applied to denoising and segmentation. *In:* Reproducible Research in Pattern Recognition: Second International Workshop, RRPR 2018, Beijing, China, August 20, 2018, Revised Selected Papers 2 (pp. 138-149). Springer International Publishing.

Vaeljaots, E., H. Lehiste, M. Kiik and T. Leemet. 2018. Soil sampling automation case study using unmanned ground vehicle. Eng. Rural Dev, 17: 982-987.

Vázquez-Arellano, M., H.W. Griepentrog, D. Reiser and D.S. Paraforos. 2016. 3-D imaging systems for agricultural applications – A review. Sensors. 16(5): 618.

Vidoni, R., R. Gallo, G. Ristorto, G. Carabin, F. Mazzetto, L. Scalera and A. Gasparetto. 2017, November. ByeLab: An agricultural mobile robot prototype for proximal sensing and precision farming. *In:* ASME International Mechanical Engineering Congress and exposition (Vol. 58370, p. V04AT05A057). American Society of Mechanical Engineers.

Virlet, N., K. Sabermanesh, P. Sadeghi-Tehran and M.J. Hawkesford. 2016. Field scanalyzer: An automated robotic field phenotyping platform for detailed crop monitoring. Funct. Plant Biol. 44(1): 143-153.

Wang, A., W. Zhang and X. Wei. 2022. A review of weed detection using ground-based machine vision and image processing techniques. Comput. Electron. Agric. 192: 106616.

Wang, Y., X. Hu, Z. Hou, J. Ning and Z. Zhang. 2018. Discrimination of nitrogen fertilizer levels of tea plant (Camellia sinensis) based on hyperspectral imaging. J. Sci. Food Agric. 98: 4659-4664.

Wendel, A. and J. Underwood. 2016. Self-supervised weed detection in vegetable crops using ground-based hyper spectral imaging. IEEE International Conference on Robotics and Automation (ICRA), 5128-5135.

World Population Prospects. 2024. Department of Economic and Social Affairs Population Division, United Nations.

Xie, D., L. Chen, L. Liu, L. Chen and H. Wang. 2022. Actuators and sensors for application in agricultural robots: A review. Machines. 10(10): 913.

Xue, J. and B. Su. 2017. Significant remote sensing vegetation indices: A review of developments and applications. J. Sens. 2017(1): 1353691.

Young, S.N., E. Kayacan and J.M. Peschel. 2019. Design and field evaluation of a ground robot for high-throughput phenotyping of energy sorghum. Precis. Agric. 20(4): 697-722.

Zhang, C. and J.M. Kovacs. 2012. The application of small unmanned aerial systems for precision agriculture: A review. Precis. Agric. 13: 693-712.

Zhang, S., H. Wang, W. Huang and Z. You. 2018. Plant diseased leaf segmentation and recognition by fusion of superpixel, K-means and PHOG. Optik, 157: 866-872.

Zheng, D., J. Rademacher, J. Chen, T. Crow, M. Bresee, J. Le Moine and S.R. Ryu. 2004. Estimating aboveground biomass using Landsat 7 ETM+ data across a managed landscape in northern Wisconsin, USA. Remote Sens. Environ. 93(3): 402-411.

Zhu, H., Y. Zhang, D. Mu, L. Bai, H. Zhuang and H. Li. 2022. YOLOX-based blue laser weeding robot in corn field. Front. Plant Sci. 13: 1017803.

CHAPTER

6

Modern Technologies for Soil Health Monitoring and Management

Dewali Roy[1*], Pooja Tamuk[1], Aditya Pratap Singh[2] and Saptarshi Mondal[3]

[1] Division of Soil Science and Agricultural Chemistry, ICAR-Indian Agricultural Research Institute, Pusa, New Delhi - 110012, India

[2] Department of Genetics and Plant Breeding, Bidhan Chandra Krishi Viswavidyalaya, Mohanpur, West Bengal - 741252, India

[3] Institute of Plant Breeding, Genetics and Genomics, University of Georgia, Griffin, 30223, USA

Introduction to Soil Health

Soil health denotes to the ecological equilibrium and the functionality of a soil and its ability to maintain a well-balanced ecosystem having a rich biodiversity above and below surface, and productivity (Cardoso et al. 2012). The FAO (Food and Agriculture Organization) defines soil health as “the continued capacity of soil to function as a vital living ecosystem that sustains plants, animals, and humans.” This definition emphasizes the importance of soil as a dynamic and living system that supports several ecosystem functions, such as nutrient cycling, water filtration, and plant growth, while also promoting biodiversity and maintaining environmental quality. The concept of soil health, which emerged from discussions around soil quality in the late 1980s and 1990s (Doran and Parkin 1994), is now increasingly linked to the ‘One Health’ framework that integrates the health of humans, animals, and the environment (Keith et al. 2016). The policymakers in India have embraced the soil health concept, and distributed soil health cards (SHC) to more than 100 million farmers (Rumpel et al. 2018).

Why We Need to Study Soil Health?

Soil is a critical natural resource that offers a vast dimension of ecosystem services essential for human well-being. The contribution of soil health towards several ecosystem services can be understood from Fig. 1. Soil role as nutrient source by

*Corresponding author: dewaliroy14@gmail.com

DOI: 10.1201/9781003536932-6

means of transformation of different nutrients, and its storage; climate mitigation; raw material for agri-based industries, pharmaceutical, plant growth promoting rhizobacteria (PGPR) and plant growth promoting fungi (PGPF); biological diversity as medium for plant and soil microorganism growth; and hydrological cycle with its role in bioremediation and filtering soil pollutant, and storage of soil water all are associated with soil health. The fauna and microbial communities within soil play a pivotal role in ecosystems by aiding in the decomposition of organic matter, facilitating nutrient cycling, and influencing different properties of soil—that influence the soil fertility and sustainability outcomes (Lavelle and Spain 2001; Rillig et al. 2018). The flow of nutrients through various trophic levels in both natural and agricultural ecosystems is largely driven by soil organisms (Fig. 1). While the role of soil in supporting agricultural production has long been recognized, the importance of conserving and enhancing the broader ecosystem services it provides such as carbon sequestration, water filtration, groundwater recharge, pathogen suppression, biological nitrogen fixation, and biodiversity preservation has only recently gained widespread attention (Kumari et al. 2023). Furthermore, the emphasis on maintaining and improving soil quality emerged later compared to concerns about water and air quality.

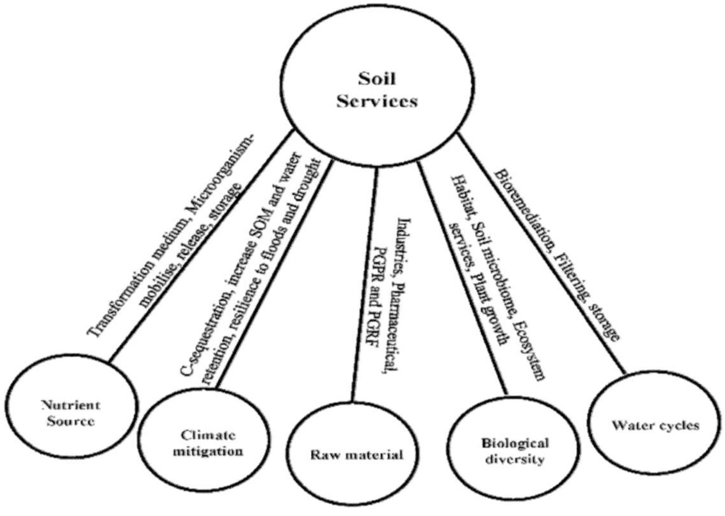

Fig. 1: Soil services in soil health and ecosystem.

As agricultural expansion nears its global limits, the urgency around ensuring the sustainability of agricultural systems has intensified. The challenge of providing food security for over seven billion people while maintaining environmental sustainability is one of the defining issues for future generations. Soil management is critical to agriculture, yet there is significant evidence of widespread soil health degradation, including issues such as erosion, declining organic matter, metal contamination, subsurface compaction, and increased salinity (Zimnicki et al. 2020; Patra et al. 2023). This degradation often occurs gradually and subtly; impacting agricultural productivity and the environment over time. In response, researchers have developed methods to assess soil health, enabling better monitoring and management to prevent degradation, as only healthy soils can perform multiple functions and support the delivery of key ecosystem services (Foley et al. 2011). Thus, soil health provides a comprehensive assessment of soil functionality.

Sustainable crop production is essential to meet the rising demand for food security, and this places a strong emphasis on soil health management (Jian et al. 2020, Karlen et al. 2019). While intensive agricultural practices have been prioritized to increase yields, many of these practices can negatively impact soil health (Norris and Congreves 2018). A focus on soil health involves transitioning from dependence on inorganic fertilizers to the use of organic amendments, recycling crop residues, reducing tillage, increasing crop diversity, and implementing erosion control strategies (Congreves et al. 2015; Guddaraddi et al. 2023). Furthermore, soil health management also requires optimizing plant-microbe interactions within the root zone to promote plant growth. This necessitates the careful regulation of the soil micro biome in terms of its abundance, activity, composition, and functional capabilities (Bünemann et al. 2018). Additionally, some of the main reasons for studying and understanding soil health are outlined below:

Water Quality

Soil functions acts both as a source and a filter for pollutants as water passes through it (Zimnicki et al. 2020). Nutrient pollution from synthetic fertilizers is a widespread problem, contributing to issues such as waterbodies eutrophication, harmful algal blooms, and the deterioration of drinking water quality (Jha et al. 2023). Approaches towards improvement of soil health and water quality involve retaining, buffering, and biologically transforming pollutants. Increasing soil organic matter aids in trapping heavy metals and toxins, while soil microbes can degrade pollutants like toluene into harmless substances (Hanson et al. 1999; Jha et al. 2024). Therefore, the presence of organic matter and microbial activity in the soil is essential for improving water quality.

Human Health

Soil health is closely connected to human health, particularly through the nutritional quality and type of crops produced. Soils that are in micronutrients are linked to lower rates of malnutrition, as higher levels of soil organic matter contribute to improved crop nutrition (Welch and Graham 2004). Furthermore, diverse soil ecosystems can enhance nutrient availability, suppress plant diseases, and influence the taste, storage, and preparation of food (Schlatter et al. 2017; Rillig et al. 2018).

Climate Change

Soil management plays a critical role in either mitigating or exacerbating climate change and affects other ecosystem services such as water quality and plant productivity (Lal 2004; Kurmi et al. 2023). However, there are trade-offs, such as the use of nitrogen fertilizers, which increase crop yields but also lead to higher emissions of nitrous oxide, a powerful greenhouse gas (Paustian et al. 2016). These challenges highlight the importance of balancing soil usage and establishing clear objectives for effective soil health management.

Conventional Approach of Soil Health Management

With growing concerns about the sustainability of agricultural systems to feed a global population of seven billion, this issue presents a significant challenge for future generations. Soil quality indicators must be carefully selected based on land use, management practices, soil characteristics, and environmental conditions. According to Doran and Safley (1997), assessing the sustainability of a system requires monitoring changes in the soil's properties, as well as its capacity to support plant growth and perform critical environmental functions. Therefore, commonly used chemical, physical, and especially biological indicators of soil health are highlighted below (Table 1). Integrating these indicators offer a more holistic and effective approach to evaluate soil health.

Table 1: Some basic soil health indicators and their measurements

Soil Health Indicator	Unit	Ideal Values Desirable in Agricultural Soil	Methods and Reference
Texture	Textural classes based on the proportion of sand, silt and clay	Loam texture having 7–27% clay, 28–50% silt and 23–52% sand ideal in general	Hydrometer method, International pipette method (Bouyoucos 1962; Gupta and Ghildyal 1998)
Bulk density	g cm^{-3}/mg m^{-3}	1.33-1.35 g cm^{-3}	Core sampler (Gupta and Ghildyal 1998)
Penetration resistance	Mega Pascal (MPa); N m^{-2}		Cone penetrometer method (Gupta and Ghildyal 1998; Herrick and Jones 2002)
Aggregate stability	Mean weight diameter (mm)		Wet sieving and dry sieving method (Yoder 1936)
Water holding capacity	%	Crop specific	Pressure plate and membrane apparatus (Richards and Fireman 1943)
Infiltration rate	mm $hour^{-1}$		Ring Infiltrometer (Johnson 1963; Gupta and Ghildyal 1998)
pH	In scale of 1–14	Neutral (6.7–7.3) pH (ideal)	Saturation soil extract or soil-water suspension (1:2 or 1:2.5) (Rao and Reddy 2013)

(Contd.)

(*Contd.*)

Soil Health Indicator	Unit	Ideal Values Desirable in Agricultural Soil	Methods and Reference
Organic matter	%	Low <0.5%, medium 0.5–0.75% and high >0.75%	Walkley and Black (1934)
Available soil nitrogen	kg ha^{-1}	Low <280 kg ha^{-1}, medium 280-560 kg ha^{-1}, high >560 kg ha^{-1}	Alkaline permanganate method (Subbiah and Asija 1956)
Available soil potassium	kg ha^{-1}	Low < 108 kg ha^{-1}, medium 108–280 kg ha^{-1}, high >280 kg ha^{-1}	Flame photometer method (Prasad et al. 2006)
Available sulfur	kg ha^{-1}		Turbidimetric method (Williams and Steinberg 1959)
Available soil Ca and Mg	Milli equivalent $liter^{-1}$ or milli equivalent 100^{-1} gram soil		Titration with EDTA (Hesse 1971)
Micronutrients (Fe, Mn, Cu & Zn)	mg kg^{-1}	Fe - 4.5 mg kg^{-1} Mn - 2.0 mg kg^{-1} Cu - 0.6 mg kg^{-1} Zn - 0.2 mg kg^{-1}	DTPA extraction and determination by AAS (Lindsay and Norvell 1978)
Microbial biomass carbon	µg microbial biomass carbon g^{-1} soil		Fumigation method (Jenkinson et al. 2004)
Dehydrogenase activity	µg TPF g^{-1} soil day^{-1}		Triphenyl tetrazolium chloride (Casida et al. 1964)
Soil respiration rate	µ mol m^{-2} s^{-1}		Closed or open dynamic system (Davidson et al. 2002)
Nitrogen fixation of microorganisms	n mole ethylene g^{-1} h^{-1}		Acetylene reductase activity (Stewart et al. 1987)
Soil chlorophyll	mg g^{-1}		Extraction using organic solvent (Nayak et al. 2004)

Modern Tools and Technologies to Monitor Soil Health

Soil Sensors and Internet of Things (IoT) Devices

Soil sensors and Internet of Things (IoT) devices are technologies increasingly used together to monitor and manage soil health in areas like agriculture, environmental monitoring, and urban planning (Satpute et al. 2024). Optical sensors, which rely on reflectance-based spectroscopy, measure the energy reflected and absorbed by various nutrient ions in the soil solution. Electrochemical sensors, such as ion-selective electrodes, produce a current or voltage output corresponding to the concentration of specific ions for soil nutrient analysis (Burton et al. 2020). IoT devices facilitate timely interventions and more efficient resource management, leading to enhanced productivity, cost savings, and environmental sustainability (Bhatnagar and Chandra 2020). IoT platforms integrate multiple soil sensors and devices, enabling comprehensive monitoring of soil health (Satpute et al. 2024). These systems collect data from various sensors, analyse it, and provide valuable insights for farmers. User-friendly interfaces assist with data visualization and interpretation, empowering farmers to make informed decisions about irrigation, fertilization, and overall soil management (Sudharson et al. 2023).

Soil Moisture Sensors

These sensors measure soil moisture content at different depths, providing real-time data that helps farmers optimize irrigation schedules and avoid both overwatering and under watering. This level of precision is crucial for efficient water management and preserving soil health (Soulis et al. 2015).

Soil pH Sensors

IoT-based pH sensors estimate soil acidity or alkalinity and helps farmers to adjust pH levels for nutrient management. Monitoring soil pH is essential for maintaining the correct nutrient balance and preventing deficiencies or toxicities (Yin et al. 2021).

Soil Temperature Sensors

Monitoring soil temperature is crucial, as it influence the seed germination, plant growth, and several soil microbial activities. This data helps farmers with decisions related to plant schedules, selecting the crop, frost-prone area management, and soil moisture (Wang et al. 2021).

Soil Nutrient Sensors

Equipped with nutrient sensors, IoT devices can estimate essential nutrient critical levels in the soil, such as nitrogen, phosphorus, and potassium. This information assists farmers in recognizing imbalances in nutrients, enabling precise and targeted administration of fertilizers. Optimized nutrient management improves crop health and reduces nutrient runoff (Rossel and Bouma 2016; Chowdhury et al. 2024 a,b; Sunil Kumar et al. 2024).

Soil Erosion Sensors

IoT-based erosion sensors detect and monitor soil erosion rates by measuring factors like soil moisture, slope, and rainfall intensity. Identifying erosion prone areas guide the farmers to implement erosion control measures and conservation practices to safeguard soil health (Liu et al. 2023).

Soil Organic Matter Sensors

IoT devices can measure soil organic matter content, a crucial indicator of soil health and fertility. Monitoring organic matter levels aid in understanding soil structure, water-holding capacity, nutrient cycling, and microbial activity. This information guides decisions regarding organic matter management and soil health enhancement (Wu et al. 2017).

Remote Sensing and Imaging Techniques

Remote sensing (RS) is the broad term of detecting and monitoring the physical characteristics of an object by measuring its reflectance and emitted radiation from a distance. RS techniques offer a direct, efficient, and timely means to assess critical soil indicators, a vital component of sustainable development. Soil health monitoring benefits significantly from RS, providing valuable insights into soil properties and conditions across expansive areas (Das et al. 2015). Within the visible near infrared (vis-NIR) spectrum (400 to 2500 nm), soil characteristics like mineralogy, organic matter, texture, and colour become discernible (Chowdhury et al. 2025). In contrast, mid-infrared (MIR) spectroscopy, spanning wavelengths from 2500 to 25,000 nm, utilizes molecular bond vibrations to generate MIR absorption curves when interacting with soil samples (Stenberg et al. 2010). MIR spectra has proven highly effective in predicting various soil physical, chemical, and biological attributes (Soriano-Disla et al. 2014).

Data Analytics and Modelling for Soil Health Assessment

Data analytics and modelling techniques are powerful tools for soil health assessment as they allow for the integration and interpretation of various data sources. Here are some commonly used data analytics and modelling approaches for soil health assessment (Bhatnagar and Chandra 2020).

Statistical Analysis

Statistical methods are essential for examining soil data, revealing patterns, trends, and connections among various soil health indicators. Descriptive statistics, correlation, regression, and multivariate analysis facilitate comprehensive soil parameter understanding.

Data Integration and Fusion

Assessing soil health typically combines diverse data sources like soil sensors, satellite images, climate data, and historical records. Data integration techniques merge these datasets for a holistic soil condition analysis.

Geostatistics

Geostatistical methods like kriging facilitate spatial interpolation of soil properties by leveraging spatial correlations among samples, enabling precise mapping for comprehensive landscape soil health assessment.

Soil Health Indices

Soil health indices integrate multiple indicators like organic matter, pH, nutrients, and compaction, into a single numerical value, offer a comprehensive evaluation of soil health. These indices can be generated using statistical methods or expert knowledge (Rinot et al. 2019).

Modelling of Soil Processes

Process-based models simulate soil dynamics, including water flow, nutrient cycling, carbon storage, and microbial activity. They aid in analysing soil health trends, assessing management effects, and conducting scenario assessments (Santra et al. 2009).

Decision Support Systems

Decision support systems combine data analytics, models, and expert insights to aid farmers in making informed choices about soil health management, encompassing irrigation, fertilization, crop rotation, and conservation practices (Chowdhury et al. 2023).

Geographic Information System (GIS)

Integration of remote sensing and imaging techniques with GIS and data analysis tools allows for spatial mapping and analysis of soil health indicators. These techniques help farmers to make informed decisions regarding irrigation, fertilization, crop selection, and soil management practices. By monitoring soil health remotely, farmers can optimize resource usage, improve crop productivity, and enhance sustainability in agricultural practices.

Application of Artificial Intelligence and Machine Learning in Soil Health Management

Machine learning methods are valuable for assessing soil health by analysing extensive datasets and constructing predictive models. These algorithms discern intricate data patterns, enabling the anticipation of soil characteristics and conditions through supervised techniques like decision trees, random forests, support vector machines, and neural networks (Mandal 2022). Some major applications of AI in soil health management are shown in Table 2.

Soil Monitoring and Analysis

AI algorithms can process data from multiple sources, including satellite images, soil sensors, and historical records, to assess soil health factors such as nutrient concentrations, moisture levels, and pH balance. This data-driven methods enable real-time tracking and prompt actions to enhance soil quality. Applying machine learning techniques to evaluate soil characteristics and forecast nutrient content plays a crucial role in smart soil management within contemporary agriculture.

Table 2: Some examples of AI application in soil health management

Sl. No.	Data Input	Algorithm	Features	References
1	8556 soil samples	ANN	Estimates SOM content and chemical attributes	Fernandes et al. 2019
2	100 soil samples	ANNSK (Artificial Neural Network Simple Kriging)	To predict SOM content	Mirzaee et al. 2016
3	369 soil samples	DT (Decision Tree), K-NN (K- Nearest Neighbour), ANN	To predict SOM content	Bouasria et al. 2021
4	90 soil samples	RF (Random Forest), CNN (Convolution Neural Network)	A smartphone application for predicting soil texture	Swetha et al. 2020
5	Various types of variables	ANFIS-GOA, ANFIS-SSA, ANFIS-s GWO, ANFIS- PSO, ANFIS-GA, ANFIS- DA	To predict soil temperature	Penghui et al. 2020
6	1700 soil sample image	CNN, Lenet, AlexNet, Vgg16	SOM, macronutrients	Pathumuthusabana and Priyadharsini 2021
7	1000 soil samples	DT, KNN, SVM	Predicts soil fertility	Rajamanickam 2021
8	Various types of variables	ANN, UAV	Predicts soil moisture	Hassan-Esfahani et al. 2015
9	Various types of variables	ANN	Predicts compression and recompression on index of soil	Kurnaz et al. 2016
10	721 soils samples	ANN	Evaluates Pedo transfer function of field capacity and permanent wilting point	Mohanty et al. 2015

Predictive Modelling

Incorporating sensor technology for real-time monitoring of soil moisture and pH levels is a key method of intelligent soil management in modern agriculture. AI models, including machine learning algorithms, can forecast soil conditions and health metrics using historical data. This enables farmers to make data-driven decisions regarding crop selection, fertilizer use, and irrigation planning, ultimately enhancing soil health and boosting productivity (Anand et al. 2024; Pradhan et al. 2024).

Precision Agriculture

Deploying AI-driven UAVs (Unmanned Aerial Vehicles) and satellites for mapping soil variability is an advanced soil management method in modern agriculture. AI-powered precision agriculture technologies enable farmers to apply inputs like water, fertilizers, and pesticides with precision, reducing waste and environmental impact. By analysing soil data and crop performance indicators, AI can optimize input usage for sustainable soil management (Mandal 2022; Adak et al. 2023; Gawande et al. 2023).

Disease and Pest Detection

Advanced plant protection technology leverages AI-powered image recognition systems to rapidly detect and diagnose early signs of plant diseases. By employing sophisticated image recognition algorithms and machine learning models, these systems enable timely and efficient implementation of control measures. AI-based image analysis can assess crop and soil images to identify symptoms of diseases, pest infestations, or nutrient deficiencies. Early detection facilitates targeted interventions, reducing reliance on broad-spectrum treatments and minimizing crop damage.

Recommendation Systems

AI-powered recommendation systems offer tailored guidance to farmers on soil management practices, taking into account specific soil conditions, crop types, and environmental factors. These personalized recommendations help optimize soil health, enhance crop yields, and improve resource efficiency.

Autonomous Farming Equipment

The deployment of autonomous robots for preventive and control measures, such as field management and crop coverage, represents a highly efficient approach to modern agricultural management. AI-driven autonomous tractors and drones can conduct tasks like soil health monitoring, precision planting, and more, with minimal human intervention. These technologies streamline operations and enable more efficient soil management practices. The integration of AI in soil health management offers significant potential to boost agricultural sustainability, productivity, and resilience. By harnessing AI, farmers and land managers can make data-informed decisions that enhance soil health, conserve resources, and ensure long-term food security (Adak et al. 2023).

Challenges in Digital Soil Health Monitoring

Digital soil health monitoring has the potential to revolutionize agriculture and improve sustainable land management practices but also certain challenges are there simultaneously to adopt them. Major challenges that can be considered according to Liakos and Panagos (2022) are:

- Standardizing the methodologies for data collection and equipment is crucial for accurate and consistent soil health monitoring to minimize errors and ensure reliable information.
- The next challenge is managing the huge volume of data and their management. Besides, monitoring soil health digitally requires extensive data on physical, chemical, and biological soil properties.
- Handling and analysing this data demands robust tools, infrastructure, and expertise as well.
- Affordable and accessible digital soil health monitoring systems are essential for widespread adoption, as their initial costs and on-going maintenance can hinder small-scale farmers and resource-constrained regions.
- Sensor accuracy and calibration: Accurate soil health sensors require proper calibration and validation due to factors like sensor drift, environmental variations, and soil differences, ensuring reliable data (Fan et al. 2022).
- Interpreting and integrating data: Analysing digital soil health data with physical, chemical, and biological aspects demand advanced analytics to uncover valuable insights and inform decision-making due to its complexity.
- Education and capacity building: Training farmers, agronomists, and land managers in soil health monitoring is crucial. It requires skilled individuals to operate systems, gather data, and communicate results effectively.

Conclusion

Digital soil health monitoring, integrating IoT technology with soil sensors and machine learning revolutionizes modern agriculture by providing real-time, precise delivering and accurate data. It enhances the understanding of soil conditions at any given moment. This enables farmers and land managers to respond swiftly to changes in soil quality, moisture, and nutrients, preventing issues before they escalate. The precision of advanced sensors reduce errors compared to traditional methods, leading to more reliable decision-making for crop management, irrigation, and fertilization. The combination of IoT technology and data analytics support better decision-making and helped farmers and land managers by providing detailed information about soil conditions, and enable them with informed choices about crop management, irrigation, fertilization, and other practices essential for maintaining healthy soil. While soil health assessment is complex, requiring a mix of data analytics and expert knowledge, ongoing advancements in machine learning enhance data interpretation, turning complex datasets into actionable insights. This supports sustainable farming practices and helps preserve soil health, ensuring resilient agricultural ecosystems for future generations.

References

Adak, S., K. Bandyopadhyay, T.J. Purakayastha, S. Sen, R.N. Sahoo, M. Shrivastava et al. 2023. Impact of contrasting tillage, residue mulch and nitrogen management on soil quality and system productivity under maize-wheat rotation in the north-western Indo-Gangetic Plains. Front. Sustain. Food Syst. 7: 1230207.

Anand, R., S.S. Nalla, S. Nasreen, M. Chowdhury, Madhusudan, B.S., A. Pandey et al. 2024. Fruit Counting and Analysis Using Artificial Intelligence Approaches. pp. 93-114. *In:* S.S. Chouhan, A. Saxena, U.P. Singh and S. Jain [eds.]. Artificial Intelligence Techniques in Smart Agriculture. Springer Nature, Singapore.

Bhatnagar, V. and R. Chandra. 2020. IoT-based soil health monitoring and recommendation system. IoT and Analytics for Agri. 2: 1-21.

Bouasria, A., K.I. Namr, A. Rahimi and E.M. Ettachfini. 2021. Geospatial assessment of soil organic matter variability at Sidi Bennour District in Doukkala Plain in Morocco. J. Ecol. Eng. 22(11): 120-130.

Bouyoucos, G.J. 1962. Hydrometer method improved for making particle size analyses of soils 1. Agron. J. 54(5): 464-465.

Bünemann, E.K., G. Bongiorno, Z. Bai, R.E. Creamer, G.D. Deyn, L. Fleskens et al. 2018. Soil quality – A critical review. Soil Biol. Biochem. 120: 105-125.

Burton, L., K. Jayachandran and S. Bhansali. 2020. Review – The "Real-Time" revolution for in situ soil nutrient sensing. J. Electrochem. Soc. 167: 037569.

Cardoso, E.J.B.N., R.L.F. Vasconcellos, D. Bini, M.Y.H. Miyauchi, C. Alcantara, P.R.L. Alves et al. 2013. Soil health: Looking for suitable indicators. What should be considered to assess the effects of use and management on soil health? Sci. Agric. 70(4): 274-289.

Casida, J.L.E., D.A. Klein and T. Santoro. 1964. Soil dehydrogenase activity. Soil Sci. 98(6): 371-376.

Chowdhury, M., E.V. Thomas, A. Jha, A. Kushwah, R. Kurmi, T.K. Khura et al. 2023. An automatic pressure control system for precise spray pattern analysis on spray patternator. Comput. Electron. Agric. 214: 108287.

Chowdhury, M., T.K. Khura, P.K. Upadhyay, R.A. Parray, H.L. Kushwaha, C. Singh et al. 2024a. Assessing vegetation indices and productivity across nitrogen gradients: A comparative study under transplanted and direct-seeded rice. Front. Sustain. Food Syst. 8: 1351414.

Chowdhury, M., T.K. Khura, R.A. Parray, H.L. Kushwaha, P.K. Upadhyay, A. Jha et al. 2024b. The use of destructive and nondestructive techniques in concrete nitrogen assessment in plants. J. Plant Nutri. 47: 2271-2294.

Chowdhury, M., T.K. Khura, R.A. Parray, P.K. Upadhyay, H.L. Kushwaha, A. Lama et al.. 2025. Multi-vegetation indices based handheld device for precise nitrogen assessment and prescription in direct seeded rice. Comput. Electron. Agric. 229: 109886.

Congreves, K.A., A. Hayes, E.A. Verhallen and E.L.L. Van. 2015. Long-term impact of tillage and crop rotation on soil health at four temperate agroecosystems. Soil Tillage Res. 152: 17-28.

Das, B.S., M.C. Sarathjith, P. Santra, R.N. Sahoo, R. Srivastava, A. Routray et al. 2015. Hyperspectral remote sensing: Opportunities, status and challenges for rapid soil assessment in India. Curr. Sci. 108(5): 860-868.

Davidson, E.A., K.V.L.V. Savage, L.V. Verchot and R. Navarro. 2002. Minimizing artifacts and biases in chamber-based measurements of soil respiration. Agric. For. Meteorol. 113(1-4): 21-37.

Doran, J.W. and T.B. Parkin. 1994. Defining and assessing soil quality. pp. 3-21. *In:* J.W. Doran, D.C. Coleman, D.F. Bezdicek & B.A. Stewart [eds.]. Defining Soil Quality for a Sustainable Environment. SSSA Special Publication No. 35. Soil Science Society of America, Madison, WI.

Doran, J.W. and M. Safley. 1997. Defining and assessing soil health and sustainable productivity. pp. 1-28. *In:* C.E. Pankhurst, B.M. Doube, V.V.S.R. Gupta [eds.]. Biological Indicators of Soil Health. CAB International, Wallingford, UK.

Fan, P., H. Yu, B. Xi and W. Tan. 2022. A review on the occurrence and influence of biodegradable microplastics in soil ecosystems: Are biodegradable plastics substitute or threat? Environ. Int. 163: 107244.

Fernandes, M.M.H., A.P. Coelho, C. Fernandes, M.F.d. Silva and C.C.D. Marta. 2019. Estimation of soil organic matter content by modeling with artificial neural networks. Geoderma. 350: 46-51.

Foley, J.A., N. Ramankutty, K.A. Brauman, E.S. Cassidy, J.S. Gerber, M. Johnston et al. 2011. Solutions for a cultivated planet. Nature. 478: 337-342.

Gawande, V., D.R.K. Saikanth, B.S. Sumithra, S.A. Aravind, G.N. Swamy, M. Chowdhury et al. 2023. Potential of precision farming technologies for eco-friendly agriculture. Int. J. Plant Soil Sci. 35(19): 101-112.

Guddaraddi, A., A. Singh, G. Amrutha, D.R.K. Saikanth, R. Kurmi, G. Singh et al. 2023. Sustainable biofuel production from agricultural residues an eco-friendly approach: A review. Int. J. Environ. Clim. Chang. 13(10): 2905-2914.

Gupta, R.P. and B.P. Ghildyal. 1998. Theory and Practice in Agrophysics Measurement. Allied Publishers. New Delhi.

Hanson, J.R., J.L. Macalady, D. Harris and K.M. Scow. 1999. Linking toluene degradation with specific microbial populations in soil. Appl. Environ. Microbiol. 65: 5403-5408.

Hassan-Esfahani, L., A. Torres-Rua, A. Jensen and M. McKee. 2015. Assessment of surface soil moisture using high-resolution multi-spectral imagery and artificial neural networks. Remote Sens. 7(3): 2627-2646.

Herrick, J.E. and T.L. Jones. 2002. A dynamic cone penetrometer for measuring soil penetration resistance. Soil Sci. Soc. Am. J. 66(4): 1320-1324.

Hesse, P.R. 1971. A Textbook of Soil Chemical Analysis. C.B.S. Publishers and Distributors Pvt. Ltd. New Delhi.

Jenkinson, D.S., P.C. Brookes and D.S. Powlson. 2004. Measuring soil microbial biomass. Soil Boil. Biochem. 36(1): 5-7.

Jha, A., M. Chowdhury and A.N. Satpute. 2023. Surface water quality forecasting using machine learning approach. pp. 293-315. *In:* C.B. Pande, M. Kumar and N.L. Kushwaha [eds.]. Surface and Groundwater Resources Development and Management in Semi-arid Region: Strategies and Solutions for Sustainable Water Management. Cham: Springer International Publishing.

Jha, A., S. Garg, S. Satpute, J.P. Singh, D. Singh, U.G Phutela et al. 2024. Isotherm and kinetic modelling for assessing the effectiveness of economical adsorbents in sewage water treatment. Indian J. Chem. Technol. 31: 438-450.

Jian, J., X. Du and R.D. Stewart. 2020. A database for global soil health assessment. Sci. Data. 7: 16.

Johnson, A.I. 1963. A field method for measurement of infiltration. Water Supply Paper (No. 1544-F). USGP Numbered Series.

Karlen, D.L., K.S. Veum, K.A. Sudduth, J.F. Obrycki and M.R. Nunes. 2019. Soil health assessment: Past accomplishments, current activities, and future opportunities. Soil Till. Res. 195: 104365.

Keith, A.M., O. Schmidt and B.J. McMahon. 2016. Soil stewardship as a nexus between ecosystem services and one health. Ecosyst. Serv. 17: 40-42.

Kumari, A., P. Ashoka, P. Tiwari, P. Sachan, A.K. Malla, A, Tripathy et al. 2023. Zero tillage lead to enhanced productivity and soil health. Int. J. Environ. Clim. Chang. 13(10): 3707-3715.

Kurmi, R., S.D. Lande, J. Kurmi, M. Chowdhury, C. Singh, and P. Kumar. 2023. Comparative study on carbon footprint assessment of rice-wheat production system. Int. J. Environ. Clim. Chang. 13(11): 390-398.

Kurnaz, T.F., U. Dagdeviren, M. Yildiz and O. Ozkan. 2016. Prediction of compressibility parameters of the soils using artificial neural network. SpringerPlus. 5: 1-11.

Lal, R. 2004. Soil carbon sequestration impacts on global climate change and food security. Sci. 304: 1623-1627.

Lavelle, P. and A. Spain. 2001. Soil Ecology. Kluwer Academics, Dordrecht, Netherlands.

Liakos, L. and P. Panagos. 2022. Challenges in the geo-processing of big soil spatial data. Land. 11(12): 2287.

Lindsay, W.L. and W. Norvell. 1978. Development of a DTPA soil test for zinc, iron, manganese, and copper. Soil Sci. Soc. Am. J. 42(3): 421-428.

Liu, J., T. Qiu, J. Peñuelas, J. Sardans, W. Tan, X. Wei et al. 2023. Crop residue return sustains global soil ecological stoichiometry balance. Glob. Chang. Biol. 29(8): 2203-2226.

Mandal, A.K. 2022. Soil degradations in the IGP of central Haryana, India – A spatial assessment for reclamation and management. Environ. Monit. Assess. 194(11): 835.

Mirzaee, S., S. Ghorbani-Dashtaki, J. Mohammadi, H. Asadi and F. Asadzadeh. 2016. Spatial variability of soil organic matter using remote sensing data. Catena. 145: 118-127.

Mohanty, M., N.K. Sinha, D.K. Painuli, K.K. Bandyopadhyay, K.M. Hati, K.S. Reddy et al. 2015. Modelling soil water contents at field capacity and permanent wilting point using artificial neural network for Indian soils. Natl. Acad. Sci. Lett. 38: 373-377.

Nayak, A.K., B. Gangwar, A.K. Shukla, S.P. Mazumdar, A. Kumar, R. Raja et al. 2004. Long-term effect of different integrated nutrient management on soil organic carbon and its fractions and sustainability of rice–wheat system in Indo Gangetic Plains of India. Field Crops Res. 127: 129-139.

Norris, C.E. and K.A. Congreves. 2018. Alternative management practices improve soil health indices in intensive vegetable cropping systems: A review. Front. Environ. Sci. 6: 50.

Pathumuthusabana, A.M. and S.S. Priyadharsini. 2021. An artificial intelligence-based evaluation of soil fertility. pp. 255-275. *In:* Handbook of Green Engineering Technologies for Sustainable Smart Cities. CRC Press.

Patra, K., C.M. Parihar, H.S. Nayak, B. Rana, D.R. Sena, A. Anand et al. 2023. Appraisal of complementarity of subsurface drip fertigation and conservation agriculture for physiological performance and water economy of maize. Agric. Water Manag., 283: 108308.

Paustian, K., J. Lehmann, S. Ogle, D. Reay, G.P. Robertson and P. Smith. 2016. Climate-smart soils. Nature 532(7597): 49-57.

Penghui, L., A.A. Ewees, B.H. Beyaztas, C. Qi, S.Q. Salih, N. Al-Ansari et al. 2020. Metaheuristic optimization algorithms hybridized with artificial intelligence model for soil temperature prediction: Novel model. IEEE Access. 8, 51884-51904.

Pradhan, N.C., M.A. Naik, M. Chowdhury, A. Kushwah, K.R. Asha, T. Dhar et al. 2024. Robotic seeding or sowing system in smart agriculture. pp. 495-520. *In:* K. Pandey, N.L. Kushwaha, C.B. Pande and K.G. Singh [eds.]. Artificial Intelligence and Smart Agriculture: Technology and Applications. Springer Nature, Singapore.

Prasad, R., P. Hobbs and Y.S. Shivay. 2006. New frontiers in phosphate fertilizers. Indian J. Ferti. 14(10): 20-26.

Rajamanickam, J. 2021. Predictive model construction for prediction of soil fertility using decision tree machine learning algorithm. INFOCOMP. J. Comput. Sci. 20(1).

Rao, A.S. and K.S. Reddy. 2013. Analysis of soils for pH, EC and available major nutrients, in methods of analysis of soils, plants, waters fertilizers and organic manures. (H.L.S. Tandon [ed.]). Fertilizer Development and Consultation Organization. New Delhi.

Richards, L.A. and M. Fireman. 1943. Pressure-plate apparatus for measuring moisture sorption and transmission by soils. Soil Sci. 56(6): 395-404.

Rillig, M.C., A. Lehmann, J. Lehmann, T. Camenzind and C. Rauh. 2018. Soil biodiversity effects from field to fork. Trends Plant Sci. 23: 17-24.

Rinot, O., G.J. Levy, Y. Steinberger, T. Svoray and G. Eshel. 2019. Soil health assessment: A critical review of current methodologies and a proposed new approach. Sci. Total Environ. 648: 1484-1491.

Rossel, R.A.V. and J. Bouma. 2016. Soil sensing: A new paradigm for agriculture. Agric. Syst. 148: 71-74.

Rumpel, C., F. Amiraslani, L.S. Koutika, P. Smith, D. Whitehead and E. Wollenberg. 2018. Put more carbon in soils to meet Paris climate pledges. Nature. 564: 32-34.

Santra, P., R.N. Sahoo, B.S. Das, R.N. Samal, A.K. Pattanaik and V.K. Gupta. 2009. Estimation of soil hydraulic properties using proximal spectral reflectance in visible, near-infrared, and shortwave-infrared (VIS–NIR–SWIR) region. Geoderma. 152(3-4): 338-349.

Satpute, A.N., K.P. Gavhane, S. Kaur, A. Jha, N.C. Pradhan and M. Chowdhury. 2024. Integration of AI and IoT in soilless cultivation to power sustainable agricultural revolution. pp. 387-411. *In:* K. Pandey, N.L. Kushwaha, C.B. Pande and K.G. Singh [eds.]. Artificial Intelligence and Smart Agriculture: Technology and Applications. Springer Nature, Singapore.

Schlatter, D., L. Kinkel, L. Thomashow, D. Weller and T. Paulitz. 2017. Disease suppressive soils: New insights from the soil microbiome. Phytopathol. 107: 1284-1297.

Soriano-Disla, J.M., I. Gómez, J. Navarro-Pedreño and M.M. Jordán. 2014. The transfer of heavy metals to barley plants from soils amended with sewage sludge with different heavy metal burdens. J. Soil Sediment. 14: 687-696.

Soulis, K.X., S. Elmaloglou and N. Dercas. 2015. Investigating the effects of soil moisture sensors positioning and accuracy on soil moisture based drip irrigation scheduling systems. Agric. Water Manag. 148: 258-268.

Stenberg, B., R.A.V. Rossel, A.M. Mouazen and J. Wetterlind. 2010. Visible and near infrared spectroscopy in soil science. Adv. Agron. 107: 163-215.

Stewart, J.W.B., I.P. O'halloran and R.G. Kachanoski. 1987. Influence of texture and management practices on the forms and distribution of soil phosphorus. Can. J. Soil Sci. 67(1): 147-163.

Subbiah, B.V. and G.L. Asija. 1956. A rapid procedure for assessment of available nitrogen in rice soils. Curr. Sci. 25: 259-260.

Sudharson, K., B. Alekhya, G. Abinaya, C. Rohini, S. Arthi and D. Dhinakaran. 2023. Efficient soil condition monitoring with IoT enabled intelligent farming solution. *In:* 2023 IEEE International Students' Conference on Electrical, Electronics and Computer Science (SCEECS). 1-6.

Sunil Kumar, T., H.M. Virdia, K.G. Patel, S. Ragi, M. Chowdhury, P. Kumar et al. 2024. Effect of summer legume residue incorporation and fertilizer regimes on rice growth, yield, and nutrient uptake. Front. Sustain. Food Syst. 8: 1467201.

Swetha, R.K., P. Bende, K. Singh, S. Gorthi, A. Biswas, B. Li et al. 2020. Predicting soil texture from smartphone-captured digital images and an application. Geoderma 376: 114562.

Walkley, A. and I.A. Black. 1934. An examination of the Degtjareff method for determining soil organic matter, and a proposed modification of the chromic acid titration method. Soil Sci. 37(1): 29-38.

Wang, B., S. An, C. Liang, Y. Liu and Y. Kuzyakov. 2021. Microbial necromass as the source of soil organic carbon in global ecosystems. Soil Biol. Biochem. 162: 108422.

Welch, R.M. and R.D. Graham. 2004. Breeding for micronutrients in staple food crops from a human nutrition perspective. J. Exp. Bot. 55(396): 353-364.

Williams, C.H. and A. Steinberg. 1959. Soil sulphur fractions as chemical indices of available sulphur in some Australian soils. Aust. J. Agric. Res. 10: 340-352.

Wu, J., Y. Zhao, H. Qi, X. Zhao, T. Yang, Y. Du et al. 2017. Identifying the key factors that affect the formation of humic substance during different materials composting. Bioresour. Technol. 244: 1193-1196.

Yin, H., Y. Cao, B. Marelli, X. Zeng, A.J. Mason and C. Cao. 2021. Soil sensors and plant wearables for smart and precision agriculture. Adv. Mater. 33(20): 2007764.

Yoder, R.E. 1936. A direct method of aggregate analysis of soils and a study of the physical nature of erosion losses. Agron. J. 28(5): 337-351.

Zimnicki, T., T. Boring, G. Evenson, M. Kalcic, D.L. Karlen, R.S. Wilson et al. 2020. On quantifying water quality benefits of healthy soils. BioSci. 70(4): 343-352.

CHAPTER
7

Crop Disease Detection and Prevention Using Artificial Intelligence

Karishma Kumari[1], Dattatray G. Bhalekar[2*], Srinivas Chappa[3], Pradeep Kumar[4], Ruchita Bhattarai[5] and Prasad Nagineni[6]

[1] Department of Agronomy, Horticulture and Plant Science, South Dakota State University, Brookings, SD, 57007, USA

[2] Department of Biological Systems Engineering, Washington State University, Prosser, WA, USA - 99350

[3] Electrical and Computer Engineering Department, State University of New York, Binghamton, Vestal, NY, 13850, USA

[4] Agricultural and Biosystems Engineering Department, South Dakota State University, Brookings, SD 57007, USA

[5] Department of Biology and Microbiology, South Dakota State University, Brookings, SD, 57007, USA

[6] College of Engineering and Computer Science, Florida Atlantic University, Boca Raton, FL 33431, USA

Introduction

Agriculture plays a vital role in the global economy while facing several challenges that could affect food security. Various anthropogenic, biotic, and abiotic factors are responsible for crop losses impacting the overall agricultural system. (Junaid and Gokce 2024). Those factors mainly include climate change, drought, salinity, soil loss, weeds, rodents, pests, and other postharvest losses. However, crop diseases have been a major challenge, significantly impacting food quality and overall productivity. The human population is projected to increase to 11 billion by 2100 (Gu et al. 2021). Therefore, crop protection has become even more critical to ensure food security for this growing global population.

Background and Importance of Crop Disease Management

Various pathogenic agents cause crop diseases, including fungi, bacteria, viruses, and pests. If these pathogens are left unmanaged, they might cause harmful disease

*Corresponding author: dattatray.bhalekar@wsu.edu

DOI: 10.1201/9781003536932-7

outbreaks that can potentially reach a global scale. Currently, insect pests are causing a loss of up to 38% of total agricultural production. Effectively managing these causal agents is important to avoid their devastating effect on the agricultural ecosystem.

Global Agricultural Challenges Due to Crop Diseases

Globally, crop diseases may reduce agricultural productivity, ranging from 20% to 40%, causing a substantial decrease in production quantity and quality. The estimated annual economic loss globally due to plant disease is up to $220 billion, with an additional loss of $70 billion due to invasive insects (Food and Agriculture Organization, USA 2023). Developing countries are most vulnerable to this issue since their economy predominantly relies on agriculture. The impact of crop disease may extend beyond the yield loss (Sunil Kumar et al. 2024). Disease outbreaks disrupt the food production and supply chain, substantially increasing food prices. The agricultural communities must invest high costs in chemicals and machinery for crop protection (Bhalekar et al. 2024a). Globally, the amount of pesticide used in agriculture was estimated to be 3.54 metric tons in 2021 (Global Pesticide Agricultural Use 1990-2021, 2024). Excessive use of chemicals may negatively affect human and animal health. The chemical residues on crops and in spray equipment are also a potential medium for human chemical exposure (Kumar et al. 2023; Guddaraddi et al. 2023; Mozzanini et al. 2024). A recent study by the United Nations estimated that fatalities due to pesticide usage are up to 200,000 annually (DuVall 2017). The off-target losses in conventional crop protection practices may impact the environment and biodiversity. Alternative chemical application techniques (Bhalekar et al. 2023, 2024b) have been researched to minimize these effects in the last few decades. However, substantial innovations are required for safe and efficient chemical applications compared to existing disease management approaches (Kurmi et al. 2023; Gawande et al. 2023).

Evolution of Disease Detection Methods

Crop health monitoring and pathogen detection are critical to reduce disease spread and ensure adequate crop protection. Conventionally, crop disease monitoring and management relied heavily on manual methods such as visual inspections. This usually includes scouting pests or inspecting the disease symptoms by experts. For example, farmers use visual disease rating keys, scales, and diagrams to monitor the severity of several crop diseases. The first standardized tool, which included five classes, was developed for severity assessment (Cobb 1892). Over last 80 years, traditional disease scales have often been used for visual disease assessment. Although manual methods provide direct disease detection, it can be subjective, inefficient, and time-consuming. Moreover, detecting a specific disease accurately requires expertise for classifying complex symptoms. The accuracy of manual detection also depends on scale type, rater, level of experience, and training (Bock et al. 2022). Presently, significant advancements from laboratory to field scale detection can be observed in modern disease detection methods. These new technologies usually offer disease detection with higher precision and accuracy. The modern methods of automatic disease detection are often divided into direct and indirect methods.

Direct methods such as polymerase chain reaction (PCR) tests are critical innovations in modern disease detection for various pathogens. PCR offers higher accuracy by rapidly amplifying the DNA sequences of the causal agents. Portable PCR instruments are often used for infield diagnostics (Schaad and Frederick 2002). However, direct methods involving destructive tissue sampling, sample processing, analysis, and real-time disease detection using this approach are still challenging.

In contrast, indirect approaches capable of disease detection based on diverse plant characteristics commonly include optical imaging techniques (Bakicha et al. 2020; Galletti et al. 2020) observing morphological changes and transpiration rates. Fluorescence and hyperspectral imaging are indirect techniques for early-stage disease detection. Powdery mildew (PM) in the symptomatic stage has been detected using optical imaging under controlled environments in various crops (Shin et al. 2021; Li et al. 2022). Researchers have also studied PM detection in the early stage using infrared (Awad et al. 2015) and hyperspectral (Zhang et al. 2012; Khan et al. 2021) imaging techniques.

Various machine vision and learning approaches have also been utilized for automated ground and remote sensing-based disease detection (Oberti et al. 2016; Mahmud et al. 2019; Abdulridha et al. 2020). A computer vision-based infield plant-level disease rating system was developed for Cercospora leaf spot, a prominent disease in sugar beet (Atoum et al. 2016). The study reported higher consistency in rating compared to human experts. Study evaluated the PM detection performance of UAV-based high-resolution RGB and multispectral sensors in an apple orchard (Chandel et al. 2021). The study reported that multiple stress indicators varied significantly in infected and non-infected apple trees. Variations in vegetation indices are closely related to changes in chlorophyll, anthocyanin, and carotenoid pigments, collectively representing disease incidence (Vescovo et al. 2012; Kyratzis et al. 2017). These studies indicate the advancements in disease monitoring in the present scenario.

Emergence of AI in Agriculture

Artificial Intelligence (AI) has recently emerged as a key technology in agriculture. The applications of AI are substantially increasing in various fields of agriculture, including biotic and abiotic stressors monitoring, weather forecasting, and overall crop health monitoring (Malkhani et al. 2024). AI technologies are also being integrated into farming practices, including soil management, crop load monitoring, and crop protection. AI algorithms can analyze large datasets more efficiently and accurately than humans. In agriculture, machine learning can be used to collect and analyze data from sources such as crop physiology and soil and weather sensors to predict crop health (Jha et al. 2023; Pradhan et al. 2024; Chowdhury et al. 2024). Various unmanned aerial and ground vehicles can be automated using AI algorithms for more precise applications. Automated tractors, sprayers, and harvesters use AI algorithms to perform activities with higher precision (Anand et al. 2024; Chowdhury et al. 2025). Moreover, AI-powered robots are used in specialty crops for pollination, weeding, blossom thinning, and harvesting applications, considerably reducing labor requirements (Satpute et al. 2024). These AI applications in agriculture assist growers in enhancing productivity and optimizing resource use efficiency.

Crop disease monitoring and management has become a significant application of AI in agriculture. AI-based machine vision, machine learning, and deep learning can reliably be used for disease detection using a non-destructive approach (Kushwah et al. 2024). Various convolutional neural network (CNN) architecture models have been trained to detect disease incidence in plant leaves, fruits, stems, and root tissues using imagery data (Kumari and Nafchi 2024). Currently, YOLO architecture has shown promising accuracy in object detection compared to various CNN-based models (Kumari et al. 2024b). YOLO is a unified object detection model that is considerably robust and simple to train (Fig. 1). Since YOLO outperforms several object detection deep learning models, it is recommended for real-time disease detection (Chandana and Ramachandra 2022).

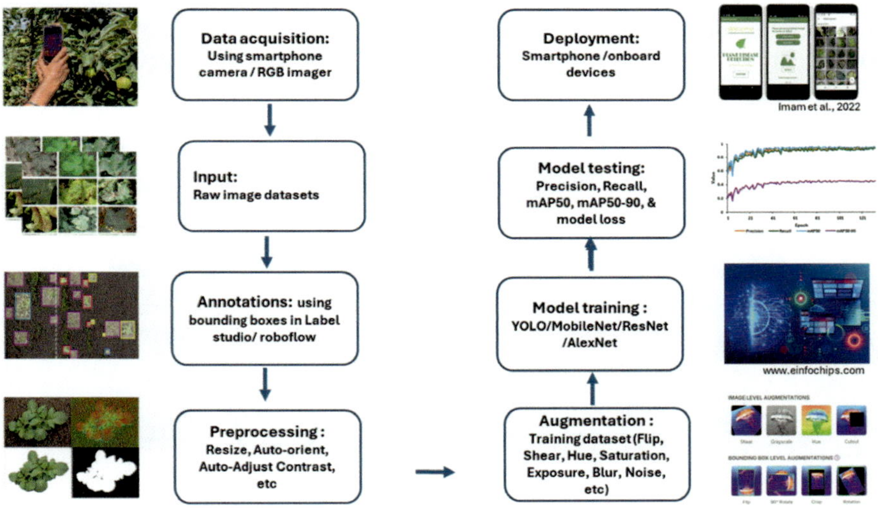

Fig. 1: Standard workflow to develop AI-based deep learning models for crop disease detection applications.

Integrating these AI models with aerial UAVs or remote sensing techniques can be used in large-scale disease monitoring. AI-based machine learning models have also been developed to predict disease incidence based on real-time weather, soil moisture, and leaf wetness data. Along with disease monitoring, AI algorithms have also been developed for targeted precision chemical applications. A machine vision-based advanced robotic system for chemical thinning has also been developed (Mu et al. 2023). This robotic system consists of a machine vision system for blossom detection, a cartesian robotic system for navigation, and an end-effector for spray applications. Laser-guided (2D LiDAR) precision sprayers have been adopted for crop protection in specialty crops (Chen et al. 2020). These applications of AI in disease management can substantially reduce chemical inputs and with disease control comparable to conventional chemical application methods. This chapter will broadly discuss various AI applications in crop disease monitoring and management.

AI Techniques for Crop Disease Detection

AI techniques have revolutionized crop disease detection by enabling rapid, accurate, and automated identification of plant diseases. Advanced models, including deep learning networks and machine learning algorithms, have demonstrated high accuracy in diagnosing various crop diseases. These methods enhance precision agriculture by facilitating early detection and reducing dependency on manual inspections. A summary of AI techniques and their performance in crop disease identification is presented in Table 1.

Machine Learning Approaches

The use of machine learning (ML) in crop disease detection has significantly improved the identification and diagnosis of plant diseases, Fig. 2. ML techniques process image data and identify patterns that correlate with specific diseases, enhancing agricultural productivity.

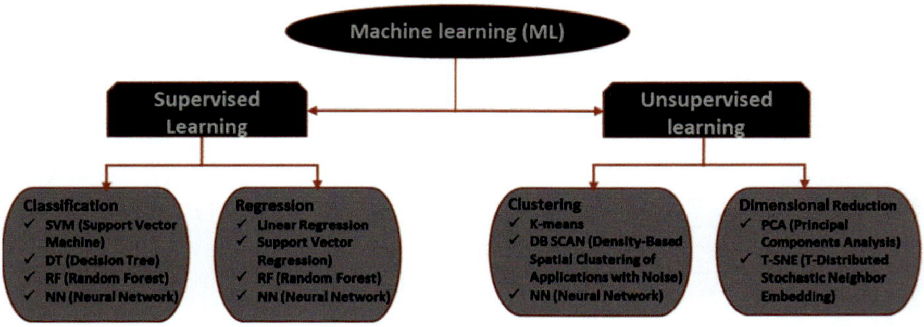

Fig. 2: Types of ML models in agriculture.

Supervised Learning

Supervised learning models, such as Support Vector Machines (SVMs) and Decision Trees, are widely used for crop disease classification. These models are trained on labeled datasets where the disease type is already known. The models then learn to differentiate between healthy and diseased crops based on the patterns present in the data. For instance, SVMs can separate different disease categories based on the extracted features, such as texture or color changes in plant leaves (Whitmire et al. 2021).

Unsupervised Learning

Unsupervised learning methods, such as clustering, have been used for pattern recognition in disease detection. These techniques do not require labeled datasets

Table 1: Overview of AI techniques for crop disease identification and their performance

Sl. No.	Crop	Disease	AI Technique/Model Used	Performance	References
1.	Tomato	Bacterial spot, early blight, and leaf mold	Deep learning networks: (AlexNet, VGG16, ResNet34, EfficientNet-b0, and MobileNetV2) and machine learning algorithms: (SVM), (kNN) and (RF)	ResNet34 achieved the best results with 99.7% accuracy, 99.6% precision, 99.7% recall, and a 99.7% F1 score.	Tan and Jiang 2021
2.	Paddy	Bacterial leaf blight, brown spot, leaf smut and leaf blast	Convolution Neural Network (CNN) based model named VGG-16 for classification	The model, trained on 4,500 healthy and unhealthy rice leaf images from Kaggle, achieved 90% accuracy.	Jangid and Sharma 2023
3.	Capsicum	Bacterial or fungal disease	Tree, Linear Discriminant, KNN and SVM	KNN and SVM performed best, with SVM achieving 100% accuracy on 62 healthy and diseased capsicum images.	Sood and Singh 2020
4.	Soybean	Reddish-brown lesions on seedling hypocotyls	Cetalatran-optimized deep KNN algorithm	The proposed method achieved 91.88% accuracy, 92.74% sensitivity, and 95.95% specificity, outperforming existing techniques.	Gaikwad and Musande 2023
5.	Cucumber	Downy mildew, anthracnose, and blight	Global-local single value decomposition (SVD) and SVM techniques	Proposed method is effective for cucumber disease recognition, offering the highest recognition rate.	Zhang and Wang 2016
6.	Basil	Leaf diseased	9 different classifiers (Decision tree, Random Forest, Support vector machine, AdaBoost, Linear models, Naives Bayes, KNN and Artificial neural networks)	Validated on 400 cases (200 healthy, 200 diseased), the system achieved 98.4% accuracy, showing promise as an effective tool for leaf disease identification.	Dhingra et al. 2019

but instead group the image data based on inherent similarities. This approach is useful when dealing with large datasets where labeling every image is impractical. Clustering can reveal natural groupings of diseased and healthy plants, facilitating early disease detection (Srivastava et al. 2020).

Deep Learning Models

Deep learning (DL) has revolutionized crop disease detection by leveraging neural networks capable of processing vast amounts of image data. DL models have been highly effective in detecting subtle disease symptoms and distinguishing between multiple types of diseases.

Convolutional Neural Networks (CNNs)

CNNs are the most prominent deep-learning models used for image-based disease detection. These networks are capable of recognizing complex patterns within images, such as minute changes in leaf texture or color, that indicate disease (Ferentinos et al. 2018). CNN architectures such as AlexNet, VGG16, and MobileNet have demonstrated high accuracy across various crop disease detection tasks, particularly when applied to large and diverse datasets (Chen et al. 2020). Moreover, CNNs can be fine-tuned using transfer learning, reducing the need for extensive training data.

Recurrent Neural Networks (RNN)

While CNNs excel in spatial data analysis, RNNs are often employed to analyze temporal data, such as changes in crop health over time. This approach is particularly useful when monitoring disease progression across seasons or in specific regions of a plant. By using RNNs, agriculturalists can gain deeper insights into the timing and development of plant diseases (Too et al. 2019).

Transfer Learning

Transfer learning involves fine-tuning pre-trained models on specific agricultural datasets. This technique has proven particularly beneficial for crop disease detection, where training a CNN from scratch would require large datasets and computational resources (Mohanty et al. 2016). Models pre-trained on general image datasets, such as ImageNet, can be adapted for crop disease classification with relatively smaller datasets, significantly improving detection accuracy and reducing computational overhead (Ramcharan et al. 2017).

Data Sources and Types

The performance of AI models in crop disease detection heavily relies on the quality and type of data used for training and testing.

Satellite Imagery, Drone-Based Imaging, and IoT Sensors

Crop disease detection often involves a variety of data sources, including satellite imagery, drone-based imaging, and data from IoT sensors. Each data source provides different levels of spatial and spectral information. For example, drone-based imaging captures high-resolution images of crops, enabling early detection of disease symptoms (Iqbal et al. 2020). Satellite imagery can be used for large-scale monitoring of crop health, providing a macroscopic view of disease spread across fields (Devaraj et al. 2019).

Importance of Annotated Datasets in Training Models

The accuracy of ML and DL models in crop disease detection is directly influenced by the availability and quality of annotated datasets. These datasets must contain well-labeled images that distinguish between various diseases, healthy plants, and background noise. Proper annotation ensures that models are trained on the most relevant features, such as disease-affected regions of leaves or fruits. The use of robust datasets is crucial for enhancing the model's ability to generalize and perform accurately on unseen data.

AI Techniques for Crop Disease Prevention

The most likely and early encounters that a crop faces during its production are disease and spread by insects, which eventually lead to destroying the entire crop by spreading if it is not taken care of on time and should be nullified at the earliest detection. To avoid the crop disease spreading, it is significant to identify the disease during the early stages to avoid low yields and rotten produce. The key source of getting disease on a crop is inclimate weather and a lack of knowledge on the health of the soil and the intended crop. Hence, in the literature, there have been prominent articles mentioning various algorithms based on artificial intelligence about crop disease identification (Tirkey et al. 2023; Kumari et al. 2024b). In this section, we will be learning about various models to understand patterns of diseases and various ways to prevent crop diseases by knowing the health of crops and soil.

Predictive Modeling

Rainfall and humidity are the primary determinants of crop disease (Matengu et al. 2023). And if the disease is left uncontrolled, it controls the crop yield severely. To understand and analyze any disease, modeling is the origin. According to published research, weather-based prediction models account for how weather patterns affect the likelihood of diseases or pest infestations, producing very accurate predictions (David 2023). Farmers have adopted recent trends in agriculture like AI and ML algorithms to predict diseases. In ML techniques, a historical dataset is used to train the system (model) to gain knowledge about the disease and predict the outbreak of the same (Shahare et al. 2023). A typical architecture is shown in Fig. 3.

Weather-Based Prediction Models

As the ML models are more inclined by previous data, this is a primary step in weather-based prediction models. Researchers gather the weather and disease outbreak data from weather stations, private organizations, field sensors, and the internet. Huge data ML models give more iterations and the best prediction. Field data is prominent in the data section of the ML architecture. The typical weather

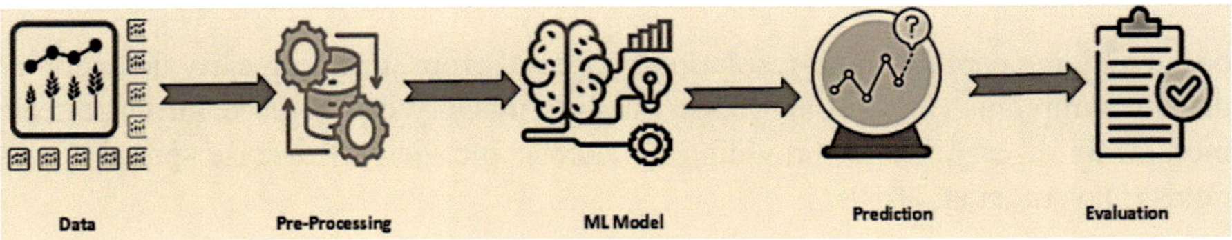

Fig. 3: ML architecture model.

data is temperature, humidity, evaporation, precipitation, wind speed, sunlight, etc. Once the researcher has the previous data, it has to be preprocessed to dump into ML algorithms because accurate data yields accurate predictions. Hence, the data sourced from various sources has to be processed by scaling, normalization, etc. ML models are meant to warn farmers about pest prediction; thus, yield loss can be mitigated. Convolutional neural networks (CNN) and long-short-term memory (LSTM) hybrid models are the most commonly used and accurate ones (Shahare et al. 2023). A brief comparison of various ML models has been summarized in (David 2023). The ML model embedded into the algorithm runs multiple iterations with the preprocessed data and yields the predicted outcome based on the type of parameter feed-in. Once the predicted data is out of the ML model, it will be evaluated with performance indexes as the final steps of the prediction model, and the usual indexes are accuracy, mean square error (MSE), root mean square error (RMSE), precision, and mean based error (MBE), and geometric mean (GM). Thus, the crop yield can be improved by estimating the disease outbreak earlier and altering the farmer.

Crop and Soil Health Monitoring for Early Intervention

As agriculture is the backbone of a country, still higher sides of farmers are not educated about the subject of agriculture. This lack of knowledge, like suitable crops for the soil and utilizing proper fertilizers on time by knowing the health of the crop. Eventually, all these lead to poor crop yields (Gotlur et al. 2024). Knowledge of soil and crops is prominent in irrigation to achieve profit out of efforts. The literature says there have been effective ways to understand the above-addressed concerns, and among them, ML-based solutions are the best fit (Kumar et al. 2020).

To know the soil or crop health, it is necessary to have a monitoring device called a sensor. Various sensors have been reported in the literature (Khanal et al. 2024). Anyway, electrical, electromagnetic, and electrochemical sensors are widely used in agriculture. All these sensors work on the principle of sensing the electrical characteristics of soil via electrodes and then correlating with corresponding soil properties like moisture, temperature, humidity, crop health, etc. Modernization is an added advantage that helps in agriculture too, by utilizing the internet of things (IoT) integrated with the advanced sensors (Rajak et al. 2023). As depicted in Fig. 4, the ecosystem consists of all the monitoring, sensor systems, and centralized computers embedded with AI-based models. The computer receives the data from all the other components and processes all the steps as shown in Fig. 4. Explained in the weather-based prediction model. By doing so, farmers will be warned at each stage of the crop and hence prevent the likely diseases, thus raising the crop yields.

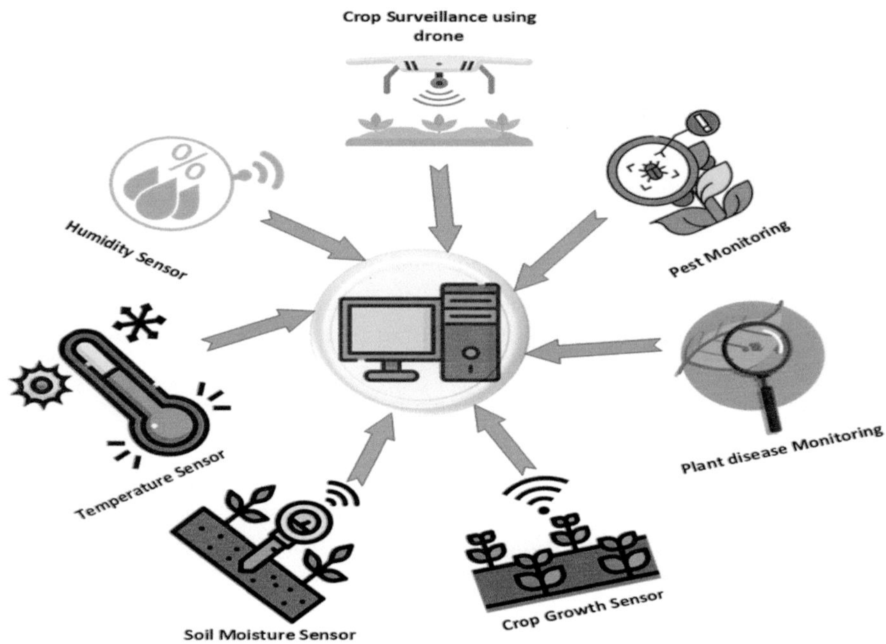

Fig. 4: IoT-based sensor and monitoring system.

Autonomous Systems and Robotics

Drones and Robots for Disease Monitoring and Treatment Application

In agriculture, the use of drones and robots for disease monitoring and treatment is growing because they provide accurate and effective results that are unmatched by conventional approaches (Abbas et al. 2023). These are applied in various ways, such as:

Disease Monitoring

Drones are used for aerial imaging with high-resolution cameras, and multispectral sensors capture detailed images of crops. These images aid in spotting early disease indicators, like color changes or irregular growth patterns. Some drones use thermal cameras to detect changes in plant temperature that might indicate stress or disease. Drone imagery is analyzed by AI systems to find trends and abnormalities, giving farmers comprehensive data on crop health. Autonomous robots equipped with cameras and sensors can move through fields, capturing close-up images and data from the ground. The detailed inspection can identify disease symptoms that the aerial view might miss.

Disease Treatment

Precision spraying in which drones can apply pesticides, fungicides, or other treatments precisely needed. This decreases the need for chemicals and lowers the possibility of disease spreading to untouched areas. Drones assist in lowering the overall amount of chemicals needed by focusing on specific areas that have been recognized as unhealthy, which is economical and environmentally beneficial.

Advantages of Using Drones and Robots

Precision: Both drones and robots can operate with high precision, targeting specific areas, and reducing the risk of overapplication.

Efficiency: They can cover large areas quickly and operate in challenging conditions, such as wet or uneven terrain, where traditional might struggle.

Early Detection: The high-resolution imaging and advanced sensors used in drones and robots allow for early detection of disease, which is crucial for effective management.

Reduced Labor: Automation of disease monitoring and treatment reduces the need for manual labor, saving time and resources.

Drones and robots have the potential to revolutionize agriculture disease surveillance and treatment, enabling previously unheard-of levels of accuracy and efficiency in crop health management. Their contribution to disease control will probably become even more crucial to contemporary farming methods as technology develops.

AI-Driven Smart Irrigation and Fertilization

AI-driven smart irrigation and fertilization systems are revolutionizing agriculture by optimizing resource use, improving crop yields, and promoting sustainability (Tace et al. 2022). AI is enhancing the critical aspects of farming, such as:

Smart Irrigation

AI-enhanced decision-making systems analyze data from soil moisture sensors, weather forecasts, and satellite imagery to determine the precise amount of water. This helps in over- or under-watering. Machine learning models predict future water needs based on historical data, weather patterns, and crop growth stages. AI-controlled irrigation systems adjust water delivery in real-time based on current soil conditions and weather forecasts. This ensures that crops receive the optimal amount of water without manual intervention. AI contributes to waste reduction and water resource conservation by optimizing water consumption with data. This is particularly important in regions facing water scarcity.

Smart Fertilization

AI analyzes the data from soil tests, crop health sensors, and historical yields to create detailed nutrient maps of fields. This helps in applying fertilizers precisely when needed. Growth Stage Optimization: AI models predict the nutritional needs of crops at various growth stages, allowing for timely and targeted fertilization. AI can forecast prospective yields and modify fertilization techniques to optimize yield by evaluating data on crop health and nutrient levels.

Integration and Synergy

AI-driven smart irrigation and fertilization systems often work together, integrating data from various resources (i.e., soil moisture, weather, crop sensors) to provide a comprehensive approach to resource management. AI technologies offer detailed analyses and reports on fertilization and irrigation efficiency, assisting farmers in making well-informed decisions and improving their methods. AI also predicts when maintenance is needed for irrigation and fertilization equipment, reducing downtime and ensuring consistent performance (Wei et al. 2024). Smart irrigation helps farmers reduce labor and operating costs while optimizing resource use, enhancing crop health, and enhancing sustainability through the use of automation and advanced data analytics.

Integration with Precision Farming

Precision farming, which combines cutting-edge technology with conventional agriculture methods to improve productivity, efficiency, and sustainability, has made great strides in agriculture with the integration of AI. Precision farming aims to address the heterogeneity found in farmed fields, seeks to maximize overall efficiency, and standardizes yields. This method opens the door to high-quality and sustainable production by combining AI with agronomic management in a synergistic fashion. The goal of precision farming is to become a decision-support tool for comprehensive and sustainable farm management (Pero et al. 2024). The overview of AI transforming precision farming is explained in detail:

Data Collection and Analysis

AI-driven systems gather and examine enormous volumes of data from numerous sources, like sensors and IoT devices, drones and satellites, and climate data.

Predictive Analytics

Artificial intelligence systems examine both past and current data to forecast crop yields, disease and pest outbreaks, and resource needs like predicting the amount of water, fertilizer, and other inputs required for optimal growth.

Precision Applications

AI helps in applying resources more efficiently for variable rate technology (VRT), which shows the adjusting of the number of seeds, fertilizers, and pesticides applied to different areas of a field based on specific needs, and automated machinery, which controls tractors and harvesters that can adapt to real-time field conditions and optimize their performance.

Decision Support Systems

AI-driven decision support systems help farmers make informed choices about customized recommendations offering tailored advice on planting strategies, crop rotation, and soil management.

Enhance Crop Management

AI helps in monitoring and managing crop health with image analysis, which analyzes images from drones or satellites to detect signs of stress or disease in crops using AI.

Sustainability and Environmental Impact

AI helps in promoting sustainable farming practices with optimized resource use. reducing waste and environmental impact by applying the right number of resources where they are needed most. Soil Health Monitoring will track the soil conditions to prevent degradation and promote long-term fertility. Thus, the combination of AI and precision farming is transforming the agriculture industry by increasing its data-drivenness, efficiency, and sustainability. These advantages will probably be further enhanced by continuing developments in AI and similar technologies, which will increase farming's productivity and resilience in the face of global challenges.

Challenges and Limitations

Data Quality and Availability

One significant challenge in applying AI for disease detection and prevention is the lack of adequate and suitable datasets (Okengwu et al. 2023). Often, the models required for detection rely on relatively large datasets. The bulk of crop disease detection operations have been carried out in perfect, controlled environments, such as when there is just one disease with a uniform backdrop. Furthermore, the level of precision gained is going to be greater than what is really attained in real-world situations since environmental circumstances are not taken into account. Moreover, categorizing images is an extremely time-consuming and demanding operation. These considerations make it very difficult to produce a complete, dependable, and effective database. Currently, six approaches exist to address the issue of missing datasets: few-shot learning, transfer learning, citizen science, data augmentation techniques, and synthetic data (Orchi et al. 2021). The majority of frequently employed databases in plant pathogen identification are cleansed and its imbalance is overlooked to focus only on developing algorithms and prevent getting sidetracked by other issues. In contrast, the distribution between categories can be distorted and imbalanced in real-life situations, varying from slightly biased to drastically unbalanced. This challenges predictive modeling, and it could be necessary to apply specialist methods, such as resampling, as most machine learning algorithms used for categorization are predicated on the idea that there are indeed a similar number of samples in every category.

Problem in Disease Classification

The form, color, and size of the symptoms of a particular disease might vary greatly depending on the stage of the disease's development, making diagnosis extremely difficult. It should be highlighted that a wide variety of disorders can manifest concurrently, making it very difficult to differentiate between specific symptoms and symptom combinations. Generally, all approaches assume that each picture represents just one illness, but in actuality, multiple illnesses may coexist with various kinds of ailments including pests and nutritional deficits. The diagnosis of the underlying cause for such multiple indications may be quite difficult because they might appear separately or visibly united. In this context, (Bock et al. 2009) noticed the co-occurrence of indications from several illnesses pointed out how this might cause problems regarding diagnosis and suggested that more developments would be needed to address this issue. Physically identical indications may result from a variety of conditions, including but not limited to illnesses, phytotoxicity, parasitic infections, as well as inadequate nutrition. Because of this, pinpointing the exact cause of an indication can be quite challenging, especially when doing so just using the spectrum of light. This makes it necessary for techniques to distinguish among indications based on minute variations (Orchi et al. 2021).

Issues with Light

In their natural habitats, crops experience significant fluctuations. As a result, a variety of elements, including wind, lighting, and other meteorological circumstances, affect photographs. Lighting problems are therefore unavoidable, but it is nearly difficult to totally eradicate the variances. However, certain efforts have been made to lessen them. Pourreza et al. (2014) created a method that uses polarizing filter set and narrow-band imaging to identify citrus Huanglongbing illness in real-time. The hardest issue to solve is reflective illumination, which is the concurrent presence of both shade and light. Although this technique probably results in some contemplation, it is possible to reduce the amount of reflected illumination by adjusting the leaf's location or the angle at which the picture is shot. In addition, since automated catches make it more challenging to minimize light issues, study found that reflected light and shades were the primary sources of mistakes when tracking Cercospora leaf spots on sugar beet plants (Zhou et al. 2015).

Ethical and Environmental Considerations

Artificial intelligence (AI) offers opportunities for environmentally friendly practices and plays a crucial role in advancing agricultural sustainability. However, if not managed properly, AI can have negative impacts on the environment (Okengwu et al. 2023). One significant concern is the carbon emissions generated by AI. According to recent research from the University of Massachusetts, training natural language processing (NLP) models can release up to 300,000 kg of carbon (Dhar 2020). In some instances, the environmental impact of AI models can be indirect. Another issue is the continuous power required to operate these systems, which contributes to environmental degradation. Additionally, the materials used to build AI systems can generate significant waste, posing environmental hazards. If improperly disposed of, these materials could reduce soil fertility and harm the ecosystem (Okengwu et al. 2023). Despite the rapid advancement of AI, there are insufficient regulations and guidelines governing its use, raising ethical and data security concerns. Improper handling of data could lead to violations of Intellectual Property Rights and the potential misuse of sensitive information (Okengwu et al. 2023). While technologies evolve and are applied systematically, the use of large datasets in agriculture could still negatively affect certain social groups, even with anonymization efforts. Agencies and corporations may make large-scale decisions that farmers may find unacceptable or detrimental. The implementation of artificially intelligent robots and their capacity to replace occupations raises further questions regarding ethics (Okengwu et al. 2023). With the advancement of agricultural technology, the need of permanent staff members has decreased. Concerns about the nature of work in future generations make sense because numerous jobs on farms anticipate the use of "intelligent" devices, which will reduce the need for people. The two problems with artificial intelligence for farming are the availability of technology and growth promotion. Innovative tools are usually only available to large organizations since they are very costly to create and employ for smaller agricultural operations and businesses. Second, it discovered that most farmers are unable to access the data, even though the majority believe they have the necessary skills to do so (Sykuta 2016).

Future Directions and Prospects

Advancements in AI for Agriculture

Emerging technologies such as generative AI and Reinforcement Learning have great potential in optimizing crop disease management. Generative AI can create synthetic datasets to address the scarcity of training data, while Reinforcement Learning offers dynamic decision-making for resource allocation and disease control strategies.

Collaborative Research and Open Data Initiatives

Importance of collaboration between academia, industry, and governments.

Sustainable and Inclusive AI Solutions

Collaboration among academia, industry, and governments is essential for the growth of AI in agriculture. Open data sharing initiatives can accelerate research, allowing models to be trained on diverse datasets and increasing the accuracy and applicability of AI tools in various agricultural environments.

Vision for the Future

Ensuring that AI technologies are accessible to all farmers, regardless of region or resources, is crucial for inclusive agricultural advancements. These technologies should be developed in a way that minimizes environmental impact while maximizing benefits for smallholder and large-scale farmers alike.

Conclusion

The advancements in Artificial Intelligence (AI) are poised to significantly transform crop disease management, offering solutions that can address both existing challenges and future demands in agriculture. Through the integration of machine learning, deep learning, and advanced data analytics, AI has enabled more accurate and timely detection of crop diseases. This has led to improved precision in disease management, reducing the reliance on traditional, resource-heavy methods such as excessive pesticide use. By using AI-powered systems like Convolutional Neural Networks (CNNs), Unmanned Aerial Vehicles (UAVs), and Internet of Things (IoT) sensors, farmers can now monitor large fields in real-time, detecting diseases at their early stages and preventing widespread crop damage. Moreover, the potential of AI goes beyond detection. Predictive models, driven by AI, can forecast disease outbreaks based on weather patterns, soil conditions, and other environmental factors, enabling farmers to act preemptively. This helps to minimize crop losses and improve yields, fostering more sustainable agricultural practices. As AI technology continues to evolve, collaborative efforts between researchers, policymakers, and farmers must be strengthened to ensure its accessibility and scalability. The development of AI-based solutions must also be inclusive, addressing the needs of small-scale farmers in developing regions. Ultimately, AI holds the promise of creating a more sustainable, efficient, and disease-free agricultural future, contributing to global food security.

References

Abbas, A., S. Jain, M. Gour and S. Vankudothu. 2021. Tomato plant disease detection using transfer learning with C-GAN synthetic images. Comput. Electron. Agric. 187: 106279.

Abbas, A., Z. Zhang, H. Zheng, M.M. Alami, A.F. Alrefaei, Q. Abbas and L. Zhou. 2023. Drones in plant disease assessment, efficient monitoring, and detection: A way forward to smart agriculture. Agro. 13(6): 1524.

Abdulridha, J., Y. Ampatzidis, P. Roberts and S.C. Kakarla. 2020. Detecting powdery mildew disease in squash at different stages using UAV-based hyperspectral imaging and artificial intelligence. Bio Eng. 197: 135-148.

Anand, R., S.S. Nalla, S. Nasreen, M. Chowdhury, Madhusudan, B.S., A. Pandey et al. 2024. Fruit Counting and Analysis Using Artificial Intelligence Approaches. pp. 93-114. *In:* S.S. Chouhan, A. Saxena, U.P. Singh and S. Jain [eds.]. Artificial Intelligence Techniques in Smart Agriculture. Springer Nature, Singapore.

Atoum, Y., M.J. Afridi, X. Liu, J.M. McGrath and L.E. Hanson. 2016. On developing and enhancing plant-level disease rating systems in real fields. Pat. Recog. 53: 287-299.

Awad, Y.M., A.A. Abdullah, T.Y. Bayoumi, K. Abd-Elsalam and A.E. Hassanien. 2015. Early detection of powdery mildew disease in wheat (*Triticum aestivum* L.) using thermal imaging technique. *In:* Intelligent Systems' 2014: Proceedings of the 7th IEEE International Conference Intelligent Systems IS'2014, September 24-26, 2014, Warsaw, Poland, Volume 2: Tools, Architectures, Systems, Applications (pp. 755-765). Springer Intl. Pub.

Bachika, N.A., N. Hashima, A. Wayayoka, H.C. Mana and M.M. Alia. 2020. Optical imaging techniques for rice diseases detection: A review. J. Agric. Food Eng. 2(1): 123-134.

Bhalekar, D.G., R.A. Parray, I. Mani, H. Kushwaha, T.K. Khura, S.K. Sarkar and M.K. Verma. 2023. Ultrasonic sensor-based automatic control volume sprayer for pesticides and growth regulators application in vineyards. Smart Agric. Technol. 4: 100232.

Bhalekar, D.G., R.A. Parray, I. Mani, T.K. Khura, S. Kumar and N.V. Sarkar. 2024a Agrochemical spray technology adoption and safety awareness assessment in crop protection in vineyard cultivation. CNS&E Journal. 1(3): 188-196, https://doi.org/10.63015/9S-2407.1.3.

Bhalekar, D.G., R.K. Sahni, M.J. Schrader and L.R. Khot. 2024b. Pneumatic spray delivery-based fixed spray system configuration optimization for efficient agrochemical application in modern vineyards. Pest Manag. Sci. 80(8): 4044-4054.

Bock, C.H., A.Z. Cook, P.E. Parker and T. R. Gottwald. 2009. Automated image analysis of the severity of foliar citrus canker symptoms. Plant dis. 93(6): 660-665.

Bock, C.H., K.S. Chiang and E.M. Del Ponte. 2022. Plant disease severity estimated visually: A century of research, best practices, and opportunities for improving methods and practices to maximize accuracy. Trop. Plant Pathol. 47(1): 25-42.

Chandana, R.K. and A.C. Ramachandra. 2022. Real-time object detection system with YOLO and CNN models: A review. arXiv preprint arXiv: 2208.00773.

Chandel, A.K., L.R. Khot and B. Sallato. 2021. Apple powdery mildew infestation detection and mapping using high-resolution visible and multispectral aerial imaging technique. Sci. Hortic. 287: 110228.

Chen, J., J. Chen, D. Zhang, Y. Sun and Y. A. Nanehkaran. 2020. Using deep transfer learning for image-based plant disease identification. Comput. Electron. Agric. 173: 105393.

Chen, L., M. Wallhead, M. Reding, L. Horst and H. Zhu. 2020. Control of insect pests and diseases in an Ohio fruit farm with a laser-guided intelligent sprayer. Hort. Technol, 30(2): 168-175.

Chowdhury, M., T.K. Khura, P.K. Upadhyay, R.A. Parray, H.L. Kushwaha, C. Singh et al. 2024. Assessing vegetation indices and productivity across nitrogen gradients: A comparative study under transplanted and direct-seeded rice. Front. Sustain. Food Syst. 8: 1351414.

Chowdhury, M., T.K. Khura, R.A. Parray, P.K. Upadhyay, H.L. Kushwaha, A. Lama and A. Kushwah. 2025. Multi-vegetation indices based handheld device for precise nitrogen assessment and prescription in direct seeded rice. Comput. Electron. Agric. 229: 109886.

Cobb, N.A. 1892. Contributions to an Economic Knowledge of Australian Rusts (Urediniæ): What Has Been Found Out in this and Other Countries Concerning Wheat Rust. Charles Potter.

Costa, J., C. Silva and B. Ribeiro. 2019. Hierarchical deep learning approach for plant disease detection. *In:* Pattern Recognition and Image Analysis: 9th Iberian Conference, IbPRIA 2019, Madrid, Spain, July 1–4, 2019, Proceedings, Springer Int. Pub Part II, 9: 383-393.

David, D. 2023. Weather based prediction models for disease and pest using machine learning: A review. Asian J. Agric. Ext. Econ. Sociol. 41(11): 334-345.

Devaraj, A., K. Rathan, S. Jaahnavi and K. Indira. 2019. Identification of plant disease using image processing technique. Int. Conf. Commun. Signal Process. IEEE: 0749-0753.

Dhar, P. 2020. The carbon impact of artificial intelligence. Nat. Mach. Intell. 423-425

Dhingra, G., V. Kumar and H.D. Joshi. 2019. A novel computer vision-based neutrosophic approach for leaf disease identification and classification. Meas. 135: 782-794.

DuVall, Eric. U.N. report estimates pesticides kill 200,000 people per year. 2017. UPI. https://www.upi.com/Top_News/World-News/2017/03/09/UN-report-estimates-pesticides-kill-200000-people-per-year/1161489037649/

Ferentinos, K.P. 2018. Deep learning models for plant disease detection and diagnosis. Comput. Electron. Agric. 145: 311-318.

Food and Agriculture Organization of the United Nations. 2023. https://www.fao.org

Gaikwad, V.P. and V. Musande. 2023. Advanced prediction of crop diseases using cetalatran-optimized deep KNN in multispectral imaging. Trait. Du. Signal. 40(3): 1093.

Galletti, P.A., M.E. Carvalho, W.Y. Hirai, V.A. Brancaglioni, V. Arthur and C. Barboza da Silva. 2020. Integrating optical imaging tools for rapid and non-invasive characterization of seed quality: Tomato (*Solanum lycopersicum* L.) and carrot (*Daucus carota* L.) as study cases. Front. Plant Sci. 11: 577851.

Gawande, V., D.R.K. Saikanth, B.S. Sumithra, S.A. Aravind, G.N. Swamy, M. Chowdhury and B.V. Singh. 2023. Potential of precision farming technologies for eco-friendly agriculture. Int. J. Plant Soil Sci. 35(19): 101-112.

Geetharamani, G. and A. Pandian. 2019. Identification of plant leaf diseases using a nine-layer deep convolutional neural network. Comput. Electr. Eng. 76: 323-338.

Global Pesticide Agricultural use 1990-2021. 2024. Statista. https://www.statista.com/statistics/1263077/global-pesticide-agricultural-use/

Gu, D., K. Andreev and M.E. Dupre. 2021. Major trends in population growth around the world. China. CDC Wkly. 3(28): 604.

Guddaraddi, A., A. Singh, G. Amrutha, D.R.K. Saikanth, R. Kurmi, G. Singh et al. 2023. Sustainable biofuel production from agricultural residues an eco-friendly approach: A review. Int. J. Environ. Clim. Chang. 13(10): 2905-2914.

Iqbal, M.A. and K.H. Talukder. 2020, August. Detection of potato disease using image segmentation and machine learning. Int. Conf. Wirel. Commun. Signal Process. Netw (WiSPNET). IEEE : 43-47.

Jangid, B. and R.S. Sharma. 2023. Rice disease detection using deep learning VGG-16 model and flask. Int. J. Sci. Res. Eng. Manag. (IJSREM): 7(03).

Jha, A., M. Chowdhury and A.N. Satpute. 2023. Surface water quality forecasting using machine learning approach. pp. 293-315. *In:* C.B. Pande, M. Kumar and N.L. Kushwaha [eds.]. Surface and Groundwater Resources Development and Management in Semi-arid Region: Strategies and Solutions for Sustainable Water Management. Cham: Springer International Publishing.

Junaid, M.D. and A.F. Gokce. 2024. Global agricultural losses and their causes. Bull. Biol. Allied Sci. Res. 2024(1): 66-66.

Karuna, G., R.R. Kumar, P. Sanjeeva, P. Deepthi, H.Y. Saeed, V. Asha and P. Praveen. 2024. Crop recommendation system and crop monitoring using IoT. *In:* E3S Web Conf. EDP Sci. Vol. 507: 01063.

Khan, I.H., H. Liu, W. Li, A. Cao, X. Wang, H. Liu and X. Yao. 2021. Early detection of powdery mildew disease and accurate quantification of its severity using hyperspectral images in wheat. Remote Sens. 13(18): 3612.

Khanal, K., G. Ojha, S. Chataut and U.K. Ghimire. 2024. IoT-based real-time soil health monitoring system for precision agriculture. Inter. Res. J. Eng. Technol. 11(7): 123-134.

Kumar, Y.J.N., V. Spandana, V.S. Vaishnavi, K. Neha and V.G.R.R. Devi. 2020. Supervised machine learning approach for crop yield prediction in the agriculture sector. *In:* 2020 5th Int. Conf. Commun. Electron. Syst. (ICCES). IEEE: 736-741.

Kumar, Y., A. Koul, R. Singla and M.F. Ijaz. 2023. Artificial intelligence in disease diagnosis: A systematic literature review, synthesizing framework and future research agenda. J. Ambient Intell. Humanized Comput. 14(7): 8459-8486.

Kumari, A., P. Ashoka, P. Tiwari, P. Sachan, A.K. Malla, A. Tripathy and M. Chowdhury. 2023. Zero tillage lead to enhanced productivity and soil health. Int. J. Environ. Clim. Chang. 13(10): 3707-3715.

Kumari, K. and A.M. Nafchi. 2024. Sustainable agriculture with AI, Machine Learning, Deep Learning, and IoT for future farming. ASABE Annual International Meeting (p. 1). ASABE.

Kumari, K., K. Won and A.M. Nafchi. 2024a. YOLOv5 deep learning model for mixed seed detection, classification, and counting. ASABE Annual International Meeting (p. 1). ASABE

Kumari, K., R. Parray, Y.B. Basavaraj, S. Godara, I. Mani, R. Kumar and H. Mirzakhaninafchi. 2024b. Spectral sensor-based device for real-time detection and severity estimation of groundnut bud necrosis virus in tomato. J. Field Robot. 42(1): 5-19.

Kurmi, R., S.D. Lande, J. Kurmi, M. Chowdhury, C. Singh and P. Kumar. 2023. Comparative study on carbon footprint assessment of rice-wheat production system. Int. J. Environ. Clim. Chang. 13(11): 390-398.

Kushwah, A., P.K. Sharma, H.L. Kushwaha, R.H. Nag, G. Carpenter, M.K. Choudhary et al. 2024. Economic evaluation of precise intelligent cauliflower harvester: A comparative study with manual harvesting. J. Sci. Res. Rep. 30(1): 33-42.

Kyratzis, A.C., D.P. Skarlatos, G.C. Menexes, V.F. Vamvakousis and A. Katsiotis. 2017. Assessment of vegetation indices derived by UAV imagery for durum wheat phenotyping under a water limited and heat stressed mediterranean environment. Front. Plant Sci. 8: 1114.

Li, Y., J. Wang, H. Wu, Y. Yu, H. Sun and H. Zhang. 2022. Detection of powdery mildew on strawberry leaves based on DAC-YOLOv4 model. Comput. Electron. Agric. 202: 107418.

Mahmud, M.S., Q.U. Zaman, T.J. Esau, G.W. Price and B. Prithiviraj. 2019. Development of an artificial cloud lighting condition system using machine vision for strawberry powdery mildew disease detection. Comput. Electron. Agric. 158: 219-225.

Malkani, P., I. Mani, P.K. Sahoo, R.A. Parray, S.S. Swain, K.R. Asha et al. 2024. Development and evaluation of a sensor enabled smart herbicide applicator for inter row crops. Smart Agric. Technol. 9: 100649.

Martinelli, F., R. Scalenghe, S. Davino, S. Panno, G. Scuderi, P. Ruisi and A.M. Dandekar. 2015. Advanced methods of plant disease detection. A review. Agron. Sustain. Dev. 35: 1-25. https://doi.org/10.1007/s13593-014-0246-1

Matengu, T.T., P.R. Bullock, M.S. Mkhabela, F. Zvomuya, M.A. Henriquez, E.R. Ojo and W.D. Fernando. 2024. Weather-based models for forecasting Fusarium head blight risks in wheat and barley: A review. Plant Pathol. 73(3): 492-505.

Mohanty, S.P., D.P. Hughes and M. Salathé. 2016. Using deep learning for image-based plant disease detection. Front. Plant Sci. 7: 1419.

Mozzanini, E., D.G. Bhalekar, M. Grella, G.A. Hoheisel, S. Samy, P. Balsari and L.R. Khot. 2024. Cleaning performance evaluation of pneumatic spray delivery based solid set canopy delivery system. J. ASABE: 0.

Mu, X., M. Hussain, L. He, P. Heinemann, J. Schupp, M. Karkee and M. Zhu. 2023. An Advanced robotic system for precision chemical thinning of apple blossoms. J. ASABE: 0.

Oberti, R., M. Marchi, P. Tirelli, A. Calcante, M. Iriti and A.N. Borghese. 2014. Automatic detection of powdery mildew on grapevine leaves by image analysis: Optimal view-angle range to increase the sensitivity. Comput. Electron. Agric. 104: 1-8.

Okengwu, U.A., L.N. Onyejegbu, L.U. Oghenekaro, M.O. Musa and A.O. Ugbari, 2023. Environmental and ethical negative implications of AI in agriculture and proposed mitigation measures. Sci. Afr. 22(1): 141-150.

Orchi, H., M. Sadik and M. Khaldoun, 2021. On using artificial intelligence and the internet of things for crop disease detection: A contemporary survey. Agric. 12(1): 9.

Pero, C., S. Bakshi, M. Nappi and G. Tortora. 2023. IoT-driven machine learning for precision viticulture optimization. IEEE. J. Sel. Top. Appl. Earth Obs. Remote Sens. 47: 2437-2447.

Pourreza, A., W.S. Lee, E. Raveh, R. Ehsani and E. Etxeberria. 2014. Citrus Huanglongbing detection using narrow-band imaging and polarized illumination. Trans. ASABE. 57(1): 259-272.

Pradhan, N.C., M.A. Naik, M. Chowdhury, A. Kushwah, K.R. Asha, T. Dhar et al. 2024. Robotic seeding or sowing system in smart agriculture. pp. 495-520. *In:* K. Pandey, N.L. Kushwaha, C.B. Pande and K.G. Singh [eds.]. Artificial Intelligence and Smart Agriculture: Technology and Applications. Springer Nature, Singapore.

Rajak, P., A. Ganguly, S. Adhikary and S. Bhattacharya. 2023. Internet of Things and smart sensors in agriculture: Scopes and challenges. J. Agric. Food Res. 14: 100776.

Ramcharan, A., K. Baranowski, P. McCloskey, B. Ahmed, J. Legg and D.P. Hughes. 2017. Deep learning for image-based cassava disease detection. Front. Plant Sci. 8: 1852.

Satpute, A.N., K.P. Gavhane, S. Kaur, A. Jha, N.C. Pradhan and M. Chowdhury. 2024. Integration of AI and IoT in soilless cultivation to power sustainable agricultural revolution. pp. 387-411. *In:* K. Pandey, N.L. Kushwaha, C.B. Pande and K.G. Singh [eds.]. Artificial Intelligence and Smart Agriculture: Technology and Applications. Springer Nature, Singapore.

Schaad, N.W. and R.D. Frederick. 2002. Real-time PCR and its application for rapid plant disease diagnostics. Can. J. Plant Pathol. 24(3): 250-258.

Shahare, K., A. Mitra, D. Naware, R. Keshri and H.M. Suryawanshi. 2023. Performance analysis and comparison of various techniques for short-term load forecasting. Energy Rep. 9: 799-808.

Shin, J., Y.K. Chang, B. Heung, T. Nguyen-Quang, G.W. Price and A. Al-Mallahi. 2021. A deep learning approach for RGB image-based powdery mildew disease detection on strawberry leaves. Comput. Electron. Agric. 183: 106042.

Sood, M. and P.K. Singh. 2020. Hybrid system for detection and classification of plant disease using qualitative texture features analysis. Procedia Comput. Sci. 167: 1056-1065.

Srivastava, A.R. and M. Venkatesan. 2020. Tea leaf disease prediction using texture-based image processing. Emerg. Res. Data Eng. Syst. Comput. Commun.: Proc. Springer Sing. CCODE 2019: 17-25.

Sunil Kumar, T., H.M. Virdia, K.G. Patel, S. Ragi, M. Chowdhury, P. Kumar et al. 2024. Effect of summer legume residue incorporation and fertilizer regimes on rice growth, yield, and nutrient uptake. Front. Sustain. Food Syst. 8: 1467201.

Sykuta, M.E. 2016. Big data in agriculture: Property rights, privacy and competition. The Int. Food Agribusiness Manag. Rev. 19(A): 57-74.

Tace, Y., M. Tabaa, S. Elfilali, C. Leghris, H. Bensag and E. Renault. 2022. Smart irrigation system based on IoT and machine learning. Energy Rep. 8: 1025-1036.

Tan, L., J. Lu and H. Jiang. 2021. Tomato leaf diseases classification based on leaf images: A comparison between classical machine learning and deep learning methods. Agri. Eng. 3(3): 542-558.

Tirkey, D., K.K. Singh and S. Tripathi. 2023. Performance analysis of AI-based solutions for crop disease identification, detection, and classification. Smart Agric. Technol. 5: 100238.

Too, E.C., L. Yujian, S. Njuki and L. Yingchun. 2019. A comparative study of fine-tuning deep learning models for plant disease identification. Comput. Electron. Agric. 161: 272-279.

Vescovo, L., G. Wohlfahrt, M. Balzarolo, S. Pilloni, M. Sottocornola, M. Rodeghiero and D. Gianelle. 2012. New spectral vegetation indices based on the near-infrared shoulder wavelengths for remote detection of grassland phytomass. Int. J. Remote Sens. 33(7): 2178-2195.

Wei, H., W. Xu, B. Kang, R. Eisner, A. Muleke, D. Rodriguez and M.T. Harrison. 2024. Irrigation with Artificial Intelligence: Problems, premises, promises. Hum. Cent. Intell. Syst. 1-19.

Whitmire, C.D., J.M. Vance, H.K. Rasheed, A. Missaoui, K.M. Rasheed and F.W. Maier. 2021. Using machine learning and feature selection for alfalfa yield prediction. AI 2021, 2: 71-88.

Zhang, J.C., R.L. Pu, J.H. Wang, W.J. Huang, L. Yuan and J. Luo. 2012. Detecting powdery mildew of winter wheat using leaf level hyperspectral measurements. Comput. Electron. Agric. 85: 13-23.

Zhang, S. and Z. Wang. 2016. Cucumber disease recognition based on Global-Local Singular value decomposition. Neurocomputing. 205: 341-348.

Zhou, R., S.I. Kaneko, F. Tanaka, M. Kayamori and M. Shimizu. 2015. Image-based field monitoring of Cercospora leaf spot in sugar beet by robust template matching and pattern recognition. Comput. Electron. Agric. 116: 65-79.

CHAPTER
8

Weed Management Strategies Employing Artificial Intelligence and Robotics

Samriddha Das, Arjun Upadhyay and Xin Sun*
Department of Agricultural and Biosystems Engineering, North Dakota State University, USA

Introduction

The perception of weeds as harmful to agricultural output arises from the way they compete with crops for crucial nutrients. However, the classification of a plant as a "weed" is subjective and context dependent. These plants pose challenges in agricultural fields by consuming essential resources like sunlight, water, and nutrients (Sunil Kumar et al. 2024). Weeds, by definition, are plants that grow in locations where they are not desired, particularly in agricultural and managed environments. They are typically characterized by their aggressive growth and reproductive strategies which allow them to thrive without human intervention (Ngugi et al. 1978). The environment can be greatly impacted by weeds, particularly in agriculture. They compete directly with crops for natural resources which can in turn lead to a reduction in crop growth and yield (Tshewang et al. 2016). They cause irregular crop maturity and make the harvesting process more difficult by disrupting the consistency of crop growth (Yu et al. 2019). Weeds can also affect the broader environment. They often serve as hosts for pests and diseases, which can then spread to crops, making them more difficult to manage (Ahmad Loti et al. 2021). Additionally, some weeds can block irrigation channels and drainage systems, further complicating farming operations. In natural environments, invasive weed species can outcompete native plants, leading to a loss of biodiversity. The problem is especially serious when weeds have no natural enemies in the area, allowing them to grow unchecked.

Weeds can have a major economic impact on agriculture. They not only reduce the crop yields by taking away resources that the crops need to grow but also their

*Corresponding author: xin.sun@ndsu.edu

DOI: 10.1201/9781003536932-8

presence often become labor intensive followed by a sudden spike in the production cost with a quality degradation (Adeux et al. 2019; Molinari et al. 2020). In a research study carried out over a period of three years, (Adeux et al. 2019), revealed that most of the weed communities resulted in significant yield losses in unweeded zones, ranging from 19% to 56%.

In a research study based on six herbaceous field crops, losses based on the yield were assessed which resulted in 23% for wheat, 40% for corn, 30% for potatoes, 37% for soybeans, 36% and 37% for cotton and rice respectively. Hence overall reduction in agricultural produce due to weed infestation has been obtained to be 34% (Oerke 2006).

Agricultural techniques whether mechanized or laborious it can be hampered by weeds, which signifies that neutralizing them is expensive in terms of both time and money (Molinari et al. 2020; Madhusudan et al. 2024). This increases the cost of farming which can reduce the profits of the farmers. Continuous competition for nutrients and changes in the soil conditions caused by certain weed species can make the soil less fertile over time, further reducing crop yields (Kushwah et al. 2024b). The cost of managing weeds can sometimes be so high that it outweighs the benefits of growing crops, especially in areas where weeds are particularly difficult to control (Bridges 2017). Thus, a crucial strategy for raising crop output is by regulating weeds.

At present, there are four basic strategies of dealing with weed regulation namely physical, mechanical, biological and chemical. These traditional methods of weed management have been the cornerstone of agricultural practices for decades. In this chapter, we address several methods and techniques of weed management practiced globally, alongside various approaches. Traditional methods for weed management have specific applicable scenarios, whereas integrated weed management enhances agricultural yield and quality and provides greater control (Scavo et al. 2020). Considering the drawbacks of conventional approaches it is imperative to integrate Artificial Intelligence and robotics into existing weeding approaches.

There are three sections to this chapter: Firstly, there is a discussion about the traditional methods and techniques for weeding with the short comings and the drawbacks in each of the methods which lead to the next part that explores the integration of Artificial Intelligence, automation as well as robotics for precise control and management of weeds, and the last part covers the challenges and future prospects along with the directions for future research. The main objectives are to (a) offer a thorough analysis of individual techniques and their shortcomings in weed management, (b) explore ongoing research in AI and robotics for site-specific weed management, (c) investigate design trends for intelligent weeding robots, UAVs, and worldwide commercial options, and (d) outline emerging areas of research for efficient technology regarding precise regulation of weeds.

Traditional Weed Management Techniques and their Limitations

Traditional weed management techniques have long been essential in agriculture for controlling unwanted plant growth in crop fields. Developed and refined over time, these methods address the challenges posed by weeds that may have a major effect on crop quality and output. Now, extensive spraying of pesticides has been the most common approach, but it often results in resource wastage and environmental pollution (Rani et al. 2021). Historically, farmers and researchers have relied on four main solutions: physical, mechanical, chemical and biological weed control. Each method offers unique benefits and comes with its own set of limitations, making it essential to understand the context in which each technique is most effective. Fig. 1 provides a diagrammatic representation of conventional weed management strategies that have traditionally been practiced.

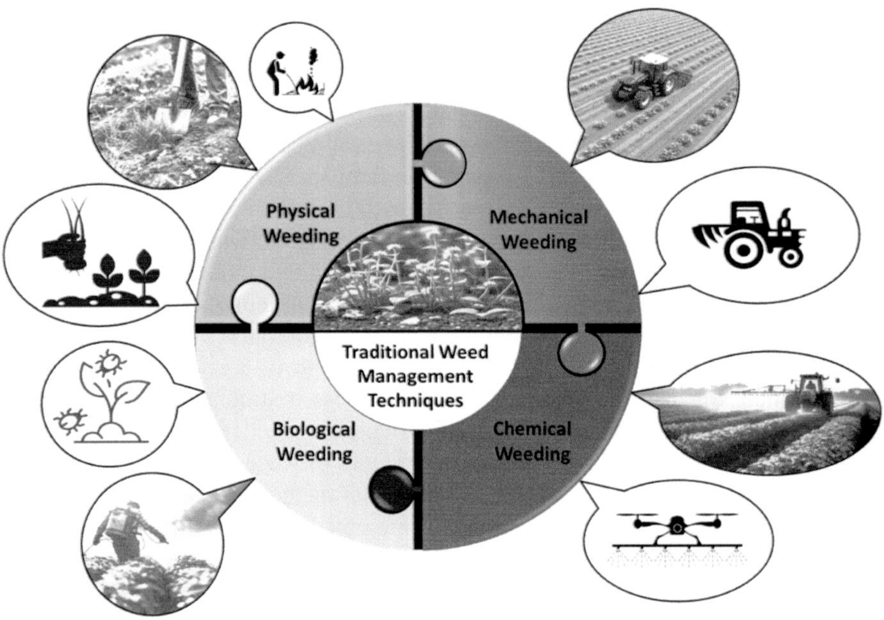

Fig. 1: Traditional weed management techniques.

Physical weed control encompasses traditional practices such as hand weeding, frequent tillage, crop rotation, late planting, mulching, and intercropping, alongside modern approaches like thermal technologies, including flame and laser weed control. Hand weeding, the oldest and most widely used method, requires minimal explanation but is primarily feasible for small-scale landholdings where labour is abundant or financial resources are limited. Frequent tillage disrupts weed growth by repeatedly disturbing the soil, while crop rotation suppresses weed populations by alternating crops, thus interrupting weed cycles in the following season (Desta 2000). Techniques like intercropping—where two crops are grown simultaneously, with one acting as a biological control for weeds—and living mulches, in which cover crops are sown before or alongside the main crop and maintained throughout the growing season, also serve to control weeds while improving soil fertility (Hatcher and Melander 2018). Modern techniques such as flame weeding employ heat to destroy plant tissues, disrupting cell membranes and halting cellular functions, effectively killing the weeds (Vincent et al. 2001). Flame weeding is particularly useful for large-scale operations where the use of chemicals is restricted, especially in organic farming, where propane burners or renewable fuels such as hydrogen are used to generate combustion temperatures reaching up to 1900°C. However, the effectiveness of flame weeding depends on the weed species, seedling size, and the method of ignition, with risks including CO_2 emissions and fire hazards (Scavo et al. 2020). Laser weed control, on the other hand, uses high-energy beams, typically powered by renewable sources such as wind, water, or solar energy, to target and destroy weed tissues with precision. This method is particularly effective against smaller weeds and minimizes the risk of damage to surrounding crops, though its efficacy can vary depending on beam diameter, and there is a potential risk of igniting dry materials in the field. Laser technology offers a non-polluting, sustainable alternative to chemical weeding, although its efficiency remains dependent on the specific conditions of use.

Another method of weeding is mechanical or conventional approach, which is widely used but can lead to significant crop injury if not carefully managed. Unsupervised application may cause physical damage to crops, resulting in stunted growth and increased vulnerability to pathogens, ultimately reducing productivity. For example, techniques like harrowing in peas and hoeing in soybeans can elevate the risk of Fusarium and Phytophthora infections in sugar beet roots (Findlay et al. 1996). Precision in mechanical weed control is essential to minimize crop damage, especially in the more challenging intra-row weed removal compared to inter-row management.

In the context of biological weed control, it is the strategical use of living organisms to manage weed populations. Biological weed control agents or simply control agents can be categorized into several groups. (Cordeau et al. 2016) classified them into macro-organisms (e.g., predators, parasitoid insects, nematodes), microorganisms (e.g., bacteria, fungi, viruses), chemical agents (e.g., pheromones), and natural substances (e.g., plant or animal extracts). For example, Sorghum bicolor extracts effectively control Echinochloa crus-galli, enhancing rice production (Dayan et al. 2009). Similarly, Brassica nigra extracts, high in glucosinolates, inhibit weed growth by converting to isothiocyanates (Tsiamis et al. 2016). Biological weed control offers an environmentally friendly, sustainable alternative to chemical herbicides, reducing reliance on synthetic options and their associated risks (Duke et al. 2000).

Despite its benefits, biological control faces challenges such as shorter environmental half-lives of bioherbicides, higher production costs, and lower efficiency compared to chemical herbicides (Singh and R.K 2018). The development of bacterial and fungal bioherbicides is further limited by high costs, reduced efficacy, and risks to non-target organisms (Osadebe et al. 2021). Additionally, safety concerns for humans and animals necessitate careful management (Bailey et al. 2014).

Herbicides perform a crucial part in chemical weed management, which is a dominating technique in contemporary agriculture for regulating weeds in farmlands. The introduction of 2,4-D in the 1940s marked the beginning of widespread herbicide use. It has now emerged as the most effective weed management strategy around the world (Llewellyn et al. 2016). Herbicides are classified by chemical family, application period (preplant, pre-emergence, postemergence), mechanism of action, and specificity (Malkani et al. 2024). Some which are light reliant inhibit photosynthesis and pigment production, while others block fatty acid or amino acid synthesis (Dodge et al. 1992).

The primary advantages of chemical weed control include its effectiveness in rapidly controlling weeds, which significantly increases crop yields and reduces labour and costs. Additionally, Farmers can control weeds without using labour-intensive manual effort due to the introduction of herbicides with distinctive activation techniques. In comparison to the liquid formulations, granular or microencapsulated herbicides provide enhanced weed control when used in no-till or reduced tillage systems (Eslami 2014).

However, Chemical weed control faces challenges, including the rise of herbicide-resistant weeds, now observed globally in 262 species across 23 of 26 herbicide action sites (Heap 2024, Engel et al. 1969). Herbicides also cause environmental pollution by contaminating soil, groundwater, and disrupting microbial communities (Curran, 2016; Liao et al. 2021). Glyphosate has been found in urban wetlands and streams, highlighting groundwater contamination risks (Hollaway et al. 2006; Gaw et al. 2008). Additionally, herbicide drift can damage nearby plants and ecosystems (Florencia et al. 2017).

Precision chemical weed control, which involves region specific spraying of herbicides, serves as a potential solution to reduce herbicide usage. Studies have shown that this method can decrease herbicide application by 34% while maintaining average crop yields like those achieved with uniform herbicide spraying (Maillot et al. 2023). Despite its advantages, chemical weed control requires careful management to reduce environmental impacts and prevent resistant weed species. As challenges grow, there is increasing demand for affordable, eco-friendly alternatives, where automation, robotics, and AI play a key role. Recent reviews have focused and highlighted technological components like imaging, weed identification, and weeding techniques and robotics (Rai et al. 2023; Upadhyay et al. 2024). This chapter has been structured to provide an overview of the recent progress and developments in various robotics and automation technologies for site specific weed management.

Integration of AI And Robotics for Precision Autonomous Weed Control

Weed control based on the integration of AI and robotics for an autonomous operation marks a significant advancement in modern agriculture. Leveraging intelligent systems, these technologies enable the development of robots which operate efficiently in complex environments. The following sections discuss the key aspects related to the integration of the navigation system with the AI based weed detection and identification followed by the implementation on the robotic platforms.

Intelligent Navigation Strategies for AI-Powered Weeding Robots

Navigation in AI-powered weeding robots is a critical feature that ensures precise movement along a predetermined path while achieving specific agricultural tasks without manual intervention. This involves the robot's capability to store real-time positional and orientation data, irrespective of its operational state. Accurate navigation is essential for efficiency, as the robot must continuously track its position and adjust its movements accordingly. Research in this area focuses on enhancing the precision of Unmanned Ground Vehicles (UGVs) (Pradhan et al. 2024) and Unmanned Aerial Vehicles (UAVs) (Alsalam et al. 2017) for agricultural applications. A key aspect of autonomous navigation is the robot's ability to detect and avoid obstacles while adapting to varying field conditions to ensure it reaches its target location effectively. These capabilities form the foundation for achieving precision in tasks such as weed management, planting, and crop monitoring (Gawande et al. 2023).

Navigation in agricultural environments poses significant challenges due to unstructured terrain and various external disturbances, such as variable lighting conditions and environmental noise. Hardware limitations also have a considerable impact, with factors like wheel slippage and actuator noise complicating navigation tasks (Ding et al. 2022). To address these issues, several sensor-based approaches have been developed, aimed at rectifying errors to improve both positioning systems and navigation precision. Among these, the most widely used system is the Global Navigation Satellite System (GNSS), which relies on precise tracking based on the global GNSS coordinate system. This technique proves particularly beneficial when operated in open environments and has been successfully utilized and incorporated in numerous research works over the years (Bechar and Vigneault 2017). However, GNSS can be unreliable in areas with signal interference or obstructions (Khan et al. 2018). Additionally, obstacles within the field of view can be effectively managed by Light Detection and Ranging (LiDAR), which senses the position of objects in real time and maps the surrounding environment (Abbas et al. 2020). This approach demonstrates stability and effectiveness in orchard environments; however, its high cost limits its applicability for agricultural vehicles intended for large-scale operations (Iqbal et al. 2020). Other popular sensors that have been in practice over the past decades include Inertial Measurement Units (IMUs), Ultra-Wideband (UWB) systems, and vision sensors (Zhang et al. 2021). Computer vision also plays a significant role in autonomous navigation, characterized by features such as intelligence, signal stability, and an extended detection range. Autonomous guidance using machine vision utilizes the sensor's ability to detect vehicle's relative position concerning landmarks and determine the vehicle's heading (Radcliffe et al. 2018). These vision systems are considered cost-effective compared to other technologies, offering large-scale applicability across various agricultural operations.

Several use cases of navigation techniques in agricultural settings demonstrate the potential of various sensor-based systems. For example, lateral guidance with GPS provides precise positioning, as evidenced in research by Kushwah et al. (2024a), which achieved an accuracy of less than 2.5° in heading compensation and an average deviation of 45.9 mm from the ideal path. Additionally, to improve autonomous navigation of off-road vehicles, Real-Time Kinematic GPS (RTK-GPS) has been incorporated with additional sensors for guidance. In an experiment conducted by (Bakker et al. 2011) a navigation system was built on RTK-DGPS platform for sugar beet crops which obtained lateral errors within –4.5 cm and 3.4 cm while operating at 0.3 m/s. Vision-based sensors, including RGB-D cameras, have gained popularity in autonomous navigation, with research contributions from Fu et al. (2020) highlighting their applicability. Additionally, thermal sensors have been explored for agricultural use (Nguyen et al. 2021). Visual odometry sensors, another promising tool for navigation, have been studied by (Galati et al. 2017), while various system-based filtering techniques such as Kalman Filters (Pradhan et al. 2024), threshold-based segmentation (Riehle et al. 2020), Particle Filters (Zhou et al. 2017), and Hough Transform approaches (Illingworth and Kittler 1988) have been applied to improve navigation accuracy. Xiuzhi et al. (2020) investigated line detection-based computation for further precision in positioning. One notable experiment by Pfrunder et al. (2017) utilized both LiDAR and RTK-GPS, resulting in 60 km of fully autonomous driving based on outdated maps, further emphasizing the reliability of multi-sensor navigation systems. Similarly, Inoue et al. (2019) employed YOLOv3 for object detection and obstacle avoidance using a camera, while RTK-GNSS was used for positioning, achieving a final position error of 0.36 m by integrating IMU and landmark data.

Despite the effectiveness of machine vision techniques, environmental factors such as varying light conditions, weather, and dust can affect performance. To overcome these challenges, adaptive algorithms and multi-sensor fusion techniques, including IMU and odometers, have been employed successfully.

Weed Identification and Localization using Machine and Vision based Smart Technologies

In the development of autonomous weeding systems, real-time detection and classification of crops and weeds are critical components following the navigation technologies discussed in the previous section. Accurate weed identification is fundamental for precise and effective weed management, as ground robotic technologies heavily depend on this capability to execute automated weeding tasks. Weed identification presents significant challenges due to the physical similarities between weeds and crops in terms of color, texture, shape, and growth patterns (Shanmugam et al. 2020). This complexity necessitates advanced detection systems capable of distinguishing subtle differences in plant characteristics to minimize errors and improve precision.

Traditionally, herbicides have been applied uniformly across fields, regardless of the presence of weeds, leading to unnecessary chemical use and environmental damage. To address this, site-specific weed management has emerged as a more sustainable and efficient approach. This method integrates sensor-based detection technologies to identify weeds in real-time and optimize herbicide application accordingly. The process involves several critical stages: first, weed detection and identification using computer vision and other sensors; second, decision-making based on the detected weed information combined with farmer expertise; third, the execution of the weeding action through automated actuators; and finally, the performance evaluation of the precision operation (López-Granados 2011). This multi-step framework ensures that herbicides or other chemical treatments are applied only where necessary, reducing environmental impact and promoting precision agriculture. Through the integration of real-time weed detection and site-specific decision-making, autonomous systems can significantly enhance the efficiency and accuracy of weeding operations, marking a substantial advancement in precision weed control. Precise localization of weeds, in addition to detection, is essential for accurate robotic actuation in weed removal. RGB-D cameras combined with computer vision algorithms enable real-time mapping of weed positions. This allows the weeding system to precisely target weeds with minimal crop disturbance, significantly improving the efficiency and accuracy of autonomous weeding operations (Raja et al. 2020).

Over the past decades, a variety of sensors have been employed for weed detection, each contributing to advances in precision agriculture. Among these, infrared spectroscopy, multi- and hyperspectral imaging, and distance sensing techniques such as LiDAR and ultrasonic sensors and RGB cameras and even mobile phone cameras (Sunil et al. 2022) are notable. These sensors are typically mounted on both aerial vehicles, such as unmanned aerial vehicles (UAVs) (Reiser et al. 2017), and ground-based vehicles (López-Granados 2011). In UAV applications, hyperspectral imaging is widely used to generate classified weed maps through remote sensing techniques over large areas. Initially, images are captured from the field, and subsequent segmentation is performed to analyze and identify specific weed-infested zones, enabling site-specific weed management (Dammer et al. 2013). For ground-based unmanned vehicles, real-time detection sensors are commonly integrated with weeding mechanisms. These systems allow for the simultaneous detection and destruction of weeds during operation (Lin 2009). Spectral sensors, which rely on reflected light intensities across different wavelengths, are often used to differentiate between crops and weeds. This spectral data is processed to produce segmented images, which are then used to generate weed maps, guiding the robotic system's weed removal actions (Peteinatos et al. 2014). Cost-effective alternatives to spectrometers, such as optoelectronic sensors, offer efficiency at lower costs. Although they operate with fewer spectral bands, mainly in the red and near-infrared regions, they are still capable of effectively distinguishing crops from weeds (Tremblay et al. 2009). However, despite their ability to differentiate between crops and weeds, these sensors are limited in their ability to distinguish between different plant species, which restricts their use in precision weed management (Taghadomi-Saberi and Hemmat 2015). While machine vision approaches have been developed to overcome some of these limitations, they too face challenges. These include sensitivity to lighting conditions, occlusion of weeds and crops, and the variability in weed density across the field (Shaner and Beckie 2014). Although these traditional approaches have been widely used for weed detection, there remain significant challenges that hinder their practical implementation. As a result, researchers are increasingly exploring the use of deep learning techniques to address these limitations and improve weed detection accuracy in real-world agricultural environments (Wang et al. 2019).

The implementation of deep learning and artificial intelligence (AI) for weed identification has been extensively illustrated in Fig. 2, which presents a comprehensive flow diagram outlining the operational workflow. Fig. 2 offers an overview of the entire process, from dataset collection to pre-processing, model training, and deployment into agricultural robots for weeding operations. Among the earliest works in this domain is that of (Petersen and Krutz 1992), who emphasized the importance of using color images rather than binary classification to enhance the accuracy of weed identification systems. Significant progress has been made in

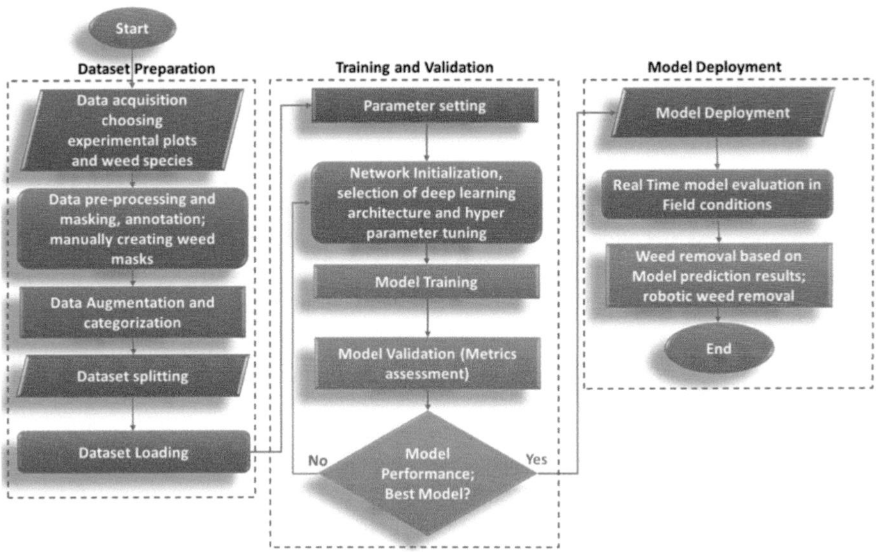

Fig. 2: End-to-end workflow of AI-driven weed control: From data processing and model validation to field deployment.

applying machine learning (ML) techniques to weed detection and identification. For instance, Hamuda et al. (2016) and Mennan et al. (2020) have explored various ML methods, including Support Vector Machines (SVMs), as demonstrated by Ahmed et al. (2012), who identified six different weed species with a precision of 97.3%. Similarly, Eide et al. (2021) employed thermal cameras to distinguish between herbicide-resistant and susceptible weed biotypes based on canopy temperature values, achieving 89% accuracy in identifying herbicide-resistant horseweed. However, further research is required to broaden this technique's applicability to other weed species. Additionally, Geographic Information System (GIS) software, such as ArcGIS Pro, has been employed for generating prescription weed maps, although the cost of such tools makes them less feasible for widespread adoption (Abd-Elrahman et al. 2021). Chen et al. (2013) proposed an innovative classification approach using binocular stereo vision, which distinguishes between crops and weeds based on their height.

Deep learning (DL) models have also played a critical role in advancing weed identification technologies. For example, (Asad and Bais 2020) utilized architectures like SegNet, UNET, VGG16, and ResNet-50 to identify weeds in canola fields. In another study, (Veeranampalayam and Sivakumar 2020) compared Faster R-CNN and Single Shot Detector (SSD) models for weed detection, concluding that Faster R-CNN outperformed SSD in terms of inference speed and Intersection over Union (IoU) results. Furthermore, (Bakshipur and Jafari 2018) compared SVM and Artificial Neural Networks (ANN) for detecting multiple weed species, reporting that the SVM model achieved a higher accuracy (95%) compared to ANN (92.92%). The introduction of YOLO (You Only Look Once) models has significantly advanced object detection and classification in both still images and real-time applications.

Researchers, including (Rai et al. 2023), have emphasized YOLO's superior performance, stability, and ability to operate at high frame rates, making it a reliable choice for real-time weed identification. Table 1 provides a detailed comparison of various weed detection, classification, and identification techniques, illustrating their strengths and limitations. Recent research has focused on optimizing deep learning models to enhance performance and reduce computational requirements. Techniques such as pruning unnecessary layers in deep learning architectures and kernel-based pruning methods have been explored (Jani et al. 2023).

Though various detection techniques, from RGB and multispectral cameras to hyperspectral imaging, offer significant advances in weed identification, challenges like cost, environmental sensitivity, and data complexity remain. To fully realize the potential of these technologies, their integration with robotic systems for automated weed removal and intervention is critical, as explored in the next section.

Robotic Implementations and Applications for Weed Identification and Removal

Building upon advancements in autonomous navigation and real-time weed detection, the next crucial phase in precision agriculture is the development of automated weed removal systems. Ground robots, deployed for this purpose, must adhere to specific operational constraints to ensure effective and safe performance. Key operational rules include maintaining the robot's movements within the predefined field borders and ensuring that weeding operations halt at the end of each row. At the headland, the robot must execute a smooth turn to identify the next row for intra-row weeding operations. Accurate field mapping prior to the operation is essential, enabling the robot to follow a preset path while avoiding obstacles in real-time. The most critical operational rule is to prevent any damage to the crops, ensuring the robot's autonomy throughout its weeding tasks (Bakker et al. 2010). In addition to these fundamental rules, several other criteria influence robotic weeding efficacy, including the accuracy of weed removal, reliable performance under varying field conditions, user-friendliness, and continuous operational capacity. For effective autonomous navigation, robots rely on prescription maps detailing specific plots within the field. Unmanned Aerial Vehicles (UAVs) have significantly enhanced the generation of such maps, especially for site-specific weed management and removal (Krishna 2018). UAVs equipped with RGB, and high-resolution cameras offer precise geographic data essential for accurate field mapping (Kavhiza et al. 2020). While the initial investment for UAV systems is substantial, given the costs associated with professional pilots, high-resolution cameras capable of precision up to 0.1 cm/px, and complex image processing, the frequency and quality of data generation make these systems highly valuable (Anderson et al. 2013). Additionally, UAVs outperform satellite imagery in unfavorable weather conditions, providing consistent and reliable data (Manfreda et al. 2018).

For weed identification, UAVs offer significant advantages over ground vehicles, particularly when operating between crop rows, due to their superior speed and responsiveness (Hassanein et al. 2018). They are capable of surveying extensive areas, providing blanket coverage of both crops and weeds, which can then be used for subsequent identification and segmentation processes (Krishna 2018). Based on

Table 1: Comparison of weed identification approaches with the technologies incorporated

Source	Sensor/Capturing Technology	Software/Algorithm/ Detection Model	Performance/Accuracy/Outcomes	Application
Jiang et al. 2023	RER-USBFHD01M-LS36 camera 1980 × 1080	YOLOv5	93.3% detection accuracy; 90%-94.5% removal rate with 1.17% injury to crops	Chemical weeding
Fawakherji et al. 2021	Multispectral camera (JAI AD-13)	Data augmentation (GANs), Bonnet network	Generate realistic multispectral images of plants, enhancing segmentation quality.	Crop/ Weed segmentation
Hussain et al. 2020	Canon PowerShot SX540 HS camera, Logitech C920 Webcam HD Pro	tiny-YOLOv3	42% reduction in liquid spray	Chemical weeding
Fatima et al. 2023	1280 × 1024 pixel camera	YOLOv5	mAP = 88%	NA
Modi et al. 2023	48 MP Sony IMX586	DarkNet53	Accuracy 99.1%, precision 99.3%	Chemical weeding
Valente et al. 2022	Phantom 3 Professional UAV's camera	MobileNet, VGG, Resnet, DenseNet, ShuffleNet, EfficientNet, MNASNet	F1-Score = 78.36% and AUROC = 93.74%	NA
Subeesh et al. 2022	Xiaomi Mi 11 × mobile device's rear camera	Alexnet, GoogLeNet, InceptionV3, Xception	InceptionV3: accuracy = 97.7%, precision = 98.5%, recall = 97.8%	Chemical weeding
Reiser et al. 2019	Sonar, electrochemical sensor for detection and laser scanner for row following	ROS-Software and laser scanner data	Tilled area for outdoor test reached a min of 57% and a max of 98% excluding unreachable areas	Mechanical weeding
Olsen et al. 2019	1920 × 1200 px camera	Inception-v3 and ResNet-50	Accuracy for average classification: 95.1% and 95.7%	Chemical weeding

Sujaritha et al. 2017	Intex IT-105 web camera, raspberry pi native camera	Color extraction algorithm, leaf texture extraction and fuzzy real time classifier	Overall weed classification accuracy of 92.9% was achieved	Mechanical weeding
Halstread et al. 2021	PATHoBot camera	Mask-RCNN	Precision = 86.5%, recall = 75.2%, F1-score = 80.4	Chemical weeding
Shapira et al. 2010	FieldSpec Pro FR spectrometer, Spectral Camera HS (Specim)	General discrimination analysis (GDA) and partial least squares regression (PLSR)	Classification for canopy spectra was 87% for heterogeneous, 95% for specific and leaf spectra was 99%	Weed detection
Genze et al. 2022	DJI Mavic 2 Pro camera	UNet-like, ResNet34	F1-score >89%	Chemical weeding
Sunil et al. 2022	Google Pixel 5 mobile camera	sVM and VGG16	Average f1-score >93%	Chemical weeding
Jeon et al. 2011	STH-MDCS2-VAR-C RGB stereo vision camera, polarizing filter on lens	ANN model and image processing algorithm	Maximum algorithm processing error 2.9%, 95.1% detection accuracy	Weed detection
Hansen et al. 2013	Multi spectra camera, GPS	Image processing and genetic algorithm for planning	Capable of solving travelling salesman problems and identification of biomass with high density	Characterization of edges of leaves in RGB images
Zou et al. 2022	RERVISION USB8MP02G	U-net	IOU 88.98%	Chemical weeding
Ofori and Omar 2021	Open-source dataset	EfficientNet and transfer learning	Accuracy = 95.44%	NA
Cai et al. 2023	DJI mavic 12 megapixels camera	ResNet50	F-score = 87.79%	Chemical weeding
Osorio et al. 2020	Mavic Pro with Parrot Sequoia multispectral camera	SVM, YOLOv3, n Mask R-CNN	F1-scores of 88%, 94%, and 94% respectively	Chemical weeding
Khan et al. 2021	DJI Spark camera	Faster R-CNN	Accuracy = 94.7%	Chemical weeding

a comprehensive review of the literature and an assessment of sensor technologies discussed in previous sections, UAVs typically operate with three primary types of cameras: RGB, multispectral, and hyperspectral cameras (Esposito et al. 2021; Chowdhury et al. 2024b).

Several practical implementations of UAVs in agriculture demonstrate their growing utility in precision farming, particularly for weed mapping and management. For instance, the Phantom 3 and Phantom 4 are small consumer drones capable of mapping up to 10 hectares of field within 20 minutes. Both models are equipped with RGB cameras that are highly effective for weed mapping applications (De Castro et al. 2017). Another notable UAV is the eBee, which is equipped with a multispectral camera for capturing detailed images of weeds, enhancing its ability to map vegetation health and density (Chowdhury et al. 2024a). This UAV is also integrated with GPS, allowing for precise georeferencing and mapping of field areas (Mesas-Carrascosa et al. 2015). In terms of weed removal, UAVs have demonstrated remarkable efficiency in pre-emergence herbicide applications, where they have been shown to achieve over 98% eradication of weeds. However, post-emergence herbicide spraying presents significant challenges, particularly due to the rise of herbicide-resistant weed variants. This resistance causes a dramatic reduction in the efficacy of UAV-based weed removal, lowering the success rate to between 10% and 70% (Mink et al. 2018). A notable experiment by Hunter III et al. (2020) utilized the DJI AGRAS MG-1, an octocopter drone, for site-specific weed detection and management. The UAV-based weed removal approach demonstrated a 19% higher effectiveness compared to ground-based broadcast applications. Furthermore, the UAV-based approach was able to treat 20% to 60% less area than the ground-based broadcast approach, highlighting the precision and efficiency of UAV spray applications. Despite this, crop injury remains a concern, particularly with post-emergence herbicide applications. Certain crops, such as wheat, have shown resilience to pre-emergence herbicides, but post-emergence applications can result in crop damage rates ranging from 10% to 20% (Chen et al. 2019). While UAVs offer several advantages in weed removal, including precision and speed, there are constraints that limit their effectiveness. One major challenge is maintaining uniformity in spraying, which is critical for both weed eradication and crop protection. The effectiveness of UAV spraying is highly dependent on various parameters, including the altitude of the drone, flight speed, nozzle shape, droplet size, application pressure, and external weather conditions such as wind speed, direction, and ambient temperature (Liao et al. 2019). These independent variables make it difficult to optimize UAV spraying for consistent weed control across diverse conditions. Thus, while UAVs excel in generating prescription maps and real-time weed mapping, they still face significant challenges in site-specific weed removal. For instance, uniform spray coverage and herbicide application remain areas of concern, and achieving an optimal balance among the various parameters affecting the spraying process is complex (Wang et al. 2020).

Despite the advancements in UAV technology, ground-based robotic systems also offer several advantages for weed removal, particularly addressing some of the critical operational parameters. Some case studies of ground-based robotic systems give a broader perspective of their operational benefits. Several robotic platforms have demonstrated the viability of autonomous weed management, showcasing the potential for improving precision, reducing herbicide use, and minimizing labor. One notable example is AgBot II, a modular robot designed with a vision-based weed detection system. During field trials, it demonstrated impressive classification accuracy, identifying 96% of plants and 92.3% of individual weed species. This system integrates a mechanical weeding tool and a precision sprayer, enabling it to efficiently target weeds while minimizing herbicide use (Bawden et al. 2017). Such platforms exemplify how combining mechanical and chemical weeding methods can enhance the effectiveness of autonomous systems in agricultural fields. In another research conducted by (Kotaniemi et al. 2024), a fully autonomous mechanical weeding robot equipped with 6 degrees of freedom robotic arm has been developed for the removal of Longleaf Dock weeds which are poisonous for cattle. The system employs GNSS based navigation system with a position accuracy of ±0.25 m and a computer vision-based weed detection using image processing and Principal Component Analysis approach for identification of larger leaves and root location with a detection accuracy of ±0.02 m. Force application of the robot is 50 N within the operational range of 50-150 N. The Remote Farming project has also made strides in the field, developing a robotic weed control system named BoniRob, specifically designed to tackle the challenges of weed control in organic farming. BoniRob became a benchmark for integrating machine vision and various sensor technologies to optimize weed control, particularly in the context of organic farming, where chemical herbicides are not an option (Lottes et al. 2017). An additional breakthrough in robotic weed control comes from Ecorobotix, which developed a solar-powered robotic vehicle capable of reducing herbicide usage by 90%. This robot operates with a performance capacity of covering 10 hectares per day, providing a sustainable and efficient alternative for large-scale farming operations (Ecorobotix 2023). Furthermore, advancements in laser weeding technology, such as the Carbon Robotics Laser Weeder, offer a non-chemical, environmentally friendly approach to weed removal. With the ability to eliminate up to 200,000 weeds per hour and achieve sub-millimeter accuracy, this platform provides an effective solution for large-scale weed control while avoiding the use of herbicides (Carbonrobotics 2023). The system's success highlights the importance of sensor fusion and vision-based technologies in improving the precision and efficiency of robotic weeding in challenging agricultural environments.

Although both Unmanned Aerial Vehicles (UAVs) and Unmanned Ground Vehicles (UGVs) have proven effective in autonomous weed management, determining the superior platform is challenging. UAVs offer significant advantages in terms of wide area coverage, efficiency, and accessibility, particularly in large fields or inaccessible areas. However, they are limited by payload capacity, susceptibility to weather conditions, and regulatory restrictions (Wang et al. 2020). In contrast, UGVs provide enhanced stability, greater payload capacity, and adaptability to different terrains, making them more suitable for precise chemical application and mechanical weed removal. However, UGVs are typically slower, have more limited coverage areas, and encounter challenges in obstacle avoidance, especially on steep slopes where terrain adaptability can be compromised. Both platforms have distinct strengths and limitations, and in many cases, one compensates for the limitations of

Table 2: Weeding robots developed over the years

Name	Target Crop	Weeding Method	Technology/Sensors	Year	Reference
Lee's Robot	Tomato	Chemical weeding	RGB color video camera (Model 2222-1340/0000, Cohu, Inc)	1999	Lee et al. 1999
Astrand's Robot	Sugar beet	Mechanical weeding	Gray level vision for row detection and colour vision for distinguishing crop form weeds	2002	Åstrand and Baerveldt, 2002
Bakker's Robot	Sugar beet	-	GPS, optical sensor, ultrasonic sensor, tactile sensor	2009	Bakker 2009
BoniRob	Carrot, sugar beet	Mechanical weeding	RGB, NIR camears and ultrasonic sensor	2014	Sellman et al. 2014
EcoRobotix	-	Chemical	RGB camera	2015	Ecorobotix 2018
AgBotII	Cotton	Chemical or mechanical weeding	ROS framework, IDS UI-1240SE 1.3MP global shutter camera	2017	Bawden et al. 2017; Hall et al. 2017
Tertill	-	Mechanical weeding	Capacitive sensors	2017	Sanchez et al. 2021
Digo	Carrot	Chemical weeding	Nvidia Jetson TK1, Omnivision 4682 4MP sensor	2018	Utstumo et al. 2018
Weed robot 1 & Weed robot 2	Paddy field	Mechanical weeding	Capacitive touch sensor, azimuth sensor	2018	Sori et al. 2018; Uchida et al. 2019
Deepfield	Sugar beet	Mechanical weeding	Classification sensor	2019	Amend et al. 2019
Naiture	Carrot	Mechanical weeding	-	2019	Naiture 2019
AVO	Rapeseed, bean, spinach, etc.	Chemical weeding	RGB camera	2020	EcoRobotix 2020

FarmWise	Broccoli, cauliflower, etc.	Mechanical weeding	-	2020	FarmWise 2020
WeLaser	Maize, sugar beet, cereals	Laser weeding	Front Stereo Camera and artifcial lighting, recognition RGB cameras	2020	Krupanek et al. 2024
Carbon's LaserWeeder	Broccoli, cauliflower, carrots, etc.	Thermal weeding	GPS, LiDAR, Front and rear parallax drive cameras, and 12 high-resolution predict cameras	2021	Carbonrobotics 2021

the other. For instance, UAVs excel in data acquisition and real-time weed mapping, while UGVs provide more stable and precise weed removal. As research continues to evolve in both domains, future efforts will likely focus on integrating the best features of both systems to create hybrid solutions that maximize efficiency, adaptability, and sustainability in weed management.

Scientific Challenges with Solutions and Future Research Directions

Despite recent advancements in autonomous weed control systems, significant technical challenges persist, particularly concerning navigation accuracy, weed detection, sensor calibration, and system affordability. Ground-based robotic platforms although they are cost efficient compared to UAVs for weed management, but they are still expensive, limiting adoption by average-scale farmers. One of the primary challenges is achieving precise navigation within agricultural environments. While RTK-GPS is the most used technology for field navigation, its accuracy is often compromised by environmental factors such as obstructions, signal interference, and satellite geometry. As a result, for precise navigation it is recommended to integrate RTK-GPS data with additional sensor such as LiDAR, RGB cameras and inertial measurement units (IMUs) for better approximation. Moreover, use of machine learning and deep learning techniques also enhances path planning accuracy (Bechar and Vigneault 2016).

In real agricultural settings, weed detection and localization are extremely difficult because of variable lighting conditions, plant occlusion, and overlapping leaves, especially in later growth stages. This makes typical image processing and machine learning (ML) algorithms ineffective in these situations. Because of these changes in the physical and spectral characteristics of leaves, it becomes more challenging to distinguish between crops and weeds, and conventional feature extraction techniques are unable to take these complexities into account. To address occlusion, techniques like active shape models and stereoscopic imaging are suggested methods based on the literature review, yet their success remains limited to specific datasets or controlled conditions. Advanced methods for deep learning (DL) have emerged as a more flexible option that enhances classification accuracy in various circumstances by allowing for independent feature learning (Anand et al. 2024). However, factors like weed density, shadows, and inconsistent lighting continue to impact detection accuracy. Expanding datasets to include more images under varied environmental conditions, alongside pre-processing methods like histogram equalization, can further enhance system robustness and adaptability in real-time applications (Rai et al. 2023).

In terms of application as discussed in the previous section, both Unmanned Aerial Vehicles (UAVs) and Unmanned Ground Vehicles (UGVs) face significant operational challenges. UAVs, while offering advantages such as wide coverage, efficiency, and accessibility, are constrained by limited payload capacity, dependence on favorable weather conditions, and stringent regulatory constraints. In contrast, UGVs provide greater terrain adaptability and payload capacity, but their limitations include slower speeds, reduced coverage area, and challenges related to weather conditions, obstacle avoidance, steep terrain operation, and operational constraints such as fuel or battery capacity. These trade-offs complicate the deployment of either system for large-scale, precision weed management (Wang et al. 2020). While significant progress has been made in autonomous weed control systems, ongoing research is crucial to further advance the field through innovation and refinement. First, navigation systems should not rely on a single sensor type. Instead, sensor fusion—integrating LiDAR, ultrasonic sensors, RTK-GPS, and RGB cameras with machine learning—should be implemented for more precise and adaptive environmental mapping. Second, the deployment of highly efficient, compact deep learning models for weed detection and identification is necessary to ensure low-cost, responsive computation. Devices such as Raspberry Pi and Nvidia Jetson computers, known for their low latency and high computational power, are ideal for running advanced algorithms on robotic platforms (Chowdhury et al. 2025). Third, the availability and accessibility of comprehensive agricultural datasets should be prioritized, enabling the development of robust models that integrate sensor data from multiple sources for more accurate weed mapping. Weed detection and management systems should move beyond limited datasets and embrace multi-sensor fusion approaches for more comprehensive and accurate identification. Additionally, integrated weed management strategies that combine mechanical and chemical methods should be explored as hybrid solutions, minimizing the reliance on herbicides while enhancing efficiency. Robotic platforms must evolve to become robust, easy to operate, and cost-effective, with a design focus on adaptability. Instead of being task-specific, future robots should be capable of performing multiple agricultural tasks, necessitating the development of a generalized robotic operating platform. Future research should also focus on the integration of Unmanned Aerial Vehicles (UAVs) and Unmanned Ground Vehicles (UGVs), utilizing a hybrid approach for efficient, precise, and site-specific weed control. This synergy between aerial and ground-based systems offers a promising direction for achieving highly effective and sustainable agricultural operations. For a comprehensive understanding and deeper insights into the discussed concepts, the review paper based on current advancement in robotic implementations for weed management by Upadhyay et al. (2024) is a must read.

Conclusion

The advancements in modern robotics and artificial intelligence (AI) for weed control have led to significant transformations in the agricultural landscape. The integration of these technologies provides promising solutions to the limitations of traditional methods, particularly in terms of efficiency, precision, and sustainability. This chapter has explored the potential of AI and robotics to replace conventional, labor-intensive, and cost-prohibitive weed management techniques. These emerging technologies are not only capable of accurately detecting, identifying, and removing weeds but also aim to enhance crop yields while minimizing crop damage. Despite the ongoing development of these innovations, substantial research efforts continue to address their broad-scale application and integration into agricultural systems. The chapter has also examined the critical challenges that these technologies face, particularly in the areas of autonomous navigation, weed identification, and system implementation, while suggesting potential future research directions to overcome these obstacles. The fusion of AI and robotics into weed management practices is set to revolutionize agriculture by providing more efficient, precise, and environmentally sustainable solutions. As these technologies mature and become more widely adopted, they will play an essential role in addressing global issues related to food security, environmental sustainability, and modern farming practices. While the path to fully realizing these advancements is still evolving, the discussions in this chapter highlight the potential of AI and robotics to drive the future of agriculture.

References

Abbas, I., J. Liu, M. Faheem, R.S. Noor, S.A. Shaikh, K.A. Solangi et al. 2020. Different sensor based intelligent spraying systems in Agriculture. Sens. Actuator A-Phys. 316: 112265.

Abd-Elrahman, A., K. Britt and T. Liu. 2021. Deep learning classification of high-resolution drone images using the ArcGIS Pro software: FOR374/FR444, 10/2021. EDIS 2021(5).

Adeux, G., E. Vieren, S. Carlesi, P. Bàrberi, N. Munier-Jolain and S. Cordeau. 2019. Mitigating crop yield ET losses through weed diversity. Nat. Sustain. 2(11): 1018-1026.

Ahmad Loti, N.N., M.R. Mohd Noor and S.W. Chang. 2021. Integrated analysis of machine learning and deep learning in chi Adão pest and disease identification. J. Sci. Food Agric. 101(9): 3582-3594.

Ahmed, F., H.A. Al-Mamun, A.H. Bari, E. Hossain and P. Kwan. 2012. Classification of crops and weeds from digital images: A support vector machine approach. J. Crop Prot. 40: 98-104.

Alsalam, B.H.Y., K. Morton, D. Campbell and F. Gonzalez. 2017. Autonomous UAV with vision-based on-board decision making for remote sensing and precision agriculture. 2017 IEEE AESS: 1-12.

Amend, S., D. Brandt, D. Di Marco, T. Dipper, G. Gässler, M. Höferlin et al. 2019. Weed management of the future. KI-Künstliche Intelligenz. 33: 411-415.

Anand, R., S.S. Nalla, S. Nasreen, M. Chowdhury, Madhusudan, B.S., A. Pandey et al. 2024. Fruit counting and analysis using artificial intelligence approaches. pp. 93-114. *In:* S.S. Chouhan, A. Saxena, U.P. Singh and S. Jain [eds.]. Artificial Intelligence Techniques in Smart Agriculture. Springer Nature, Singapore.

Anderson, K. and K.J. Gaston. 2013. Lightweight unmanned aerial vehicles will revolutionize spatial ecology. Front. Ecol. Environ. 11(3): 138-146.

Andújar, D., V. Rueda-Ayala, H. Moreno, J.R. Rosell-Polo, A. Escolá, C. Valero et al. 2013. Discriminating crop, weeds and soil surface with a terrestrial LIDAR sensor. Sensors. 13(11): 14662-14675.

Asad, M.H. and A. Bais. 2020. Weed detection in canola fields using maximum likelihood classification and deep convolutional neural network. Inf. Process. Agric. 7: 535-545.

Åstrand, B. and A.J. Baerveldt. 2002. An agricultural mobile robot with vision-based perception for mechanical weed control. Autonomous Robots. 13: 21-35.

Bailey, K.L. 2014. The bioherbicide approach to weed control using plant pathogens. pp. 245-266. *In:* Integrated Pest Management. Academic Press.

Bakker, T., H. Wouters, K. Van Asselt, J. Bontsema, L. Tang, J. Müller et al. 2008. A vision-based row detection system for sugar beet. Comput. Electron. Agric. 60: 87-95.

Bakker, T. 2009. An Autonomous Robot for Weed Control: Design, Navigation and Control. Wageningen University and Research.

Bakker, T., K. van Asselt, J. Bontsema, J. Müller and G. van Straten. 2011. Autonomous navigation using a robot platform in a sugar beet field. Biosyst. Eng. 109: 357-368.

Balappa, B., A. Lokesh and N.M. Babu. 2021. Development of semiautomated sowing machine for multiple crops. Mater. Today. 42: 1317-1320.

Barnhart, I.H. 2024. Use of artificial intelligence to locate and treat weeds in Midwestern United States corn (*Zea mays*) and soybean (*Glycine max*) cropping systems. Ph.D. Thesis, Kansas State University.

Bawden, O., J. Kulk, R. Russell, C. McCool, A. English, F. Dayoub et al. 2017. Robot for weed species plant-specific management. J. Field Robot. 34: 1179-1199.

Bechar, A. and C. Vigneault. 2017. Agricultural robots for field operations. Part 2: Operations and systems. Biosyst. Eng. 153: 110-128.

Bridges, D.C. 2017. Ecology of weeds. pp. 19-34. *In:* Handbook of Weed Management Systems. Routledge.

Cai, Y., F. Zeng, J. Xiao, W. Ai, G. Kang, Y. Lin et al. 2023. Attention-aided semantic segmentation network for weed identification in pineapple field. Comput. Electron. Agric. 210: 107881.

Carbonrobotics. 2021. Carbonrobotics. [Online] Available: https://carbonrobotics.com/.

Chen, Y., X. Jin, L. Tang, J. Che, Y. Sun and J. Chen. 2013. Intra-row weed recognition using plant spacing information in stereo images. *In:* 2013 Kansas City, Missouri, July 21-24, 2013. ASABE, p. 1.

Chowdhury, M., R. Anand, T. Dhar, R. Kurmi, R.K. Sahni and A. Kushwah. 2024a. Digital insights into plant health: Exploring vegetation indices through computer vision. pp. 7-30. *In:* S.S. Chouhan, U.P. Singh and S. Jain [eds.]. Applications of Computer Vision and Drone Technology in Agriculture 4.0. Springer Nature, Singapore.

Chowdhury, M., T.K. Khura, R.A. Parray, H.L. Kushwaha, P.K. Upadhyay, A. Jha et al. 2024b. The use of destructive and nondestructive techniques in concrete nitrogen assessment in plants. J. Plant Nutri. 47: 2271-2294.

Chowdhury, M., T.K. Khura, R.A. Parray, P.K. Upadhyay, H.L. Kushwaha, A. Lama and A. Kushwah. 2025. Multi-vegetation indices based handheld device for precise nitrogen assessment and prescription in direct seeded rice. Comput. Electron. Agric. 229: 109886.

Christensen, S., H.T. Søgaard, P. Kudsk, M. Nørremark, I. Lund, E.S. Nadimi et al. 2009. Site-specific weed control technologies. Weed Res. 49: 233-241.

Coleman, G.R., A. Stead, M.P. Rigter, Z. Xu, D. Johnson, G.M. Brooker et al. 2019. Using energy requirements to compare the suitability of alternative methods for broadcast and site-specific weed control. Weed Technol. 33(4): 633-650.

Cordeau, S., M. Triolet, S. Wayman, C. Steinberg and J.-P. Guillemin. 2016. Bioherbicides: Dead in the water? A review of the existing products for integrated weed management. J. Crop Prot. 87: 44-49.

Curran, W.S. 2016. Persistence of herbicides in soil. J. Crop. Soil. 49: 16-21.

Dammer, K.H., J. Intress, H. Beuche, J. Selbeck and V. Dworak. 2013. Discrimination of Ambrosia artemisiifolia and Artemisia vulgaris by hyperspectral image analysis during the growing season. Weed Res. 53: 146-156.

Dayan, F.E., C.L. Cantrell and S.O. Duke. 2009. Natural products in crop protection. Bioorg. Med. Chem. 17: 4022-4034.

De Castro, A.I., J.M. Peña, J. Torres-Sánchez, F. Jiménez-Brenes and F. López-Granados. 2017. Mapping *Cynodon dactylon* in vineyards using UAV images for site-specific weed control. Advan. Ani. Biosci. 8: 267-271.

Desta, K. 2000. Weed control methods used in Ethiopia. Animal Power for Weed Control. Technical Centre for Agricultural and Rural Cooperation (CTA), Wageningen, The Netherlands.

Devine, M., S.O. Duke and C. Fedtke. 1992. Physiology of herbicide action. PTR Prentice Hall, Englewood Cliffs, New Jersey, USA. 441 pp. Exp. Agric. 29: 524.

Duke, D. and R. Romagni. 2000. Natural products as sources of herbicides: Current status and future trends. Weed Res. 40: 99-111.

EcoRobotix. 2018. Autonomous robot weeder. Available at: https://www.ecorobotix.com/en/autonomous-robot-weeder/.

EcoRobotix. 2020. Avo robot désherbant autonome. Available at: https://www.ecorobotix.com/fr/avo-robot-desherbant-autonome/.

Ecorobotix. 2021. Passez au désherbage intelligent avec Ecorobotix. Available at: https://www.ecorobotix.com/en/. Accessed 12 Dec 2021.

EcoRobotix. 2023. Weeding robotic platform: AVO, autonomous robot. Available at: https://ecorobotix.com/en/avo/.

Eide, A., Y. Zhang, C. Koparan, J. Stenger, M. Ostlie, K. Howatt et al. 2021. Image-based thermal sensing for glyphosate-resistant weed identification in greenhouse conditions. Comput. Electron. Agric. 188: 106348.

Engel, R. and R.D. Ilnicki. 1969. Turf weeds and their control. Turfgrass Science 14: 240-287.

Eslami, S. 2014. Weed management in conservation agriculture systems. *In:* Recent Advances in Weed Management (pp. 87-124). Springer, New York, NY, USA.

Esposito, M., M. Crimaldi, V. Cirillo, F. Sarghini and A. Maggio. 2021. Drone and sensor technology for sustainable weed management: A review. Chem. Biol. Technol. Agric. 8: 1-11.

FarmWise. 2020. Available at: https://farmwise.io/.

Fatima, H.S., I. ul Hassan, S. Hasan, M. Khurram, D. Stricker and M.Z. Afzal. 2023. Formation of a lightweight, deep learning-based weed detection system for a commercial autonomous laser weeding robot. Appl. Sci. 13: 3997.

Fawakherji, M., C. Potena, A. Pretto, D.D. Bloisi and D. Nardi. 2021. Multi-spectral image synthesis for crop/weed segmentation in precision farming. Robot. Auton. Syst. 146: 103861.

Fernández-Quintanilla, C., J.M. Peña, D. Andújar, J. Dorado, A. Ribeiro, F. López-Granados et al. 2018. Is the current state of the art of weed monitoring suitable for site-specific weed management in arable crops? Weed Res. 58(4): 259-272.

Findlay, S., M. Carreiro, V. Krischik and C.G. Jones. 1996. Effects of damage to living plants on leaf litter quality. Ecol. Appl. 6: 269-275.

Florencia, F.M., T. Carolina, B. Enzo and G. Leonardo. 2017. Effects of the herbicide glyphosate on non-target plant native species from Chaco Forest (Argentina). Ecotoxicol. Environ. Saf. 144: 360-368.

Fu, L., F. Gao, J. Wu, R. Li, M. Karkee and Q. Zhang. 2020. Application of consumer RGB-D cameras for fruit detection and localization in field: A critical review. Comput. Electron. Agric. 177: 105687.

Galati, R., G. Reina, A. Messina and A. Gentile. 2017. Survey and navigation in agricultural environments using robotic technologies. *In:* 2017 14th IEEE International Conference on Advanced Video and Signal Based Surveillance (AVSS), pp. 1-6. IEEE.

Gaw, S., M.E. Close and M.J. Flintoft. 2008. Fifth national survey of pesticides in groundwater in New Zealand. N.Z. J. Mar. Freshw. Res. 42: 397-407.

Gawande, V., D.R.K. Saikanth, B.S. Sumithra, S.A. Aravind, G.N. Swamy, M. Chowdhury and B.V. Singh. 2023. Potential of precision farming technologies for eco-friendly agriculture. Int. J. Plant Soil Sci. 35(19): 101-112.

Genze, N., R. Ajekwe, Z. Güreli, F. Haselbeck, M. Grieb and D.G. Grimm. 2022. Deep learning-based early weed segmentation using motion-blurred UAV images of sorghum fields. Comput. Electron. Agric. 202: 107388.

Gerhards, R., D. Andujar Sanchez, P. Hamouz, G.G. Peteinatos, S. Christensen and

C. Fernandez-Quintanilla. 2022. Advances in site-specific weed management in agriculture – A review. Weed Res. 62: 123-133.

Hajjaj, S.S.H. and K.S.M. Sahari. 2016. Review of agriculture robotics: Practicality and feasibility. *In:* 2016 IEEE International Symposium on Robotics and Intelligent Sensors (IRIS), pp. 194-198. IEEE.

Hall, D., F. Dayoub, T. Perez and C. McCool. 2017. A transplantable system for weed classification by agricultural robotics. *In:* 2017 IEEE/RSJ International Conference on Intelligent Robots and Systems (IROS), pp. 5174-5179. IEEE.

Halstead, M., A. Ahmadi, C. Smitt, O. Schmittmann and C. McCool. 2021. Crop agnostic monitoring driven by deep learning. Front. Plant Sci. 12: 786702.

Hamuda, E., M. Glavin and E. Jones. 2016. A survey of image processing techniques for plant extraction and segmentation in the field. Comput. Electron. Agric. 125: 184-199.

Hansen, K.D., F. Garcia-Ruiz, W. Kazmi, M. Bisgaard, A. la Cour-Harbo, J. Rasmussen et al. 2013. An autonomous robotic system for mapping weeds in fields. IFAC Proc. Vol. 46(10): 217-224.

Hassanein, M. and N. El-Sheimy. 2018. An efficient weed detection procedure using low-cost UAV imagery system for precision agriculture applications. The Int. Arch. Photogramm. Remote Sens. Spat. Inf. Sci. 42: 181-187.

Hatcher, P.E. and B. Melander. 2003. Combining physical, cultural, and biological methods: Prospects for integrated non-chemical weed management strategies. Weed Res. 43: 303-322.

Heap, I.J.A.M. 2024. The International Survey of Herbicide Resistant Weeds. Available online: https://weedscience.org/Home.aspx (accessed 2 February 2024).

Hollaway, K., R.S. Kookana, D. Noy, J. Smith and N. Wilhelm. 2006. Crop damage caused by residual acetolactate synthase herbicides in the soils of south-eastern Australia. Aust. J. Exp. Agric. 46: 1323-1331.

Hu, K., Z. Wang, G. Coleman, A. Bender, T. Yao, S. Zeng et al. 2024. Deep learning techniques for in-crop weed recognition in large-scale grain production systems: A review. Precis. Agric. 25(1): 1-29.

Hu, L., H. Liu, J. He, G. Chen, Z. Wang and C. Wang. 2023. Research progress and prospect of intelligent weeding robot. J. S. China Agric. Univ. 44: 34-42.

Hunter III, J.E., T.W. Gannon, R.J. Richardson, F.H. Yelverton and R.G. Leon. 2020. Integration of remote-weed mapping and an autonomous spraying unmanned aerial vehicle for site-specific weed management. Pest Manag. Sci. 76: 1386-1392.

Hussain, N., A.A. Farooque, A.W. Schumann, A. McKenzie-Gopsill, T. Esau, F. Abbas et al. 2020. Design and development of a smart variable rate sprayer using deep learning. Remote Sens. 12(24): 4091.

Illingworth, J. and J. Kittler. 1988. A survey of the Hough transform. Comput. Vis. Graph. Image Process. 44: 87-116.

Inoue, K., Y. Kaizu, S. Igarashi and K. Imou. 2019. The development of autonomous navigation and obstacle avoidance for a robotic mower using machine vision technique. IFAC-PapersOnLine. 52: 173-177.

Iqbal, J., R. Xu, S. Sun and C. Li. 2020. Simulation of an autonomous mobile robot for LiDAR-based in-field phenotyping and navigation. Robotics. 9: 46.

Jani, M., J. Fayyad, Y. Al-Younes and H. Najjaran. 2023. Model compression methods for YOLOv5: A review. arXiv preprint arXiv:2307.11904.

Jeon, H.Y., L.F. Tian and H. Zhu. 2011. Robust crop and weed segmentation under uncontrolled outdoor illumination. Sensors. 11(6): 6270-6283.

Jiang, W., L. Quan, G. Wei, C. Chang and T. Geng. 2023. A conceptual evaluation of a weed control method with post-damage application of herbicides: A composite intelligent intra-row weeding robot. Soil Tillage Res. 234: 105837.

Kavhiza, N.J., V. Vvedenskiy, A. Behzad, M. Bayat, M.H. Kargar and M. Zargar. 2020. Weed mapping technologies in discerning and managing weed infestation levels of farming systems. Res. Crops. 21(1): 93-98.

Khan, N., G. Medlock, S. Graves and S. Anwar. 2018. GPS guided autonomous navigation of a small agricultural robot with automated fertilizing system. SAE Technical Paper (No. 2018-01-0031).

Khan, S., M. Tufail, M.T. Khan, Z.A. Khan and S. Anwar. 2021. Deep learning-based identification system of weeds and crops in strawberry and pea fields for a precision agriculture sprayer. Precis. Agric. 22(6): 1711-1727.

Koo, G., K. Kim, J.Y. Chung, J. Choi, N.Y. Kwon, D.Y. Kang et al. 2017. Development of a high precision displacement measurement system by fusing a low-cost RTK-GPS sensor and a force feedback accelerometer for infrastructure monitoring. Sensors. 17(12): 2745.

Kotaniemi, J., N. Känsäkoski, E. Halbach and T. Heikkilä. 2024. A weeding robot for seedling removal. *In:* 2024 20th IEEE/ASME International Conference on Mechatronic and Embedded Systems and Applications (MESA), pp. 1-8. IEEE.

Krishna, K.R. 2018. Agricultural drones: A peaceful pursuit. Apple Academic Press.

Krupanek, J., P.G. de Santos, L. Emmi, M. Wollweber, H. Sandmann, K. Scholle et al. 2024. Environmental performance of an autonomous laser weeding robot – A case study. Int. J. Life Cycle Assess. 1: 1-32.

Kushwah, A., A. Chouriya, V.K. Tewari, C. Gupta, M. Chowdhury, P. Shrivastava and P. Bhagat. 2024a. A novel embedded system for tractor implement performance mapping. Cogent Eng. 11(1): 2311093.

Kushwah, A., P.K. Sharma, H.L. Kushwaha, B.B. Sharma, A.K. Shrivastava, R.H. Nag et al. 2024b. Hybrid taguchi-grey relational analysis approach for optimizing cutter operational parameters in selective cauliflower harvesting. Sci. Rep. 14(1): 29709.

Le, T., J.G.O. Gjevestad and P.J. From. 2019. Online 3D mapping and localization system for agricultural robots. IFAC-PapersOnLine 52: 167-172.

Lee, W.S., D.C. Slaughter and D.K. Giles. 1999. Robotic weed control system for tomatoes. Precis. Agric. 1: 95-113.

Liao, H., X. Li, Q. Yang, Y. Bai, P. Cui, C. Wen et al. 2021. Herbicide selection promotes antibiotic resistance in soil microbiomes. Mol. Biol. Evol. 38: 2337-2350.

Liao, J., Y. Zang, X. Luo, Z. Zhou, Y. Lan, Y. Zang et al. 2019. Optimization of variables for maximizing efficacy and efficiency in aerial spray application to cotton using unmanned aerial systems. Int. J. Agric. Biol. Eng. 12: 10-17.

Lin, C. 2010. A support vector machine embedded weed identification system. Ph.D. Thesis, University of Illinois at Urbana-Champaign.

Llewellyn, R., D. Ronning, M. Clarke, A. Mayfield, S. Walker and J. Ouzman. 2016. Impact of weeds in Australian grain production. Grains Research and Development Corporation, Canberra, ACT, Australia.

López-Granados, F. 2011. Weed detection for site-specific weed management: Mapping and real-time approaches. Weed Res. 51: 1-11.

Lottes, P., M. Hörferlin, S. Sander and C. Stachniss. 2017. Effective vision-based classification for separating sugar beets and weeds for precision farming. J. Field Robot. 34: 1160-1178.

Machleb, J., G.G. Peteinatos, B.L. Kollenda, D. Andújar and R. Gerhards. 2020. Sensor-based mechanical weed control: Present state and prospects. Comput. Electron. Agric. 176: 105638.

Madhusudan, B.S., H.L. Kushwaha, A. Kumar, R.A. Parray, S.S. Swain, M. Chowdhury et al. 2024. Critical assessment of furrow openers and operational parameters for optimum performance under conservation tillage. Sci. Rep. 14(1): 20928.

Maillot, T., J.B. Vioix and N. Colbach. 2023. Site-specific herbicide spraying can control weeds as well as full spraying in the long-term: A simulation study. Comput. Electron. Agric. 214: 108338.

Malkani, P., I. Mani, P.K. Sahoo, R.A. Parray, S.S. Swain, K.R. Asha et al. 2024. Development and evaluation of a sensor enabled smart herbicide applicator for inter row crops. Smart Agric. Technol. 9: 100649.

Manfreda, S., M.F. McCabe, P.E. Miller, R. Lucas, V. Pajuelo Madrigal, G. Mallinis et al. 2018. On the use of unmanned aerial systems for environmental monitoring. Remote Sens. 10(4): 641.

Mennan, H., K. Jabran, B.H. Zandstra and F. Pala. 2020. Non-chemical weed management in vegetables by using cover crops: A review. Agronomy 10(2): 257.

Mesas-Carrascosa, F.J., J. Torres-Sánchez, I. Clavero-Rumbao, A. García-Ferrer, J.M. Peña, I. Borra-Serrano et al. 2015. Assessing optimal flight parameters for generating accurate multispectral orthomosaics by UAV to support site-specific crop management. Remote Sens. 7: 12793-12814.

Mink, R., A. Dutta, G.G. Peteinatos, M. Sökefeld, J.J. Engels, M. Hahn et al. 2018. Multi-temporal site-specific weed control of Cirsium arvense (L.) Scop. and Rumex crispus L. in maize and sugar beet using unmanned aerial vehicle-based mapping. Agric. 8(5): 65.

Modi, R.U., M. Kancheti, A. Subeesh, C. Raj, A.K. Singh, N.S. Chandel et al. 2023. An automated weed identification framework for sugarcane crop: A deep learning approach. Crop Prot. 173: 106360.

Molinari, F.A., A.M. Blanco, M.R. Vigna and G.R. Chantre. 2020. Towards an integrated weed management decision support system: A simulation model for weed-crop competition and control. Comput. Electron. Agric. 175: 105597.

Naiture. 2019. Naiture: Innovative solutions for sustainable agriculture. Retrieved from https://www.naiture.org/.

Ngugi, D., P.K. Karau and W. Nguyo. 1978. East African Agriculture. 282. London: Macmillan.

Nguyen, T.X.B., K. Rosser and J. Chahl. 2021. A review of modern thermal imaging sensor technology and applications for autonomous aerial navigation. J. Imaging. 7: 217.

Oerke, E.C. 2006. Crop losses to pests. J. Agric. Sci. 144(1): 31-43.

Ofori, M. and O. El-Gayar. 2021. An approach for weed detection using CNNs and transfer learning. *In:* 2021 54th Hawaii International Conference on System Sciences (HICSS), pp. 888-895.

Olsen, A., D.A. Konovalov, B. Philippa, P. Ridd, J.C. Wood, J. Johns et al. 2019. DeepWeeds: A multiclass weed species image dataset for deep learning. Sci. Rep. 9(1): 2058.

Osadebe, V., N. Dauda, A. Ede, G. Chimdi and B. Echezona. 2021. The use of bioherbicides in weed control: Constraints and prospects. Afr. J. Agric. Tech. 21: 37-54.

Osorio, K., A. Puerto, C. Pedraza, D. Jamaica and L. Rodríguez. 2020. A deep learning approach for weed detection in lettuce crops using multispectral images. AgriEngineering 2: 471-488.

Petersen, P.E.H. and G.W. Krutz. 1992. Automatic identification of weed seeds by color machine vision. Seed Sci. Technol. 20: 193-208.

Pfrunder, A., P.V. Borges, A.R. Romero and G. Catt. 2017. Real-time autonomous ground vehicle navigation in heterogeneous environments using a 3D LiDAR. *In:* 2017 IEEE/RSJ International Conference on Intelligent Robots and Systems (IROS), pp. 2601-2608. IEEE.

Pradhan, N.C., M.A. Naik, M. Chowdhury, A. Kushwah, K.R. Asha, T. Dhar et al. 2024. Robotic seeding or sowing system in smart agriculture. pp. 495-520. *In:* K. Pandey, N.L. Kushwaha, C.B. Pande and K.G. Singh [eds.]. Artificial Intelligence and Smart Agriculture: Technology and Applications. Springer Nature, Singapore.

Radcliffe, J., J. Cox and D.M. Bulanon. 2018. Machine vision for orchard navigation. Comput. Ind. 98: 165-171.

Rai, N., Y. Zhang, B.G. Ram, L. Schumacher, R.K. Yellavajjala, S. Bajwa et al. 2023. Applications of deep learning in precision weed management: A review. Comput. Electron. Agric. 206: 107698.

Raja, R., T.T. Nguyen, D.C. Slaughter and S.A. Fennimore. 2020. Real-time weed-crop classification and localisation technique for robotic weed control in lettuce. Biosyst. Eng. 192: 257-274.

Rani, L., K. Thapa, N. Kanojia, N. Sharma, S. Singh, A.S. Grewal et al. 2021. An extensive review on the consequences of chemical pesticides on human health and environment. J. Clean Prod. 283: 124657.

Reddy, N.V., A. Reddy, S. Pranavadithya and J.J. Kumar. 2016. A critical review on agricultural robots. Int. J. Mech. Eng. Technol. 7(4): 183-188.

Reid, J.F., Q. Zhang, N. Noguchi and M. Dickson. 2000. Agricultural automatic guidance research in North America. Comput. Electron. Agric. 25: 155-167.

Reiser, D., E.S. Sehsah, O. Bumann, J. Morhard and H.W. Griepentrog. 2019. Development of an autonomous electric robot implement for intra-row weeding in vineyards. Agric. 9: 18.

Reiser, D., J.M. Martín-López, E. Memic, M. Vázquez-Arellano and S. Brandner. 2017. 3D imaging with a sonar sensor and an automated 3-axes frame for selective spraying in controlled conditions. J. Imaging. 3: 9.

Riehle, D., D. Reiser and H.W. Griepentrog. 2020. Robust index-based semantic plant/background segmentation for RGB-images. Comput. Electron. Agric. 169: 105201.

Sanchez, J. and E.R. Gallandt. 2021. Functionality and efficacy of Franklin Robotics' Tertill™ robotic weeder. Weed Technol. 35: 166-170.

Scavo, A. and G. Mauromicale. 2020. Integrated weed management in herbaceous field crops. Agronomy. 10: 466.

Shahbazi, N., M.B. Ashworth, J.N. Callow, A. Mian, H.J. Beckie, S. Speidel et al. 2021. Assessing the capability and potential of LiDAR for weed detection. Sensors. 21: 2328.

Shaner, D.L. and H.J. Beckie. 2014. The future for weed control and technology. Pest Manag. Sci. 70: 1329-1339.

Shanmugam, S., E. Assunção, R. Mesquita, A. Veiros and P.D. Gaspar. 2020. Automated weed detection systems: A review. KnE Eng. 271: 271-284.

Shapira, U., I. Herrmann, A. Karnieli and J.D. Bonfil. 2010. Weeds detection by ground-level hyperspectral data. Theory Pract. 38: 27-33.

Singh, R.K. 2018. Integrated weed management. Indian Farming 35.

Slaughter, D.C., D.K. Giles and D. Downey. 2008. Autonomous robotic weed control systems: A review. Comput. Electron. Agric. 61: 63-78.

Sori, H., H. Inoue, H. Hatta and Y. Ando. 2018. Effect of a paddy weeding robot in wet rice culture. J. Robot. Mechatron. 30: 198-205.

Subeesh, A., S. Bhole, K. Singh, N.S. Chandel, Y.A. Rajwade, K.V.R. Rao et al. 2022. Deep convolutional neural network models for weed detection in polyhouse-grown bell peppers. Artif. Intell. Agric. 6: 47-54.

Sujaritha, M., S. Annadurai, J. Satheeshkumar, S.K. Sharan and L. Mahesh. 2017. Weed detecting robot in sugarcane fields using fuzzy real-time classifier. Comput. Electron. Agric. 134: 160-171.

Sun, H., D.C. Slaughter, M.P. Ruiz, C. Gliever, S.K. Upadhyaya and R.F. Smith. 2010. RTK GPS mapping of transplanted row crops. Comput. Electron. Agric. 71: 32-37.

Sunil Kumar, T., H.M. Virdia, K.G. Patel, S. Ragi, M. Chowdhury, P. Kumar et al. 2024. Effect of summer legume residue incorporation and fertilizer regimes on rice growth, yield, and nutrient uptake. Front. Sustain. Food Syst. 8: 1467201.

Sunil, G.C., Y. Zhang, C. Koparan, M.R. Ahmed, K. Howatt and X. Sun. 2022. Weed and crop species classification using computer vision and deep learning technologies in greenhouse conditions. J. Agric. Food Res. 9: 100325.

Taghadomi-Saberi, S. and A. Hemmat. 2015. Improving field management by machine vision – A review. Agric. Eng. Int.: CIGR Journal 17(3).

Tremblay, N., Z. Wang, B.L. Ma, C. Belec and P. Vigneault. 2009. A comparison of crop data measured by two commercial sensors for variable-rate nitrogen application. Precis. Agric. 10: 145-161.

Tshewang, S., B.M. Sindel, M. Ghimiray and B.S. Chauhan. 2016. Weed management challenges in rice (Oryza sativa L.) for food security in Bhutan: A review. J. Crop Prot. 90: 117-124.

Tsiamis, K., E. Gervasini, F. D'amico, I. Deriu, S. Katsanevakis and T. Backeljau. 2016. The EASIN Editorial Board: Quality assurance, exchange, and sharing of alien species information in Europe. Manag. Biol. Invasions. 7: 321-328.

Tsouros, D.C., S. Bibi and P.G. Sarigiannidis. 2019. A review on UAV-based applications for precision agriculture. J. Info. 10: 349.

Uchida, H. and T. Yamano. 2019. Development of a remote-control type weeding machine with stirring chains for a paddy field. *In:* 22nd International Conference on Climbing and Walking Robots and the Support Technologies for Mobile Machines, CLAWAR.

Underwood, J. and A. Wendel. 2016. Self-supervised weed detection in vegetable crops using ground-based hyperspectral imaging. *In:* 2016 IEEE International Conference on Robotics and Automation (ICRA), pp. 5128-5135. IEEE.

Upadhyay, A., Y. Zhang, C. Koparan, N. Rai, K. Howatt, S. Bajwa et al. 2024. Advances in ground robotic technologies for site-specific weed management in precision agriculture: A review. Comput. Electron. Agric. 225: 109363.

Utstumo, T., F. Urdal, A. Brevik, J. Dørum, J. Netland, Ø. Overskeid et al. 2018. Robotic in-row weed control in vegetables. Comput. Electron. Agric. 154: 36-45.

Valente, J., S. Hiremath, M. Ariza-Sentís, M. Doldersum and L. Kooistra. 2022. Mapping of Rumex obtusifolius in nature conservation areas using very high-resolution UAV imagery and deep learning. Int. J. Appl. Earth Obs. Geoinf. 112: 102864.

Veeranampalayam Sivakumar, A.N., J. Li, S. Scott, E. Psota, A.J. Jhala, J.D. Luck et al. 2020. Comparison of object detection and patch-based classification deep learning models on mid-to-late-season weed detection in UAV imagery. Remote Sens. 12: 2136.

Vincent, C., B. Panneton and F. Fleurat-Lessard. 2001. Physical Control Methods in Plant Protection. Springer Science & Business Media, Berlin/Heidelberg, Germany.

Wang, A., W. Zhang and X. Wei. 2019. A review on weed detection using ground-based machine vision and image processing techniques. Comput. Electron. Agric. 158: 226-240.

Wang, J., Y. Lan, S. Wen, A.J. Hewitt, W. Yao and P. Chen. 2020. Meteorological and flight altitude effects on deposition, penetration, and drift in pineapple aerial spraying. Asia-Pac. J. Chem. Eng. 15: e2382.

Li, X., X. Peng and H. Fang. 2020. Navigation path detection of a plant protection robot based on the RANSAC algorithm. Nongye Jixie Xuebao/Trans. of the Chinese Soc. of Agricultural Machinery. 51(9): 40-46.

Xue, J., L. Zhang and T.E. Grift. 2012. Variable field-of-view machine vision-based row guidance of an agricultural robot. Comput. Electron. Agric. 84: 85-91.

Yin, X., Y. Wang, Y. Chen, C. Jin and J. Du. 2020. Development of an autonomous navigation controller for agricultural vehicles. Int. J. Agric. Eng. 13(4): 70-76.

Yu, J., S.M. Sharpe and N.S. Boyd. 2019. Fumigants alone or in combination with herbicide for weed management in bell pepper (*Capsicum annuum*). J. Crop Prot. 118: 31-35.

Zhang, Q., Q. Chen, Z. Xu, T. Zhang and X. Niu. 2021. Evaluating the navigation performance of multi-information integration based on low-end inertial sensors for precision agriculture. Precis. Agric. 22: 627-646.

Zhou, X., H.B. Zheng, X.Q. Xu, J.Y. He, X.K. Ge, X. Yao et al. 2017. Predicting grain yield in rice using multi-temporal vegetation indices from UAV-based multispectral and digital imagery. ISPRS J. Photogramm. Remote Sens. 130: 246-255.

CHAPTER
9

Smart Irrigation Techniques and Water Conservation for Climate-Resilient Food Systems

Olipa Nyazambe Lungu[1*], Never Mujere[2], Joyce Lepetu[3] and Elly Siakasasa[4]

[1] Department of Land Management, The University of Zambia, P.O. Box 32379, Lusaka, Zambia

[2] Department of Geography Geospatial Sciences and Earth Observation, University of Zimbabwe, P.O. Box MP 167, Mt Pleasant, Harare

[3] Faculty of Natural Resources, Botswana University of Agriculture and Natural Resources

[4] GLYMO Enterprises and FutureSeeds Z LTD – Agribusiness, Chongwe, Lusaka, Box 37250

Introduction

Almost 40% of global food production comes from irrigated fields which account for nearly 24% of cropped lands. Correspondingly, irrigated agricultural production accounts for 90% of humanity's consumptive water use (Mehtra et al. 2024). While irrigation is prominent in most developed agrarian communities, the history of its development dates back to the early civilization in Egypt and Mesopotamia in 6,000 B.C. when farmers were diverting water from the Nile River to irrigate riparian crop fields all year-round (Angelakis et al. 2020, Sojka et al. 2022). Given the importance of irrigated farming in building climate-resilient food systems, lots of national, regional, and international policies have been crafted to develop water resources to mitigate the vagaries of global climate change in agriculture (IIED and HIVOS 2017; OECD 2021; Andrews et al. 2023; GP 2023 a,b,c). Therefore, the objective of this book chapter is to highlight the characteristics of a resilient food system, its requirements to foster this resilience, and techniques available to achieve profitability, efficiency and effectiveness in agriculture. This is achieved by taking into consideration the importance of smart irrigation technologies for water conservation and climate resilience and constraints to uptake of the technologies and research gaps in smart farming.

*Corresponding author: olipa.lungu@unza.zm

DOI: 10.1201/9781003536932-9

A resilient food system can face climatic shocks and continue providing adequate, nutritious, and safe food for everyone. Achieving this demands that the tendency of mono-cropping be replaced by a wide range of crops cultivated and consumed in such a system. This increases the chance of survival, for instance, a mixture of drought-tolerant, upland crops, fruit trees, and livestock using sustainable farming practices such as approaches embedded in climate-smart agriculture (CSA). CSA encourages minimum soil disturbance to maintain soil organic matter which is key to soil fertility, and it involves integrated management for pests and diseases, through intercropping, crop rotations, and reduced chemical pesticide use. This approach reduces environmental degradation while it optimizes land, water, and available nutrient use (FAO 2010; FAO 2013).

Irrigation notably contributes to food security and resilient food systems (Abamagal 2019; AERC 2020; Mume et al. 2023; Jetana and Tesfahun 2023; Kapari et al. 2023; Mupaso et al. 2024). Food security entails that communities have easy access to enough, safe and nutritious food, (FAO 2015) and can be achieved by having strong food systems. Cities heavily rely on the periphery for their food supplies. Furthermore, in most southern African communities, cities are fed by mostly smallholders from the peri-urban and surrounding rural areas, who are often ill-prepared for climatic shocks. Thus, building resilience in these communities saves lives by ensuring the growing population has enough food (UN 2019) through continued agriculture production.

Agriculture is both a victim and contributor to greenhouse gas emissions (because of mechanisation, livestock and crop production), thus leading to global warming (IPCC 2023), while it accounts for a considerable portion of water consumption in conservation farming methods. As water resources become strained, there is increasing pressure to adopt sustainable and efficient water management strategies to keep the agriculture sector effective for food systems' integrity (Jha et al. 2023; Patra et al. 2023; Jha et al. 2024).

Robust food systems are not only anchored on good production practices but on a strong linkage of the entire food value chain from production to consumption. Thus, equitable access to finance by the entire food value chain actors including consumers cannot be overemphasized. Deliberate gender equality policies must be included and enforced in sustainable food systems, to support both men and women engaged in the food value chain, and shifts from rainfed to irrigated production must be made across the region (Guddaraddi et al. 2023). Sadly, most southern African countries are still heavily dependent on rainfed production, for instance, Zambia cultivates less than 14% of its arable land of 45 million hectares of which only 5.7% percent is irrigated (GRZ 2021). Similarly, of the estimated irrigable area in Zimbabwe (550 000 hectares), only 40% has been developed, though this includes also non-functional and under-performing systems, (Nhundu and Mushunje 2010), while Botswana presents a unique scenario with most of its irrigation development being biased towards livestock production, making Botswana one of the countries with the least-developed irrigation sectors in the Southern Africa Development Community (SADC) (WAVES 2014).

Irrigation development is widely promoted using underground and surface water resources. These traditional surface gravity-fed flood canals, drip, and sprinkler irrigation use fixed schedules and manual labour (Muleke et al. 2022) and are inefficient, costly, and time-intensive. Water delivery is also often uneven, resulting in overwatering, waste of water and harm to plant health (Durmus et al. 2024). Due to adverse impacts of climate change, the irrigation sector requires development of alternative innovative solutions and technologies to get more crops per drop of water. Alternative irrigation systems utilise modern automated or digital tools to deliver water more efficiently, save water and reduce costs (Kapari et al. 2023; Pradhan et al. 2024). These automated irrigation technologies are either known as smart, climate-smart, or intelligent irrigation among other terms.

In the smart irrigation, the use of drones brings several benefits, such as improved efficiency in the use of inputs, cost reduction, increased productivity and profitability (Chowdhury et al. 2024). Drones are able to collect real-time data and provide accurate information about plant health, allowing growers to make data-based decisions. Additionally, drones are capable of covering large areas of land in high resolution quickly and efficiently, allowing farmers to monitor their crops in real time and identify problem areas that need immediate attention. With drones, farmers can scout for pest infestation, observe water-stress and prevent disease-stress in real time (Chowdhury et al. 2025). Nutrient-deficiencies are easily noted using drones and solutions are provided to extensionists and farmers (Talaviya et al. 2020; Chowdhury et al. 2023; Satapathy et al. 2024), while spraying chemicals with drones through careful preparation of the mixture, choice of equipment and definition of flight plans (Alarcão and Nuñez 2024) achieves improved crop management.

Smart Irrigation

Smart irrigation is essentially an advanced water management technology designed to optimize irrigation systems by using data-driven tools. In smart irrigation, water is applied precisely based on plant needs and environmental conditions, thereby, reducing wastage from over-irrigation and runoff, ultimately conserving water resources (McCartney and Lacombe 2011). Some smart irrigation systems are connected to weather forecasting services, allowing them to adjust irrigation schedules based on predicted rainfall, evaporation rates, and temperature changes. By adapting to changing weather patterns, smart irrigation maximizes water efficiency, and minimizes water waste, making it a valuable tool against water scarcity, food insecurity, and unsustainable water management practices (Malkhani et al. 2024).

Smart irrigation techniques in food systems ensure climatic resilience, efficient water supply and reduce wastage through excess water supply in run-off or by evaporation, thereby also reducing erosion, which then promotes good plant growth and improves yield. By automation, smart irrigation techniques increase efficiency and increase crop survival chances even during poor rainy seasons. Economically, this means, resource maximization is increased thus unlocking more resources for other uses, and also reducing pressure on energy for production (Muleke et al. 2022). Water resources not wasted in irrigation, become more available for groundwater recharge and other competitive ecosystem uses thereby fostering strong food systems, environmental protection and food security or sustainable growth. Smart irrigation and water conservation techniques are integral to climatic resilience of food systems for sustainable development and addressing water scarcity challenges, achieving zero hunger, and reducing poverty (Abamagal 2019; Anand et al. 2024). The smart irrigation system integrates various technologies such as sensor technology, automatic control technology and computer technology. It uses Internet of Things, mobile internet, and remote sensing monitoring of crops using platforms such as drones and satellites. Remotely-controlled valve switches provide plants with the right amount of water according to soil type and prevailing weather conditions. Unlike traditional irrigation controllers that operate on a preset programmed schedule and timers, smart irrigation controllers monitor weather, soil conditions, evaporation and plant water use to automatically adjust the watering schedule to actual conditions of the site.

Types of Smart Irrigation Techniques

Smart Irrigation techniques exist in the following main broad categories as:

Sensor-Based Irrigation Systems

Sensor-based systems which utilize soil moisture sensors, weather stations, and crop sensors to monitor environmental conditions and plant water requirements in real time. These are useful in the implementation of automated irrigation scheduling algorithms based on sensor data to deliver the right amount of water at the right time to crops (Ghareeb et al. 2023).

Drip Irrigation and Micro-Irrigation Systems

Which deliver water directly to the root zone of plants with minimal evaporation and runoff and can be fitted with smart controllers, flow meters, and soil moisture sensors for increased precision in water delivery and monitoring. They are generally renowned for conserving water, increasing crop yields, and reducing production costs (Muleke et al. 2022).

Precision Irrigation Technologies

Precision irrigation techniques, including variable rate irrigation (VRI), site-specific irrigation, and precision application methods use of GPS technology, remote sensing, and GIS mapping to customize irrigation treatments based on soil characteristics, topography, and crop water requirements, and maximize water use efficiency by minimizing over and under-watering (Rao et al. 2019; Ghareeb et al. 2023; Gawande et al. 2023).

Evapotranspiration (ET) Monitoring and Control

This includes the integration of ET-based irrigation controllers with weather forecasting models and soil moisture sensors to adjust irrigation schedules in real-time (Grabow et al. 2013).

Remote Monitoring and Control Systems

Remote monitoring and control systems for irrigation infrastructure, such as pumps, valves, and emitters, using IoT technology and wireless communication networks work to increase water use efficiency and reduce water wastage. They can be controlled remotely through smartphone applications or internet interfaces, allowing clients to display screens and modify their irrigation schedules from anywhere (Rao et al. 2019). This is beneficial to busy farmers who might not have the time to control their irrigation systems manually. Automated systems can also deliver indicators and notifications, which include notifications as soil moisture content is low or there is a problem with the system, allowing for proactive control.

Smart Irrigation System Components

Smart irrigation systems have powerful functions and can realize automatic water-saving irrigation and subsequently increase the cost-effectiveness of a production unit. Their compositions are relatively complex, consisting of monitoring part, control part, transmission part and terminal platform. They optimize available water by effectively and efficiently distributing it to plants and crops. Hence, timely application of inputs. Such a system can automatically start and stop water pumps on the irrigation site based on the soil moisture content determined by the moisture sensor and the ultrasonic sensor measuring the water level in reservoirs (Ogidan et al. 2019). Smart irrigation reduces wastage and promotes more responsible water use in both agricultural and urban landscapes, amidst many competing needs. Embracing smart irrigation technologies helps alleviate water scarcity concerns, conserve water for future generations, and mitigate the environmental impacts of water misuse (Satpute et al. 2024). Fig. 1 shows the smart irrigation system comprising data acquisition (sensor), microcontroller, wireless communication, cloud, user interface, actuators and irrigation control. Each of these components can be used in IoT devices.

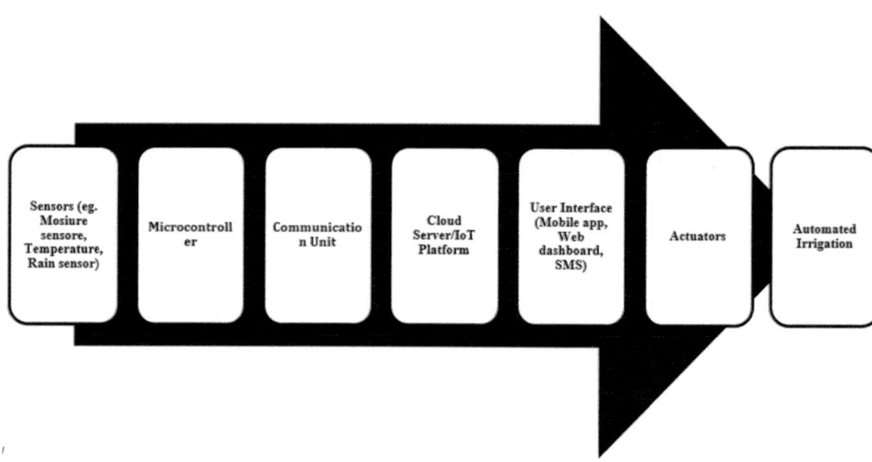

Fig. 1: Smart irrigation system components.

Controllers and Sensors

Smart irrigation controllers and sensors work together to optimize water usage. Unlike traditional irrigation controllers that operate on a preset programmed schedule and timers, smart irrigation controllers monitor weather, soil conditions, evaporation, and plant-water use to automatically adjust the watering schedule to actual conditions of the site (Singh et al. 2022). Controllers serve to manage the irrigation schedule and duration. They can be programmed to ensure precise water delivery and offer remote access for convenient management through electronic devices, thereby reducing labor costs and increasing efficiency of production systems.

Sensors provide real-time data on environmental factors useful for decision-making on watering frequency based on weather conditions, soil moisture, and specific plant requirements. Soil moisture sensors, rain sensors, weather stations, and flow sensors are used to determine situation-specific precise irrigation needs of crops (Ghareeb et al. 2023). For example, rain sensors prevent overwatering in regions with irregular rainfall, while soil moisture sensors enable real-time adjustments of water supply for optimal watering. Weather stations sensors provide intelligent irrigation adjustments based on forecasts, and flow sensors aid water conservation efforts by detecting leaks. By utilizing real-time data and automation, smart irrigation ensures that plants receive the right amount of water at the right time, minimizing water wastage and enhancing overall irrigation efficiency (Singh et al. 2022; Ghareeb et al. 2023).

Soil moisture sensor is a device that directly measures whether the soil is dry and differs according to given soil types. A probe is inserted into the same depth as the root of the plant, and the sensors in different areas are used with the nearest soil controller to control irrigation in different locations. The management platform can set high and low alarm values for soil sensors in different regions. Once the soil moisture exceeds the alarm value, the controller closes the water valve, and when the soil moisture falls below the alarm value, the controller opens the water valve (Obaideen et al. 2022).

Evaporation and leaf wetness sensors are devices related to weather monitoring. Their values reflect loss of water from soil surface. Farmers can significantly reduce irrigation water by monitoring leaf moisture, and analyze the amount of water in the soil determining whether plants lack water or there is too much water. While the evaporation sensor measures loss of water from soil surface, the leaf wetness sensor measures the loss of water from plant leaves.

Precipitation sensor controls the watering schedules by incorporating rain and snow amounts. The rain and snow sensor sends a signal to the controller to close the water valve in time after receiving it.

The wind speed sensor interrupts the irrigation cycle if the wind speed exceeds a certain threshold as watering during windy times reduces the uniformity of plant irrigation distribution and results in wastage.

Benefits and Constraints of Using Smart Irrigation Systems

Smart irrigation systems provide a more efficient, environmentally-friendly, and cost-effective approach to watering, making them an excellent investment for both residential and agricultural properties but maybe expensive for smallholder farmers who are the main producers.

Benefits of Smart Irrigation Systems

Notable benefits of smart irrigation systems include

Water Conservation: Smart irrigation systems use real-time data and automation to optimize water usage, ensuring that plants receive the right amount of water. By avoiding overwatering and minimizing water waste, these systems contribute to water conservation efforts, helping to preserve valuable water resources.

Cost Savings: By reducing water consumption, a smart irrigation system can significantly lower water bills for smallholders. The precise irrigation scheduling also reduces the need for manual labor and maintenance, resulting in potential long-term cost savings. This trend resulted in increased income among marginalized households in Malawi (Nkhata 2014).

Improved Plant Health: Smart irrigation systems deliver water based on the specific needs of plants and adjust irrigation schedules according to changing environmental conditions. This promotes healthier and more robust plant growth, as they receive adequate hydration and avoid stress caused by poor water supply (Ghareeb et al. 2023).

Remote Control and Monitoring: Many smart irrigation systems can be controlled and monitored remotely through smartphones or computers. This allows users to adjust settings, check system performance, and respond to changing conditions from anywhere, adding convenience and flexibility to the irrigation process, thus promoting adaptability to decision-making (Elufioye et al. 2024).

Integration with Weather and Site-specific Data: Smart irrigation systems can be connected to weather forecasting services, allowing them to adjust watering schedules based on predicted rainfall and other weather patterns. This integration ensures that the system adapts to changing conditions, further enhancing water efficiency (Rao et al. 2019).

Increases Climate Resilience: By promoting sustainable water use and maintaining plant health, these systems help smallholders withstand the impact of climate change and water scarcity. For farmers, this means more reliable crop production and reduced vulnerability to extreme weather events (Patel et al. 2023; Suna et al. 2023).

Additionally, smart irrigation systems use sensors and weather facts to optimize watering schedules, and they can modify watering times and portions based mostly on real-time facts, ensuring that the plants gather the right amount of water at the right time. This precision leads to large water as well as financial savings, which is crucial in the regions with water scarcity.

Disadvantages of Smart Irrigation Systems

While smart irrigation systems offer numerous benefits, they also have some potential disadvantages such as:

High Investment Cost: Smart irrigation systems typically involve higher upfront costs compared to traditional irrigation methods. The expense of purchasing and installing sensors, controllers, and other components can be a barrier for some users, especially for smallholders (Pandey and Pandey 2023).

Technical Complexity: Smart irrigation systems require a certain level of technical know-how to set up and maintain effectively. Programing and troubleshooting the system may be challenging for some users, leading to the need for professional assistance, which could add to the overall cost.

Reliability and Malfunction: Like any electronic system, smart irrigation technology can experience technical glitches, software errors, or sensor malfunctions. In some cases, these issues might result in inadequate watering or even overwatering, potentially impacting plant health and water efficiency (O'Shaughnessy et al. 2019).

Power Dependency: Smart irrigation systems typically require a reliable power source to operate, which might not be readily available in remote or off-grid locations. Power outages or disruptions can affect the system's functionality and lead to temporary watering interruptions.

Coupled with Poor Network Connectivity: This power challenge can lead to delayed information access, which can disrupt the decision-making process.

Data Privacy and Security: Smart irrigation systems gather and process data, which may include sensitive information about the property and its owners. Ensuring data privacy and protecting the system from potential cyber threats becomes crucial but can pose challenges if not adequately addressed (Ghareeb et al. 2023).

Despite these disadvantages, the continuous advancement of smart irrigation technology and its potential for water conservation and efficient irrigation make it a promising solution for sustainable water management.

Smart Irrigation Solutions with IoT

Internet of Things (IoT), smartphone tools, and sensors are technologies that enable farmers to know the exact status of their field conditions, including soil temperature, amount of water required, weather conditions, etc. IoT can be thought of as an extension of the current internet to all devices that can communicate with electronic equipment and are linked to the internet, making devices user-friendly and easy to handle (Ogidan 2019). Correspondingly, IoT is linked to automation of all areas of agriculture and farming processes in order to make the entire process more productive and efficient. The use of sensors is also important for farmers to better understand their crops, reduce the environmental impacts and conserve resources.

Internet of Things

The use of Internet of Things (IoT) sensors and smart irrigation systems can potentially alter the way farmers maximize their agribusinesses, optimize water and land resources use, save energy and ensure smart irrigation practices, and ultimately increase productivity. The use of IoT sensors and smart irrigation systems allows producers to instantaneously monitor soil moisture levels, precipitation, temperature and related weather conditions, useful to determine crop water requirements and schedule irrigation and fertigation practices (Bhardwaj and Sharma 2024). It promotes easy monitoring of the entire food system by monitoring production conditions such as pest and weed control, storage and post-harvest handling conditions, and provides real-time low-cost information on market access. Through the use of IoTs, it becomes feasible to attain detailed information regarding the weather which eventually leads to assist in developing methods that could support the process of irrigation in the long run as shown in Fig. 2.

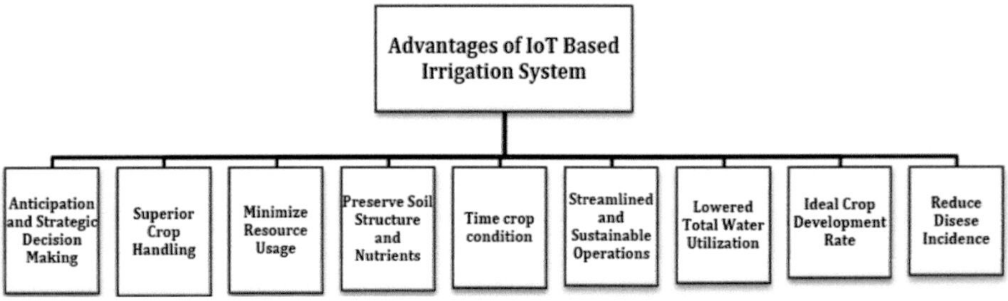

Fig. 2: Benefits of using IoT in irrigation systems.

Artificial Intelligence and Machine Learning (Ml) in Smart Irrigation Systems

Artificial intelligence (AI) allows for computers and robots to perform human activities and this has immense potential to enhance climate change resilience in agriculture. With the availability of data from satellite imagery, weather stations and automated unmanned devices or drones, AI, has become very essential in precision agriculture and are important in strengthening food systems. Most of the AI is developed in developed countries and tested in the developing countries, where there is still a big challenge of data scarcity (Bhardwaj and Sharma 2024) and satellite data is still developing (Lungu et al. 2020).

Utilization of machine learning algorithms to analyze spectral data and identify early signs of drought, heat stress, pest infestation, and disease outbreaks in crops through the incorporation of crop monitoring systems with automated alerts and decision support tools to facilitate timely interventions and adaptive management practices plays a pivotal role in improving livelihoods by increasing information on production, market access and product demands. This application of precision-agriculture techniques enabled by AI to optimize resource use, including water, in response to climate variability have been dotted across the African continents with projects in Malawi, Tanzania, Uganda, and Kenya. In Malawi, a project is building on artificial intelligence (AI) for soil moisture and nutrient-monitoring under irrigated agriculture among smallholder farmers, academic and agriculture experts (Chafuwa 2017), while the artificial intelligence for agriculture and food systems innovation research network is one of the pilots in AI for irrigation development (Feed the Future Innovation Lab for Small Scale Irrigation 2023 a, b).

Similarly, ML and AI using IoT sensors and smart irrigation systems to monitor soil moisture levels, weather conditions, and crop water requirements in real-time minimize nutrient runoff, and reduce environmental impact in agriculture (Obasi et al. 2024) remain cardinal to foster resilient food systems. Robots, Drones, IoTs, and Chatbots (Automated customer services) which are useful providers of real-time advice on plant and soil conditions for irrigation management are also built on the same principles and are gradually being integrated in farming systems. Subsequently, many sub-Sahara African farmers are educated to use smart technologies to increase production, but there is very slow uptake of the technologies.

Constraints to Uptake of Smart Irrigation Techniques and Water Conservation

While farmers are determined and committed to tackling the adverse effects of climate change, several hindrances affect their decision and level of adaptation and adoption of smart irrigation techniques and water conservation. The challenges faced by smallholder farmers particularly include:

Inadequate Information about Water Conservation

Smallholder farmers in remote areas of southern Africa often lack the education and skills needed to operate smart irrigation systems. This, coupled with limited access to information on water conservation, weather forecasts, and extension services, exacerbates the challenge of their low uptake of such technologies.

Lack of Finances and Incentives to Promote Uptake

Some irrigation systems tend to be expensive for an average smallholder so lack of financial incentives such as removal of custom duty on irrigate equipment, subsidies and grants or small loans for smallholders to promote water conservation and irrigation (Kapari et al. 2023). While individual farmers strive to access funds, lack of initiative of formation of cooperatives or farmer bodies that allow access to funds, and sometimes failure to agree, or hindrances to such institutional support further limits their access to funds (Feed the Future Innovation Lab for Small Scale Irrigation 2023a, c).

Land Tenure Systems

Installation of some water conservation infrastructure as dams, reservoirs, and some smart irrigation systems as such weather, moisture and temperature sensors require secure facilities which can be costly at times. In places where land tenure systems are unstable and short-term, the insecurity hinders installation and adoption of automated system (Hosaena 2013).

Gender

Some communities still do not promote women and youth ownership and leadership at farm level. This limits their access to land and sometimes apportions them only a small piece of land which is not sustainable to host the smart irrigation systems needed to support food system resilience, as well as foster food security (Theis et al. 2018; Benkele et al. 2018). It's important to ensure that men, women, and different social groups have equal opportunity to participate in and benefit from irrigation through deliberate gender-inclusive policies (Mujere 2016).

Conflicts

Wars and regional instability constrain farmers from investing in smart irrigation systems. Therefore, peace and stability are an essential need to foster smart farming.

Limited Investment in Water Conservation and Smart Irrigation Systems

Investment in water conservation and smart irrigation systems must be backed with corresponding output in high value crops which should cover operating costs. Therefore, if the market and output is not maximized, farmers will not invest in smart systems due to limited finances (FAO 2013).

Inadequate Government Policy Support

When Government policies are void of promoting tax exceptions for investing in smart irrigation and water conservation practices, this constrains the uptake of these techniques. Governments can learn from successes of others and have deliberate inclusions of crop monitoring systems which will improve stress detection (Lungu et al. 2020) and foster resilient food systems.

The Role of Smart Irrigation in Achieving the Sustainable Development Goals (SDGS)

Irrigation addresses various SDGs for food security and reducing poverty. However, smart irrigation indicates a wider range of SDGs for industry innovation and provides a more responsible consumption and production for food security. It has a direct influence on the current progress in SDGs. The role of irrigation in achieving the following SDGs: SDG 1: No Poverty, SDG 2: Zero Hunger, SDG 3: Good Health and Well-being, SDG 6: Clean Water and Sanitation, SDG 7: Affordable and Clean Energy when coupled with solar or wind systems, SDG 8: Decent Work and Economic Growth, SDG 9: Industry, Innovation and Infrastructure, SDG 11: Sustainable Cities and Communities through efficient use of smart irrigation systems, SDG 12: Responsible Consumption and Production by lowering the amount of waste generated, SDG 13: Climate Action SDG 14: Life Below Water by Moderating amount of water used compared to traditional irrigation systems, and, SDG 15: Life on Land by creating a reliable food supply and increasing the quality and quantity of the farm production (enhancing the agricultural yields, and ultimately altering the rainfall patterns, is an urgent cause for action towards promotion of smart irrigation across the globe. Achieving the SDGs (particularly food security) is key to most of the developmental actions and frameworks. To ensure this, forming a base for leveraging the different goals and value chain actors remains critical (UNFSS 2021).

Water Conservation Practices

While irrigation is vital for building resilient food systems, water resource conservation is highly imperative to its success, and involves water management. Water management could be considered as the management of soil moisture to make sure that the optimum level and quantity of water is applied at the right time. Effective water management is extremely crucial for the agricultural sector as it could decrease cost and augment crop production. Water management is also critical as it allows producers in the agricultural sector to manage the resources and perform related activities effectively. One of the most vital steps in good irrigation water management is regarded as knowledge of basic soil, crop water requirements, and available water. Sometimes times tradeoffs have to be made in investment of water resources (Chafuwa 2017) With limited knowledge and information, it would be not possible to explore how the irrigation activities have to be managed and controlled in adverse conditions and this may eventually lower the overall performance.

Irrigation water scarcity due to increasing competition among users has called for innovative solutions in the agricultural sectors along with other industries. Rainwater harvesting, small-scale community man-made dams, reservoirs, and mulching to conserve water and improve (*in-situ*) soil moisture retention, encompass the various water conservation practices with high potential to foster climate-resilient food systems (Mati et al. 2005; FAO 2013). Approaches such as climate-smart agriculture (CSA) in farming practices that encourage zero or reduced tillage, water conservation practices, use of organic matter, and improved crop varieties in production, enhance water conservation and unlock potential for increased climate resilience. Conservation Agriculture (CA) is an ecological approach to regenerative sustainable agriculture and ecosystem management based on the practical application of context-specific and locally adapted three interlinked principles of: (i) Continuous no or minimum mechanical soil disturbance (no-till seeding/planting and weeding, and minimum soil disturbance with all other farm operations including harvesting); (ii) permanent maintenance of soil mulch cover (crop biomass, stubble and cover crops); and (iii) diversification of cropping system (economically, environmentally and socially adapted rotations and/or sequences and/or associations involving annuals and/or perennials, including legumes and cover crops). These essential practices are combined and enhanced with other complementary practices of integrated crop, soil, nutrient, water, pest, labour, energy and land management practices to generate and sustain optimum performance of productivity and ecosystem societal services (FAO 2013).

Water management is critical for various reasons and one of the other reasons could be identified as ensuring optimum work efficiency level. Through the use of water management, it may become possible to ensure that agricultural crops receive proper amounts of water in dry areas and in periods of scarce rainfall (Mujere and Muhoyi 2022).

Rainwater harvesting, dam and reservoirs constructions, use of smart irrigation practices and managing water flow processes (Feed the Future Innovation Lab for Small Scale Irrigation. 2023a, c) are some of the most effective ways of water management in agriculture though they mainly depend on high-level country implementation and management efficiency. Zambia, Botswana, Zimbabwe have committed themselves to ensure infrastructure development in dams and irrigation to foster food security (Scoones 2019, GoB 2021, GRZ 2021).

Some in-situ water conservation practices employed by smallholders include:

Crop Rotation and Diversification

Rotation of crops and introduction of drought-resistant varieties to mitigate risks associated with climate variability and optimize water usage.

Mulching

Application of organic mulches like straw or compost to the soil surface to reduce evaporation, maintain soil moisture, and suppress weed growth.

Conservation Tillage

Adoption of minimum tillage or zero-till practices to improve soil structure, reduce water runoff, and enhance water infiltration.

With all these, there is need for capacity-building especially at higher education level in smart farming practices (such as climate-smart agriculture) together with the integration and use of climate information systems that promote efficient natural resource management (Chinene et al. 2023, 2024).

Research Gaps and Future Directions

There is limited information on AI-driven irrigation optimisation algorithms, and there's need for development of portable user-friendly sensors, and advanced access to low-cost IoT technologies. Although remote sensing has improved soil property measurements, more research is needed for calibration, sensor fusion, artificial intelligence, validation, and machine learning applications to enhance accuracy and applicability. Thus, future directions for advancing smart irrigation technologies, should include the development of AI-driven optimization algorithms, sensor miniaturization, and low-cost IoT solutions. There are several opportunities for research, innovation, and collaboration to overcome adoption challenges and accelerate the uptake of smart irrigation practices, which must be backed with deliberate investment conditions fostering irrigation development, precision farming, crop monitoring, climate-resilience and power supply using renewable energy sources.

Conclusion

Water resources available for agriculture are rapidly depleting due to increasing population and effects of climate change. For this reason, innovative solutions are needed to use less water more whilst ensuring that crop water demands are met. One of such solution can be in the form of smart irrigation. Smart irrigation comprises digital systems such as internet of things (IoTs) which are efficient and cost-effective in the long run. Smart irrigation systems based on machine learning and artificial intelligence gather information on soil moisture and other parameters using several IoT devices and sensors. These sensors can detect the moisture content of the soil, temperature, humidity, and even water level. They can also capture information that is important in collaboration between farmers, researchers, extensionists, policymakers, and technology providers to create climate-resilient food systems and ensure food security and sustainable development.

Furthermore, the information contained in this chapter can assist researchers, policy-makers, and farmers, especially smallholders, to better understand irrigation techniques and provide an adequate approach as would be sufficient to carry out irrigation-related activities. The Chapter also reveals several concerns on data privacy and cost of irrigation as well as the use of IoT in marginalized communities. Therefore, future directions must be directed to solve such concerns, while improving use of these smart technologies in developing areas.

Acknowledgments

The authors acknowledge the Climate Smart Agriculture Youth Network Global (GCSAYN) for providing a platform for collaboration in the SADC region and beyond.

References

Abamagal, A.W. 2019. Impact of small-scale irrigation on smallholder farmer livelihood improvement in the case of Damota Gale district, Wolaita zone, SNNPR, Ethiopia. JEDS. 11(5): 23-32.

Alarcão, J.C. and D.N.C. Nuñez. 2024. The use of drones in agriculture. 4.0. Braz. J. Scie. 3(1): 1-13.

Anand, R., S.S. Nalla, S. Nasreen, M. Chowdhury, Madhusudan, B.S., A. Pandey et al. 2024. Fruit counting and analysis using artificial intelligence approaches. pp. 93-114. *In:* S.S. Chouhan, A. Saxena, U.P. Singh and S. Jain [eds.]. Artificial Intelligence Techniques in Smart Agriculture. Springer Nature, Singapore.

Andrews, E., B.A. Sanderson, Food Policy Alliance Cymru, A.S. Bellamy, S. Buckland-Jones, R. Evans et al. 2023. Putting food in the driver's seat: Aligning food-systems policy to advance sustainability, health, and security. Front. Sustain. Food Syst. 7: 1204194.

Angelakıs, A.N., D. Zaccaria, J. Krasilnikoffal, S. Salgot, S. Miguel, M. Bazza et al. 2020. Irrigation of world agricultural lands: Evolution through the millennia. Water. 12(5): 1285.

Bekele, R.D, N. Lefore, R.S. Meinzen-Dick and C. Ringler. 2018. Considering gender when promoting small-scale irrigation technologies: Guidance for inclusive irrigation interventions. IFPRI-REACH Project Note. Washington, DC.

Bhardwaj, S. and R. Sharma. 2024. Artificial intelligence in irrigation. Res. Jr. Agril. Sci. 15(3): 672-673.

Bunbury, J., J. Cooper, R. Hoath, S. Ikram, C. Johnston and T. Schneider. 2023. The Egyptian Nile: Human Transformation of an Ancient River. 43-47.

Chafuwa C. 2017. Priorities for irrigation investment in Malawi. Intl Food Policy Res Inst. Washington, D.C.

Chinene, V.R.N., E. Phiri, A.M. Mweetwa, J. Yengwe, M.C. Sinda, O.N. Lungu et al. 2023. The Nationwide assessment for the needs and pathways for integration of Climate Smart Agriculture (CSA) – Climate Information Services (CIS) into the Curricula of Zambian Higher Education Institutions. AICCRA Report. AICCRA, Lusaka.

Chinene, V.R.N., E. Phiri, A.M. Mweetwa, J. Yengwe, M.C. Sinda, O.N. Lungu et al. 2024. Nationwide assessment of needs and pathways for integration of CSA – CIS into the Curricula of Zambian Higher Education Institutions. AICCRA Validation report. AICCRA. Lusaka.

Chowdhury, M., E.V. Thomas, A. Jha, A. Kushwah, R. Kurmi, T.K. Khura et al. 2023. An automatic pressure control system for precise spray pattern analysis on spray patternator. Comput. Electron. Agric. 214: 108287.

Chowdhury, M., R. Anand, T. Dhar, R. Kurmi, R.K. Sahni and A. Kushwah. 2024. Digital insights into plant health: Exploring vegetation indices through computer vision. pp. 7-30. *In:* S.S. Chouhan, U.P. Singh and S. Jain [eds.]. Applications of Computer Vision and Drone Technology in Agriculture 4.0. Springer Nature, Singapore.

Chowdhury, M., T.K. Khura, R.A. Parray, P.K. Upadhyay, H.L. Kushwaha, A. Lama and A. Kushwah. 2025. Multi-vegetation indices based handheld device for precise nitrogen assessment and prescription in direct seeded rice. Comput. Electron. Agric. 229: 109886.

Durmuş, Y., G. Arslansoy and T.K. Gürbüz. 2024. Impact of smart irrigation systems and water management on climate change. Proc. of the BIO Web of Conf. 85: 01057.

Elufioye, O.A., C.U. Ike, O. Odeyemi, F.O. Usman and N.Z. Mhlongo. 2024. AI-driven predictive analytics in agricultural supply chains: A review: assessing the benefits and challenges of AI in forecasting demand and optimizing supply in agriculture. CSITRJ. 5(2): 473-497.

FAO. 2010. Climate-Smart Agriculture: Policies, Practices and Financing for Food. FAO. Rome.

FAO. 2013. Climate Smart Agriculture Sourcebook. Food and Agriculture Organization. Rome.

Feed the Future Innovation Lab for Small-Scale Irrigation. 2023a. Report on climate change, water resources, and irrigation sustainability. Brief Water_INTERACTIVE_022123.

Feed the Future Innovation Lab for Small-Scale Irrigation. 2023b. Report on improving smallholder women farmers' access to finance for small scale irrigation technologies. Brief Financing_INTERACTIVE_030723.

Gawande, V., D.R.K. Saikanth, B.S. Sumithra, S.A. Aravind, G.N. Swamy, M. Chowdhury and B.V. Singh. 2023. Potential of precision farming technologies for eco-friendly agriculture. Int. J. Plant Soil Sci. 35(19): 101-112.

Ghareeb, A.Y., S.K. Gharghan, A.H. Mutlag, and R. Nordin. 2023. Wireless sensor network-based artificial intelligent irrigation system: Challenges and limitations. J. Tech. 5(3): 26-31.

GoB. 2021. Botswana National Food Systems Transformation Pathways. Gov. Botswana.

GP. 2023a. Food Systems and Diets: A handbook of essential policies. Global Panel on Agriculture Food Systems for Nutrition. Global Panel. London.

GP. 2023b. Food Systems and Planetary Goals: Two Inseparable Policy Agendas. Global Panel. London.

GP. 2023c. Policy Tool Global Panel on Agriculture Food Systems for Nutrition. Global Panel. London.

Grabow, G., I. Ghali, R. Huffman, G. Milller, D. Bowman and A. Vasanth. 2013. Water application efficiency and adequacy of ET-based and soil-moisture-based irrigation controllers for Turfgrass irrigation. J. Irrig. Drain. 139: 113-123.

GRZ. 2021. Zambia National Food Systems Transformation Pathways. Gov. Zambia.

Guddaraddi, A., A. Singh, G. Amrutha, D.R.K. Saikanth, R. Kurmi, G. Singh et al. 2023. Sustainable biofuel production from agricultural residues an eco-friendly approach: A review. Int. J. Environ. Clim. Chang. 13(10): 2905-2914.

Hosaena, G. 2013. Tenure (in) security and agriculture investment of smallholder farmers in Mozambique. MozSSP Working Paper 5. Intl Food Policy Res Inst. Washington D.C.

Jha, A., M. Chowdhury and A.N. Satpute. 2023. Surface water quality forecasting using machine learning approach. pp. 293-315. *In:* C.B. Pande, M. Kumar and N.L. Kushwaha [eds.]. Surface and Groundwater Resources Development and Management in Semi-arid Region: Strategies and Solutions for Sustainable Water Management. Cham: Springer International Publishing.

Jha, A., S. Garg, S. Satpute, J.P. Singh, D. Singh, U.G Phutela et al. 2024. Isotherm and kinetic modelling for assessing the effectiveness of economical adsorbents in sewage water treatment. Indian J. Chem. Technol. 31: 438-450.

Kapari M., S. Hlophe-Ginindza, L. Nhamo and S. Mpandeli. 2023. Contribution of smallholder farmers to food security and opportunities for resilient farming systems. Front. Sustain. Food Syst. 7: 1149854.

Lungu, O.N., L.M. Chabala and C. Shepande. 2020. Satellite-based crop monitoring and yield estimation – A review. J. Agric. Sci. 13: 180-194.

Malkani, P., I. Mani, P.K. Sahoo, R.A. Parray, S.S. Swain, K.R. Asha et al. 2024. Development and evaluation of a sensor enabled smart herbicide applicator for inter row crops. Smart Agric. Technol. 9: 100649.

McCartney, M.J. Robyn and G. Lacombe. 2016. Building climate resilience through smart water and irrigation management systems. *In:* Experiences from Southeast Asia (pp. 41-65). UK: Routledge.

Mehta, P., S. Siebert, M. Kummu, Q. Deng, T. Ali, M. Marston et al. 2024. Half of twenty-first century global irrigation expansion has been in water-stressed regions. Nat. Water. 2: 254-261.

Mujere, A. and H. Muhoyi. 2022. Estimating soil moisture remote sensing in Zimbabwe: A review. *In:* Soil-Water, Agriculture, and Climate Change Exploring Linkages (pp. 79-91). Springer Int Publishing.

Mujere, N. 2016. Gendered vulnerability and adaptation to climate change. pp. 171-184 *In:* promoting climate change awareness through environmental education. L. Wilson and C. Stevenson. [eds.]. Hershley, PA: IGI Global, USA.

Muleke, A., M.T. Harrison, M. Yanotti and M. Battaglia. 2022. Yields gains of irrigated crops in Australia have stalled: The dire need for adaptation to increasingly volatile weather and market conditions. CRSUST. 4: 100192.

Mume, I.S., H.M. Jema and A.O. Muhammed. 2023. Impact of small-scale irrigation on the livelihood and resilience of smallholder farmers against climate change stresses: Evidence from Kersa district, eastern Oromia, Ethiopia, Heliyon. 9(8): 18976.

Neupane, J. and W. Guo. 2019. Agronomic basis and strategies for precision water management: A review. J. Agron. 9: 87.

Nkhata, R. 2014. Does irrigation have an impact on food security and poverty: Evidence from Bwanie Valley Irrigation Scheme in Malawi. MasSSP Working Paper 4. Intl Food Policy Res Inst. Washington, DC.

O'Shaughnessy, S.A., S.R. Evett, P.D. Colaizzi, M.A. Andrade, T.H. Marek, D.M. Heeren et al. 2019. Identifying advantages and disadvantages of variable rate irrigation – An updated review. Appl. Eng. Agric. 35(6): 837-852.

Obaideen, K., A.A Bashria, M.N.A. Yousef, C.T. Yong, M. Montaser, J. Hadi et al. 2022. An overview of smart irrigation systems using IoT. Energy Nexus. 7: 100124.

Obasi, S.N., V.A. Tenebe, C.C. Obasi, G.E. Jokthan, E.A. Adjei and E.R. Keyagha. 2024. Harnessing artificial intelligence for sustainable agriculture: A comprehensive review of African apllications in spatial analysis and precision agriculture. BDA. 6(1): 06-18.

OECD. 2021. Making better policies for food systems. OECD Publishing, Paris.

Ogidan, O., A. Onile and O. Adegboro. 2019. Smart irrigation system: A water management procedure. Agric. Scie. 10: 25-31.

Pandey, P.C. and M. Pandey. 2023. Highlighting the role of agriculture and geospatial technology in food security and sustainable development goals. J. Sustain. Dev. 31 (5): 3175-3195.

Patra, K., C.M. Parihar, H.S. Nayak, B. Rana, D.R. Sena, A. Anand et al. 2023. Appraisal of complementarity of subsurface drip fertigation and conservation agriculture for physiological performance and water economy of maize. Agric. Water Manag., 283: 108308.

Pradhan, N.C., M.A. Naik, M. Chowdhury, A. Kushwah, K.R. Asha, T. Dhar et al. 2024. Robotic seeding or sowing system in smart agriculture. pp. 495-520. *In:* K. Pandey, N.L. Kushwaha, C.B. Pande and K.G. Singh [eds.]. Artificial Intelligence and Smart Agriculture: Technology and Applications. Springer Nature, Singapore.

Rao, C.S., K. Kareemulla, P. Krishnan, G.R.K. Murphy, P. Ramesh, P.S. Ananthan et al. 2019. Agro-ecosystems based sustainability indicators for climate resilient agriculture in India: A conceptual framework. Ecol. Indic. 105: 621-633.

Satapathy, B.M.J., J. Anshuman and P. Kruparani. 2024. Drone technology in agriculture: An overview. Intl. J. E-Magazine Sci. 5: 159-164.

Satpute, A.N., K.P. Gavhane, S. Kaur, A. Jha, N.C. Pradhan and M. Chowdhury. 2024. Integration of AI and IoT in soilless cultivation to power sustainable agricultural revolution. pp. 387-411. *In:* K. Pandey, N.L. Kushwaha, C.B. Pande and K.G. Singh [eds.]. Artificial Intelligence and Smart Agriculture: Technology and Applications. Springer Nature, Singapore.

Scoones, I., F. Murimbarimba and J. Mahenehene. 2019. Irrigating Zimbabwe after land reform: The potential of farmer-led systems. WaA. 12(1): 88-106.

Singh, A., R. Mehrotra, V.D. Rajput, P. Dmitriev, A.K. Singh, P. Kumar and A.K. Singh. 2022. Geoinformatics, artificial intelligence, sensor technology, big data: Emerging modern tools for sustainable agriculture. *In:* Sustainable Agriculture Systems and Technologies (pp. 295-313).

Sojka, R.E., D.L. Bjorneberg and J.A. Entry. 2022. Irrigation: An Historical Perspective Encyclopedia of Soil Science. Marcel Dekker Inc. Idaho, USA.

Talaviya, T., D. Shah, N. Patel, H. Yagnik and M. Shah. 2020. Implementation of artificial intelligence in agriculture for optimization of irrigation and application of pesticides and herbicides. Artif. Intell. Agric. 4: 1016.

Theis, S., R.D. Bekele, N. Lefore, R.S. Meinzen-Dick and C. Ringler. 2018. Considering gender when promoting small-scale irrigation technologies: Guidance for inclusive irrigation interventions. Intl. Food Policy Res. Inst. Washington, DC.

UN. 2019. United Nations Collaborative Framework for food systems – A multi stakeholder pathway for sustainable food systems. UNEP. Rio, Brazil.

UNFSS. 2021. Policy Brief: Governance of food systems transformation. United Nations. New York, USA.

WAVES. 2014. Policy Brief: Irrigation and Water Resources in Botswana. Department of Water Affairs and the Centre for Applied Research, Botswana.

CHAPTER
10

Supply Chain Optimization in Agribusiness with Artificial Intelligence

Subrata Gorain[1*], Suman Dutta[2] and Sougata Roy[3]

[1] Department of Agricultural Economics, Palli Siksha Bhavana (Institute of Agriculture), Visva-Bharati (Central University), Sriniketan, West Bengal - 731236, India

[2] Ramakrishna Mission Vivekananda Educational and Research Institute, Kolkata, West Bengal - 700103, India

[3] Senior Research Associate (Agronomist), International Potato Centre (CGIAR-CIP), Peru

Introduction

The agricultural sector faces increasing pressure to address the food needs of a rapidly growing global population, which is projected to exceed 9 billion by 2050. This growing demand, coupled with limited arable land, unstable market conditions, and climate variability, underscores the importance of efficient supply chain management to ensure global food security. Agribusinesses are actively seeking innovative solutions to streamline their supply chains and navigate the complexities inherent in agriculture, such as seasonality, perishability, and logistical challenges. In this context, Artificial Intelligence (AI) has emerged as a transformative force, revolutionizing agricultural supply chains through advanced technologies like machine learning, predictive analytics, and optimization techniques. AI offers the potential to enhance efficiency, mitigate operational risks, and support sustainability in agribusinesses worldwide (FAO 2022; Kumar et al. 2023).

The rapid evolution of artificial intelligence (AI) has led to transformative changes across various sectors, including agribusiness. Today's agricultural supply chains must navigate unprecedented challenges, such as climate variability, resource scarcity, market volatility, and a growing global population projected to exceed 9 billion by 2050. These factors exert significant pressure on food systems to maintain efficiency, sustainability, and resilience. In this context, AI has become a strategic tool for optimizing complex agribusiness supply chains, enabling enhanced productivity, reduced waste, and streamlined decision-making (Islam

*Corresponding author: subrata.gorain@visva-bharati.ac.in

DOI: 10.1201/9781003536932-10

et al. 2023; Wang et al. 2024). AI applications in supply chain management (SCM) for agribusiness extend beyond traditional optimization techniques, incorporating advanced predictive analytics, real-time data processing, and machine learning (ML) to manage demand forecasting, inventory levels, and logistics with greater precision. For instance, recent studies highlight the success of AI-based demand forecasting in minimizing waste and adapting to fluctuating market needs (Pradhan et al. 2024). These models, powered by algorithms such as Long Short-Term Memory (LSTM) and Autoregressive Integrated Moving Average (ARIMA), provide agribusinesses with accurate demand predictions, thereby aligning production schedules with market demands and reducing losses from perishable goods (Zhao and Zheng 2023, Zhang Li and Liu 2023).

Furthermore, the integration of blockchain with AI has gained traction in enhancing traceability and transparency, essential in a field where consumer trust and compliance with food safety standards are critical. Blockchain's decentralized ledger, coupled with AI's predictive analytics, allows agribusinesses to maintain a transparent record of product origins, production processes, and supply chain transitions, reducing fraud and building consumer trust (Nie and Yang 2023; Islam et al. 2023). Blockchain smart contracts are also reshaping transaction automation in agribusiness, simplifying supply chain interactions between suppliers, farmers, and retailers (Dong and Gai 2023). Logistics optimization, a critical component of agribusiness SCM, is another area where AI-driven solutions have shown substantial progress. Modern AI algorithms for route optimization consider real-time factors like traffic, fuel costs, and weather, thereby reducing transportation costs and ensuring products reach markets swiftly. These advancements are particularly valuable for the agribusiness sector, where the perishability of goods necessitates rapid and efficient transport solutions. Studies demonstrate that AI-enabled logistics management systems are helping agribusinesses lower operational costs while enhancing their environmental sustainability by minimizing fuel consumption and emissions (Aljohani 2023, Goyal and Patel 2022).

While the benefits of AI in agribusiness supply chains are considerable, recent research also highlights some challenges. For instance, the adoption of AI technologies by smallholder farmers in developing regions remains limited due to high initial costs and limited infrastructure. However, initiatives for subsidizing AI adoption and improving digital infrastructure have been proposed to enable wider implementation and equitable access across regions (Antle and Capalbo 2023; Chavas and Roumasset 2024; Satpute et al. 2024). Moreover, as AI technologies advance, the focus has broadened to incorporate sustainable practices, such as reducing resource wastage through predictive models and promoting environmentally conscious agricultural techniques (Kong et al. 2022; Lamb et al. 2024). Recent advancements in AI-driven solutions for agribusiness supply chains present substantial opportunities for efficiency and sustainability. As these technologies continue to develop, the role of AI in overcoming logistical, operational, and environmental challenges within agribusiness will likely grow, paving the way for a more resilient and adaptive food system.

Overview of AI in Agribusiness Supply Chains

AI technologies have become integral to managing the complexity of modern agribusiness supply chains. These technologies are applied across various stages, from farm production to market distribution, aiming to optimize each phase and enhance decision-making capabilities. For instance, AI-powered models help farmers and agribusinesses analyze historical data, real-time environmental conditions, and market trends, leading to improved yield forecasts and greater supply chain agility (Smith and Li 2023). Optimization techniques, such as linear and non-linear programming, are utilized to streamline logistical operations, reduce transportation costs, and maximize resource efficiency (Perez and Singh 2022). As AI technologies advance, their role in agribusiness supply chains grows due to developments in the Internet of Things (IoT), big data analytics, and cloud computing, which facilitate real-time data collection and decision-making (World Bank 2023). Historically, AI adoption in agriculture lagged due to infrastructure limitations in rural areas. However, technological advancements in satellite imaging, IoT sensors, and mobile connectivity are closing this gap, enabling smallholder farmers in low-income countries to leverage AI. North America and Europe have led the way in implementing AI-driven technologies, with emerging markets in Asia and Latin America now starting to adopt AI to address region-specific challenges such as water scarcity, market volatility, and logistical inefficiencies (FAO 2022, World Bank 2023).

Importance of Supply Chain Optimization in Agribusiness

The complexity of agricultural supply chains, characterized by the perishability of goods, fluctuating demand, and seasonal production cycles, makes optimization crucial for agribusinesses. AI-driven optimization techniques facilitate better demand forecasting, inventory management, and real-time monitoring of supply chain processes. AI models using machine learning algorithms predict future demand patterns with notable accuracy by analysing historical sales data, consumer behaviour, and market trends (Perez and Singh 2022; Smith and Li 2023; Chowdhury et al. 2024). This precision helps align production and distribution strategies with market demands, reducing overproduction and minimizing waste while addressing the risks associated with unsold perishable goods (FAO 2022). Additionally, AI technologies, including deep learning and reinforcement learning, address logistical challenges within the agricultural supply chain, particularly in transportation route optimization. Intelligent routing algorithms consider real-time traffic conditions, weather forecasts, and fuel consumption to minimize delivery times and costs (Kumar and Patel 2023). AI-driven inventory optimization systems also ensure stock levels are appropriately managed, reducing shortages and overstock situations (World Bank 2023). These advancements enhance the resilience of agribusiness supply chains, especially when confronted with disruptions from climate change, geopolitical events, or economic shocks.

Relevance to Sustainability and Competitiveness

Integrating AI into agribusiness supply chains aligns with broader sustainability goals. AI optimizes resource usage and minimizes waste, contributing to global

Table 1: Global impact of AI in agribusiness supply chains

Sl. No.	Region	AI Applications	Challenges Addressed	Outcomes	References
1.	North America	Predictive analytics for demand forecasting, optimization of transportation routes, AI-driven crop management systems	Perishability, climate change, labour shortages	Improved efficiency, reduced operational costs, better resource allocation	FAO 2022
2.	Europe	Machine learning for inventory optimization, blockchain for supply chain traceability, AI-based sustainability models	Regulatory compliance, food safety, market volatility	Enhanced traceability, increased transparency, better compliance with sustainability standards	World Bank 2023
3.	Asia	AI-powered irrigation systems, drone-based data collection, real-time monitoring of crop health and soil conditions	Water scarcity, fluctuating demand, inadequate infrastructure	Improved resource utilization, higher crop yields, reduced water consumption	Kumar and Patel 2023
4.	Africa	AI in smallholder farm management, predictive maintenance of agricultural machinery, digital platforms for market access	Lack of data infrastructure, limited access to finance, low technology adoption	Increased productivity, better market access, higher profitability for smallholder farmers	FAO 2022
5.	Latin America	AI in logistics optimization, machine learning for pest control, AI-based market analysis	Supply chain disruptions due to climate change, price volatility	Reduced food waste, optimized transportation, faster response to market shifts	Smith and Li 2023
6.	Global Initiatives	AI-powered blockchain for food traceability, AI-driven sustainability assessment frameworks, integration of IoT sensors with AI for precision agriculture	Traceability, environmental degradation, supply chain inefficiencies	Increased food security, better alignment with sustainability goals, improved collaboration across supply chain stakeholders	Perez and Singh 2022

efforts to reduce environmental degradation and combat climate change (FAO 2022). AI-powered precision agriculture systems enable more efficient management of water use, fertilizer application, and pesticide distribution, reducing the environmental footprint of farming activities (Smith and Li 2023). In supply chains, AI enhances sustainability by optimizing logistics, cutting transportation-related emissions, and preventing food waste through precise demand forecasting and inventory management (Kumar and Patel 2023). Furthermore, AI integration is essential for maintaining competitiveness in the global agribusiness market. Companies using AI can swiftly adapt to market changes, reduce operational costs, and improve customer satisfaction. AI systems that automate tasks such as crop monitoring, equipment maintenance, and inventory management reduce labour costs and boost operational efficiency (Perez and Singh 2022). Moreover, AI's real-time analysis of market trends and consumer preferences allows businesses to adjust their strategies, maintaining a competitive edge in a rapidly evolving global marketplace (Smith and Li 2023). The adoption of AI for traceability and transparency, including through blockchain technology, also enhances the reputation and credibility of agribusinesses, supporting long-term competitiveness (Kumar and Patel 2023). The global impact of AI in agribusiness supply chains are shown in Table 1.

Objective of the Chapter

In the face of complex challenges such as climate variability, resource limitations, and market volatility, optimizing agribusiness supply chains has become crucial for sustaining agricultural productivity and meeting global food demand. The primary objective of this chapter is to investigate how artificial intelligence (AI) can serve as a transformative tool for agribusiness supply chains, driving improvements in efficiency, resilience, and environmental sustainability. Specifically, this chapter examines AI applications in areas such as demand forecasting, inventory management, logistics, and traceability through blockchain integration.

Through an in-depth exploration of recent innovations and case studies, this chapter seeks to provide practical insights into AI-driven solutions that can help agribusiness stakeholders enhance decision-making, reduce operational costs, and mitigate risks associated with perishability and fluctuating demand. Additionally, it aims to highlight the opportunities AI presents for addressing pressing issues such as food waste, resource efficiency and supply chain transparency. This chapter is intended to guide agribusiness professionals, researchers, and policymakers in understanding the strategic value of AI in building adaptive and sustainable food systems, ultimately contributing to more resilient supply chains that can support future agricultural needs.

Challenges in Agricultural Supply Chains

Agricultural supply chains face several challenges, including seasonality, perishability, and market volatility (Table 2). Each of these factors introduces significant complexity into supply chain management. Addressing these issues is crucial to improving efficiency, reducing costs, and minimizing losses. AI

Table 2: Key challenges in agricultural supply chains and AI-driven solutions

Sl. No.	Challenge	Impact	Mathematical Model	AI-driven Solution
1.	Seasonality	Fluctuations in production and demand	Linear programming for production optimization	AI-based demand forecasting (ARIMA, LSTM) for dynamic production scheduling
2.	Perishability	Rapid degradation of products	Exponential decay function $Q(t) = Q_0 e^{-\lambda t}$	Predictive analytics for optimal storage and reducing spoilage
3.	Market Volatility	Unpredictable changes in prices and demand	Dynamic pricing $R = \sum_{t=1}^{T} p_t \cdot D(p_t)$	Reinforcement learning and dynamic pricing models for real-time price adjustments
4.	Logistical Complexities	Delays and inefficiencies in transportation	Vehicle routing problem (VRP)	AI-based route optimization and intelligent routing algorithms to minimize costs and delivery times
5.	Inventory Management	Mismatches in supply and demand	Inventory cost *Minimization* $C = h.I(t) + s.Q(t)$	AI-driven inventory management for just-in-time (JIT) operations and reducing excess inventory

technologies, combined with advanced mathematical models, offer robust solutions for overcoming these challenges.

Seasonality and Production Cycle Optimization

Seasonality significantly impacts agricultural supply chains due to fluctuating production volumes and varying demand patterns across seasons. The irregular nature of agricultural production cycles, driven by environmental factors such as weather, creates unpredictability in the supply chain. AI models, such as time series forecasting (e.g., ARIMA and LSTM networks), can predict future demand based on historical data and external variables. This demand forecasting can be integrated into production planning using mathematical optimization models, such as linear programming, to allocate resources efficiently:

$$\textit{Maximize: } Z = \sum_{i=1}^{n} P_i \cdot x_i \qquad \text{(i)}$$

$$\textit{Subject to: } \sum_{i=1}^{n} a_{ij}\, x_i \leq b_j,\ \forall \in \{1,2,\ldots,m\} \qquad \text{(ii)}$$

Where, P_i is the profit per unit of product i, x_i is the quantity of product i to produce, a_{ij} represents the amount of resource j required for product i and b_j is the available quantity of resource *j*. AI-based demand forecasting D_t, which predicts future demand based on market and environmental factors, can be expressed as:

$$D_t = f(X_t, X_{t-1}, \ldots, X_{t-n}) + \epsilon_t \quad \text{(iii)}$$

Where X_t represents input variables (e.g., weather data, market trends), and ϵ_t is the error term. By leveraging AI technologies for real-time demand forecasting, agribusinesses can optimize resource allocation and production schedules.

Perishability and Inventory Management

Perishability is one of the most significant challenges in agricultural supply chains. The rapid degradation of perishable goods, such as fruits, vegetables, and dairy products, leads to high levels of waste if not properly managed. AI-driven inventory management systems use predictive analytics to optimize storage times, reducing spoilage. The decay rate of perishable goods can be modelled mathematically as an exponential decay function. If Q_0 represents the initial quantity of a product and λ is the decay rate, the quantity remaining after time t is given by:

$$Q(t) = Q_0 e^{-\lambda t} \quad \text{(iv)}$$

AI-based inventory management systems can minimize holding costs and spoilage by optimizing storage based on real-time data. The objective is to minimize the total cost C of holding and spoilage:

$$\textit{Minimize: } C = h.I(t) + s.Q_0 (1 - e^{-\lambda t}) \quad \text{(v)}$$

Where, h is the holding cost per unit, $I(t)$ is the inventory level at time t and s is the spoilage cost per unit. Through the integration of AI and predictive models, agribusinesses can better coordinate supply and demand, thereby minimizing spoilage and improving overall inventory management.

Market Volatility and Price Optimization

The agricultural market is highly volatile due to fluctuations in demand, prices, and supply, driven by unpredictable factors such as weather conditions, geopolitical events, and consumer preferences. AI-based dynamic pricing models allow businesses to adjust prices in real time in response to changing market conditions. These models use reinforcement learning and other AI techniques to find the optimal pricing strategy that maximizes revenue.

The dynamic pricing problem can be formulated as:

$$\textit{Maximize: } R = \sum_{t=1}^{T} p_t \,.\, D(p_t) \quad \text{(vi)}$$

Where, R is the total revenue over time, p_t is the price at time t and $D(p_t)$ is the demand as a function of price. AI models analyze historical sales data, market conditions, and external factors to predict optimal prices, ensuring that businesses can remain competitive while maximizing profitability.

The challenges posed by seasonality, perishability, and market volatility in agricultural supply chains are complex and multifaceted. However, AI technologies combined with mathematical optimization models offer robust solutions to mitigate these challenges. Through the use of predictive analytics, real-time data processing, and dynamic optimization, agribusinesses can enhance supply chain efficiency, reduce waste, and increase profitability. By integrating these AI-driven solutions, agribusinesses can remain resilient in the face of market fluctuations, unpredictable production cycles, and other external factors that influence the global agricultural supply chain.

Artificial Intelligence Technologies in Supply Chain Optimization

Optimizing supply chains is a complex endeavour that requires careful coordination of multiple variables, such as demand forecasting, inventory management, and transportation planning. The adoption of Artificial Intelligence (AI) technologies has significantly improved decision-making processes and overall productivity in supply chain operations (Belhadi et al. 2022). Various AI-powered optimization techniques, including heuristic and metaheuristic approaches as well as linear and non-linear programming, play a pivotal role in achieving these advancements.

Linear and Non-Linear Programming

Linear and non-linear programming are mathematical optimization methods that help identify the best possible solutions for systems with various constraints and objectives. Linear programming (LP) focuses on maximizing or minimizing a linear objective function while adhering to specific linear constraints (Kalwar et al. 2023). LP is commonly used in supply chain management to tackle issues such as optimizing inventory levels, reducing transportation costs, and improving manufacturing schedules. Its simplicity makes it computationally efficient, particularly for large-scale problems with linear relationships between constraints and variables (Taşkıner and Bilgen 2021). Non-linear programming (NLP), in contrast, handles problems involving non-linear relationships between variables (Ali et al. 2021). This approach is crucial in supply chain contexts where factors such as capacity limitations, demand fluctuations, and non-linear cost structures need to be addressed. NLP is especially useful in optimizing more intricate, real-world models, such as those aiming to minimize costs while accounting for variable demand and dynamic lead times (Jackson et al. 2024). In rapidly evolving environments, AI-powered solvers are increasingly being used to manage both LP and NLP problems, offering more accurate and adaptable solutions.

Despite their utility, LP and NLP face specific challenges in supply chain optimization. For LP, scalability can be an issue as larger supply chains with more variables and constraints become computationally intensive (Fan et al. 2024). Additionally, LP's reliance on deterministic inputs limits its ability to account for the uncertainties that often exist in real-world supply chains, such as fluctuating market conditions or economies of scale. NLP, while powerful, is computationally demanding, particularly for complex supply chain networks that involve non-linear equations (Zhou et al. 2022). Furthermore, creating accurate non-linear models requires deep mathematical expertise and an in-depth understanding of supply chain dynamics. NLP techniques are also prone to converging on local optima rather than identifying global solutions, posing additional challenges.

Heuristic and Metaheuristic Methods

Heuristic and metaheuristic approaches offer flexible, approximate solutions to supply chain challenges, particularly when dealing with large, complex systems where exact optimization techniques such as LP and NLP become computationally expensive (Gholian et al. 2023). Heuristic methods rely on rules-based solutions, providing quick but often sub-optimal answers. In inventory management, for instance, heuristics can involve setting reorder points based on historical demand data to balance holding costs and stockouts (Momenitabar et al. 2022).

Metaheuristic methods take heuristics a step further by incorporating intelligent search algorithms that explore a broader solution space. Techniques such as Particle Swarm Optimization (PSO), Simulated Annealing, and Genetic Algorithms (GA) are commonly employed in supply chain optimization (Chan et al. 2020; Vishnu et al. 2021; Weinberg et al. 2023). These methods aim to avoid local optima by balancing exploration and exploitation, making them particularly effective for solving complex supply chain problems such as network design, facility placement, and vehicle routing. AI-enhanced metaheuristics are highly adaptable, making them suitable for dynamic environments that require real-time adjustments. Despite their advantages, heuristic and metaheuristic methods are not without their drawbacks. Heuristic techniques, while fast, often fail to provide optimal solutions, especially for complex problems (Gholian et al. 2023; Momenitabar et al. 2022). They also tend to rely heavily on historical data, limiting their ability to adapt to rapidly changing supply chain conditions. Metaheuristics, like PSO and GA, can also be computationally intensive, particularly when solving large-scale, multi-dimensional problems (Chan et al. 2020; Vishnu et al. 2021). Furthermore, the effectiveness of metaheuristics is sensitive to parameter settings, such as population size or mutation rates, which can significantly influence performance. While these methods offer approximate solutions, they may not always provide the level of accuracy required for mission-critical decisions.

Table 3 presents a comprehensive list of AI algorithms commonly employed in agribusiness supply chains, detailing their specific applications and associated benefits in areas such as demand forecasting, inventory management, and logistics optimization.

As outlined in Table 3, various AI algorithms offer targeted solutions across the supply chain, from optimizing transportation routes to predicting yield accurately. Each algorithm contributes distinct advantages, depending on the specific requirements of agribusiness applications.

Table 3: AI Algorithms and their applications in agribusiness supply chains

Sl. No.	Algorithm	Specific Applications	Description of Application	References
1.	Linear Programming (LP)	Resource allocation, transportation cost minimization	Optimizes supply chain logistics by minimizing costs while meeting resource constraints. Widely used for inventory and transportation management.	Kumar and Patel 2023; Perez and Singh 2022.
2.	Long Short-Term Memory (LSTM)	Demand forecasting, yield prediction	Provides accurate forecasts by learning time-based dependencies in data, useful for demand fluctuations and crop yield predictions.	Zhao and Zheng 2023; Zilberman et al. 2022.
3.	Reinforcement Learning	Dynamic pricing, adaptive logistics	Optimizes pricing and routing in real time, adjusting based on market conditions and minimizing perishable product losses.	Mohamed-Iliasse et al. 2020; Tirkolaee et al. 2021.
4.	Support Vector Machines (SVM)	Classification of crop health, market trend analysis	Analyzes patterns to classify crops based on health status or market trends, improving crop management and market strategies.	Castaldo and Del Pino 2022; Udokwu et al. 2023
5.	Genetic Algorithms (GA)	Route optimization, facility location planning	Solves complex logistics problems by generating optimal solutions for transportation routes and facility placements.	Chan et al. 2020; Gholian et al. 2023.
6.	K-means Clustering	Crop categorization, customer segmentation	Groups data into clusters for targeted marketing, customer segmentation, and crop classification based on various characteristics.	Kong et al. 2022; Smith and Li 2023
7.	Random Forest	Risk assessment, yield prediction	Enhances yield forecasting and assesses risks by creating decision trees based on historical agricultural data.	Schroeder and Lodemann 2021; Liu et al. 2023

8.	Simulated Annealing	Supply chain network design, resource optimization	Refines network design by minimizing costs associated with location and distribution within supply chains.	Vishnu et al. 2021; Weinberg et al. 2023.
9.	Neural Networks (NN)	Crop yield forecasting, disease detection	Recognizes patterns in complex agricultural data for forecasting and early disease detection.	Tirkolaee et al. 2021; Pandey et al. 2023
10.	Particle Swarm Optimization (PSO)	Inventory control, transportation scheduling	Optimizes inventory levels and reduces delivery times by determining the most efficient distribution routes.	Momenitabar et al. 2022; Aljohani 2023

Machine Learning Applications in Agribusiness and Supply Chain Management

Machine learning (ML) has revolutionized supply chain management, particularly in agribusiness, by improving decision-making, operational efficiency, and predictive capabilities (Tirkolaee et al. 2021). ML algorithms, including supervised, unsupervised, and reinforcement learning, offer distinct advantages depending on the task at hand. Supervised learning, for instance, is ideal for demand forecasting and inventory optimization. By training models on labeled data, businesses can anticipate future demand and adjust their supply chain operations accordingly (Mohamed et al. 2020). Unsupervised learning, which does not rely on pre-labeled data, is particularly useful for identifying patterns and anomalies in supply chain processes. In agriculture, unsupervised learning can detect unexpected fluctuations in crop yields or delivery schedules, helping to identify inefficiencies (Schroeder and Lodemann 2021). Reinforcement learning is well-suited for dynamic decision-making environments, where systems learn optimal policies through trial and error. This is particularly useful for real-time logistics optimization, autonomous agricultural machinery management, and adaptive pricing strategies.

Real-Time Data Processing and Decision-Making

Real-time data processing is crucial for effective supply chain management, particularly in agriculture where conditions can change rapidly (Brintrup et al. 2022). The ability to process vast amounts of real-time data ranging from weather forecasts and satellite imagery to IoT device inputs allows businesses to make more informed decisions. For example, real-time data from crop monitoring systems can optimize irrigation schedules, pesticide applications, and harvesting times, ensuring the efficient use of resources while minimizing waste. In logistics, real-time cargo tracking combined with machine learning algorithms can optimize delivery routes, reduce fuel costs, and prevent delays caused by unforeseen circumstances like traffic or weather disruptions (Aljohani 2023).

Predictive analytics, a core component of ML, allows agribusinesses to anticipate future trends and take proactive measures. By leveraging algorithms such as support vector machines, decision trees, and neural networks, businesses can forecast demand, crop yields, and market prices with greater accuracy (Udokwu et al. 2023). These predictive models enable better planning, resource allocation, and production optimization. For instance, predictive models trained on historical weather data can help farmers plan planting and harvesting schedules more effectively. Scenario analysis, another important ML-driven tool, allows businesses to explore "what-if" scenarios and assess the impact of various factors – such as shifts in demand or regulatory changes – on supply chain performance (Pandey et al. 2023). While machine learning offers tremendous potential, it also comes with challenges. Supervised learning, for example, relies heavily on labeled data, which can be costly or difficult to obtain in agricultural contexts (Brintrup et al. 2022). Unsupervised learning may struggle to interpret complex or noisy data, leading to inaccurate conclusions. Additionally, reinforcement learning, while powerful for dynamic environments, can be slow and inefficient in rapidly changing agricultural settings. Integrating large datasets from IoT sensors, satellite imagery, and weather forecasts also requires significant computational resources, which may not be readily available in all agricultural contexts (Pandey et al. 2023). In conclusion, while AI technologies hold great promise for optimizing supply chains in agribusiness, their successful implementation depends on overcoming key challenges related to data availability, computational power, and model accuracy.

Artificial Intelligence-Driven Solutions for Supply Chain Management

Artificial intelligence (AI) has become a key driver for optimizing supply chains in agriculture, where unpredictable factors like seasonality, fluctuating market conditions, and environmental challenges make operations highly complex. AI offers solutions such as predictive maintenance, intelligent logistics systems, and enhanced demand forecasting, enabling agribusinesses to make better use of their resources, reduce operational costs, and increase productivity.

Demand Forecasting Models

In agricultural supply chain management, demand forecasting plays a crucial role, helping to reduce waste, avoid stock shortages, and optimize production schedules (Abolghasemi et al. 2020). AI-driven demand forecasting models outperform traditional methods due to their ability to handle complex datasets and learn from them continuously to improve accuracy. These AI models analyze a wide range of data, including historical sales figures and market trends, to generate reliable demand predictions (Feizabadi 2022). Historical sales data sheds light on past consumer behaviour, including seasonal trends and sales patterns (Zilberman et al. 2022). In contrast, market trends capture larger shifts in consumer preferences, economic shifts, and industry developments. External factors such as weather, commodity prices, and governmental policies significantly influence agricultural demand. AI algorithms, for instance, can predict crop yields based on weather forecasts, thereby aiding in supply planning and inventory management. AI can also analyze large volumes of unstructured data from sources like social media, market research, and news reports to predict changes in consumer preferences.

The primary AI techniques used in demand forecasting include time series analysis and regression models. Time series analysis is effective in detecting trends and seasonal variations in historical data, allowing businesses to forecast future demand and adjust production and inventory accordingly. Advanced time series models such as LSTM (Long Short-Term Memory) networks (Pacella and Papadia 2021) and ARIMA (Autoregressive Integrated Moving Average) (Waqas et al. 2023) are particularly well-suited to industries like agribusiness, where demand can fluctuate greatly. Meanwhile, regression models analyze the relationships between multiple variables, helping businesses understand how factors like weather, pricing, and customer behaviour influence demand. These models provide actionable insights for better decision-making by quantifying these relationships.

Predictive Maintenance Systems

In addition to demand forecasting, AI plays a critical role in predictive maintenance, where it is used to analyze operational data from machinery, such as temperature, pressure, and vibrations (Lorenc et al. 2021). Machine learning algorithms detect patterns that indicate potential equipment failure or a drop in performance. For example, in large-scale farming operations that rely on tractors, harvesters, and irrigation systems, AI can predict when a machine is likely to malfunction, allowing managers to address the issue before it causes costly downtime (Laskar et al. 2024).

AI-driven predictive maintenance systems are more efficient than traditional maintenance approaches, which often rely on fixed schedules. Instead, AI adjusts maintenance schedules dynamically based on the actual condition and usage patterns of the equipment, reducing unnecessary repairs and maximizing the utilization of resources. This helps ensure smooth supply chain operations by minimizing the risk of unexpected breakdowns (Akintokunbo and Arimie 2021). For instance, if AI predicts that a harvesting machine will fail during a peak season, preventive maintenance can be scheduled during downtime, thus avoiding disruptions in production and delivery schedules.

Route Optimization for Transportation

Efficient transportation is essential in the agricultural sector, particularly given the perishable nature of many products (Muzylyov et al. 2020). AI-driven routing algorithms help optimize delivery routes by analyzing real-time traffic conditions, road quality, and weather data. By selecting the quickest and most fuel-efficient routes, these algorithms reduce transportation costs, shorten travel times, and ensure that products reach their destinations in optimal condition. This is particularly important for delivering fresh produce, where quick transportation is vital to preserving quality (Bhatia and Bhat 2020).

AI's benefits in route optimization extend beyond selecting the best route. These systems can also reduce operational costs by analyzing factors like fuel prices, vehicle availability, and load capacity. For example, AI might recommend grouping shipments or adjusting delivery routes to avoid congestion, thus saving time and money. AI systems can even adapt to unforeseen disruptions like road closures or accidents by providing real-time updates and alternative routes, minimizing delivery delays. In highly competitive markets, where timely deliveries are critical, AI-enabled route optimization helps agribusinesses stay efficient and cost-effective. AI-driven solutions are transforming supply chain management in agriculture by improving demand forecasting, predictive maintenance, and route optimization, ensuring that agribusinesses can stay competitive while maximizing efficiency.

Integration of Emerging Technologies in Supply Chain Management

The agricultural sector is under significant strain to adjust to the demands of a growing global population and the impacts of climate change. In this context, Artificial Intelligence (AI) has become an essential tool for refining agricultural supply chains. According to Leong et al. (2023), the integration of AI with the Internet of Things (IoT) known as AIoT is transforming the sector by enhancing resource management, streamlining production processes, and reducing reliance on manual labour. AIoT goes beyond simple automation, facilitating advancements in precision agriculture, predictive analytics, and decision support systems. Singh (2023) further underscores the role of AI in driving digital transformation within supply chains, particularly in demand forecasting, real-time logistics, and inventory management. Despite challenges related to data privacy and potential workforce impacts, AI's capacity to lower costs, boost efficiency, and fortify supply chain resilience makes it a crucial element for future growth in agriculture.

Blockchain Technology in Supply Chain Management

Originally developed for cryptocurrencies like Bitcoin, blockchain technology has now been adopted across various sectors, including supply chain management. Blockchain creates a decentralized, immutable, and time-stamped ledger that records transactions across a distributed network. Utilizing encryption, peer-to-peer networks, and consensus mechanisms, blockchain ensures data integrity and eliminates the need for intermediaries (Berdik et al. 2021). Although initially associated with financial applications, blockchain is now recognized for its potential in healthcare, energy, smart cities, and notably, supply chain management (Wüst and Gervais 2018).

In agriculture, blockchain is revolutionizing transparency and traceability within supply chains. By implementing a decentralized ledger, blockchain enables meticulous tracking of products from the farm to the consumer, enhancing food safety, verifying organic or sustainably produced items, and fostering consumer trust (Storer et al. 2014). Furthermore, blockchain-powered smart contracts can automate transactions between farmers, suppliers, and retailers, streamlining processes and bolstering supply chain reliability. The benefits of integrating blockchain into agricultural supply chains include enhanced visibility and traceability. Pandey et al. (2023) highlight that blockchain provides real-time visibility into every transaction from production to delivery, crucial for identifying inefficiencies and maintaining transparency. This capability helps ensure the integrity of the supply chain. Additionally, as noted by Patil et al. (2023), blockchain's ability to document each step of the supply chain creates a permanent and tamper-proof record. This is especially important in agriculture for verifying product origin and compliance with regulatory standards. Blockchain's transparency reduces fraud risk, builds consumer trust, and strengthens overall supply chain reliability.

AI in Demand Forecasting and Consumer Behaviour

AI has significantly enhanced demand forecasting accuracy, thereby improving agricultural supply chain management. Advanced AI models can analyze extensive data sets to provide precise forecasts, which are essential in the unpredictable agricultural sector. For instance, Pal (2023) illustrates how AI can enhance Just-In-Time (JIT) inventory systems by integrating neural networks with traditional statistical methods, leading to more accurate demand forecasts and fewer inventory discrepancies. This is particularly beneficial in agriculture for managing perishable goods and minimizing waste. Additionally, Terrada et al. (2022) explore deep learning techniques like ARIMA and LSTM models, which use historical data to better predict consumer behaviour and trends, optimizing the agricultural supply chain.

AI's application in agricultural forecasting extends beyond consumer demand. Aderibigbe et al. (2023) discusses how machine learning (ML) techniques, including deep learning and big data analytics, offer superior performance in forecasting energy use, with potential applications in agricultural forecasting. These AI-driven predictive analytics enhance resource use and adapt to market shifts, improving supply chain efficiency and resilience. Although AI adoption in agriculture is still developing compared to other industries, Rattan (2023) acknowledges its vast potential. Jonnalagadda (2021) demonstrates AI's role in predicting crop yields through geospatial mapping for soybean farming, highlighting AI's impact on precision agriculture. AI's integration into demand forecasting and consumer behaviour analysis brings substantial benefits, improving market prediction accuracy, supporting resource optimization, and reducing waste, leading to a more sustainable and efficient agricultural sector. As AI technologies advance, their influence on agriculture is set to grow, driving further innovations and improvements in the industry.

Socio-Economic and Environmental Impacts

The integration of artificial intelligence (AI) in agriculture has been rapidly advancing, fuelled by progress in machine learning (ML), deep learning (DL), computer vision, and robotics. These technologies significantly improve traditional agricultural practices by analysing extensive datasets to identify patterns, predict outcomes, and automate processes. For example, ML algorithms process data from satellite imagery, weather forecasts, and soil sensors to provide actionable insights for crop management (Klerkx et al. 2019).

Improved Resource Utilization

AI-driven models significantly enhance crop management by precisely forecasting weather, predicting pest infestations, and optimizing irrigation schedules, leading to reduced waste and increased yields. Deep learning (DL) algorithms, skilled at recognizing intricate patterns, aid in detecting plant diseases from images, enabling early intervention and minimizing crop losses. In livestock management, AI leverages sensors and computer vision to monitor animal health and behaviour, thereby improving welfare and productivity (Maohua 2021). The integration of AI with Internet of Things (IoT) devices in precision farming optimizes resource use and reduces environmental impacts by facilitating site-specific crop management. AI technologies also improve the efficiency of agricultural machinery; autonomous tractors and drones perform planting and harvesting tasks with enhanced precision. Additionally, AI-powered robots are employed for labour-intensive tasks such as weeding and harvesting, which increases efficiency and reduces labour costs. Furthermore, AI contributes to better supply chain management by analysing market trends and optimizing distribution, ensuring the timely delivery of fresh produce. Overall, AI fosters a comprehensive approach to farm management, offering real-time data and informed decision-making that boost productivity and sustainability.

Environmental Advantages of AI Technology

AI technology offers notable environmental advantages in agriculture for supply chain management. In the United States, a precision farming project showcased AI's capability to optimize resource use and boost crop yields. Collaborating with a major agricultural research university, the project integrated satellite imagery, soil sensors, and machine learning algorithms to provide actionable insights. AI-driven precision agriculture significantly reduced resource consumption—water usage decreased by approximately 43%, fertilizer application dropped by 40%, and pesticide usage was cut by 60% (Ahmed and Khan 2024). Simultaneously, crop yields improved by 25%, demonstrating AI's effectiveness in enhancing both productivity and sustainability. These improvements not only reduced costs but also minimized environmental impacts, making AI technology a cost-effective investment.

In Australia, an AI-driven livestock management system was implemented on a large dairy farm. The system, which combined computer vision and machine learning, monitored animal health and behaviour through real-time image analysis. It detected health issues early, leading to a 30% reduction in veterinary costs and a 20% decrease in mortality rates (Akhigbe et al. 2021). Additionally, the AI optimized feeding schedules, improving feed efficiency by 15% and enhancing milk production. Overall, the farm experienced a 25% increase in net income within the first year of AI adoption (Lamb et al. 2020). These case studies illustrate AI's potential to enhance both environmental sustainability and economic performance in agriculture. Maintaining soil health is essential for sustainable farming, and AI-driven soil analysis tools are playing a crucial role in this aspect. These tools assess key soil parameters such as nutrient balance, pH levels, and moisture content, providing data-driven recommendations for soil amendments and crop rotation strategies. AI-based monitoring has proven highly effective, reducing nutrient imbalances by 60%, controlling soil erosion by an equivalent percentage, and significantly increasing organic matter content by 75% (Ahmed and Khan 2024). This proactive approach ensures long-term soil fertility, prevents land degradation, and supports higher crop productivity, reinforcing AI's role in sustainable agricultural practices.

Economic Advantages of AI

The AI platform employed machine learning to predict demand and manage inventory by analysing historical sales data, weather patterns, and market trends for optimal harvest and transport timings. It also used IoT devices to monitor produce quality during transportation and storage, allowing real-time logistics adjustments. By optimizing resource allocation and reducing inefficiencies, AI has helped lower production costs significantly. Compared to traditional farming methods, AI-assisted techniques have reduced production expenses by 40%, with costs decreasing from $1,500 per hectare to $900 per hectare. Additionally, the integration of AI in farm management has resulted in a substantial increase in crop yield, with output rising from 5 tons per hectare under conventional practices to 7.5 tons per hectare, marking a 50% improvement. Furthermore, AI adoption has contributed to higher profitability for farmers. Through better market access, reduced intermediary costs, and data-driven decision-making, profit margins have increased considerably. Farmers using AI-enhanced strategies have reported profit margins of 35%, compared to 20% under traditional approaches, reflecting a 75% overall improvement (Ahmed and Khan 2024). These economic gains highlight AI's potential to transform agricultural productivity while ensuring financial sustainability for farmers.

Case Studies and Practical Applications in Agribusiness

Agribusiness supply chains are becoming more complex due to factors such as globalization, climate change, and market fluctuations. To address these challenges, Artificial Intelligence (AI) has emerged as a crucial tool for optimizing processes and improving decision-making in agriculture. The following case studies showcase the effective application of AI in agribusiness and highlight key lessons learned from these implementations.

Case Study: AI for Crop Yield Optimization in India

India, a country where agriculture plays a vital role in the economy, has implemented AI to boost crop yields. One notable initiative is an AI-based sowing application launched by the Indian government in collaboration with the International Crops Research Institute for the Semi-Arid Tropics (ICRISAT). This application uses machine learning algorithms to analyze weather data and soil health, providing farmers with optimal sowing dates. The project, implemented in Andhra Pradesh, led to a 30% increase in crop yields for groundnut and maize (Jha 2019).

Case Study: AI for Supply Chain Optimization at Walmart

Walmart has integrated AI with blockchain technology to streamline its food supply chain. This system enhances traceability and transparency, tracking food products from farm to retail shelves. By reducing the time to trace food origins from days to mere seconds, Walmart ensures food safety and minimizes waste (Tian 2017; Kamilaris et al. 2017). Additionally, AI-powered predictive analytics help Walmart manage inventory more efficiently, reducing overstock and food wastage by 15%.

Case Study: Smart Farming and IoT Integration in the Netherlands

The Netherlands, a leader in agricultural innovation, has adopted AI-powered precision farming. In collaboration with Wageningen University and Research (WUR), Dutch farmers use AI integrated with Internet of Things (IoT) sensors to monitor soil moisture, plant health, and weather conditions in real time. This system has resulted in a 20% increase in crop yields for farmers growing high-value crops like tomatoes and peppers (Singh et al. 2022).

Lessons Learned from Case Studies

These case studies reveal several key insights for the successful integration of AI in agribusiness:

Data Quality and Integration

High-quality, real-time data is essential for AI models to generate accurate insights. Collaborative data-sharing platforms between farmers and institutions play a critical role in ensuring AI systems work effectively (Antle and Capalbo, 2018).

Scalability

AI solutions must be scalable to accommodate both small and large farms. Scalable systems can address diverse agricultural challenges across different regions and operational scales (Chavas and Roumasset 2021).

Cross-Sector Collaboration

Strong partnerships between governments, technology providers, and agribusinesses are crucial. Such collaborations provide the necessary technical and financial support for large-scale AI deployment (Tsolakis et al. 2014).

These case studies illustrate how AI can significantly enhance efficiency, transparency, and productivity in agribusiness, paving the way for more sustainable agricultural practices.

Policy Implications and Future Directions

AI integration in agribusiness offers tremendous potential, but its successful implementation relies on strategic policy support, international collaboration, and

Table 4: Policy Recommendations for AI integration in agribusiness

Sl. No.	Policy Area	Recommendation	Expected Impact
1.	Data Governance	Establish data privacy frameworks and encourage open data-sharing platforms	Improved trust, more accurate AI insights
2.	Financial Subsidies	Provide financial aid for smallholder farmers adopting AI technologies	Increased AI adoption, especially among smallholders
3.	Education and Capacity-Building	Invest in training programs for farmers on digital and AI technologies	Enhanced skills, greater technology utilization
4.	Climate-Resilient Agriculture	Focus AI research on climate adaptation strategies	Improved farm resilience to climate change
5.	Blockchain and AI Integration	Promote blockchain integration for transparency and traceability in supply chains	Enhanced food safety, reduced fraud

infrastructural investments. This section explores policy implications and outlines future directions to promote AI adoption in the agricultural sector (Table 4).

Policy Interventions for AI Integration in Agribusiness

Data Governance and Privacy

The effectiveness of AI in agribusiness is heavily dependent on the availability and sharing of data. Governments must implement robust policies that address data privacy and security while promoting open data sharing among farmers, researchers, and technology providers (Lobry and Weitz 2019). A framework for the ownership and ethical use of data generated by AI systems is crucial for encouraging farmer participation and trust.

Subsidies for AI Adoption

Governments in both developed and developing countries should consider providing financial subsidies for farmers adopting AI technologies. The high upfront costs of AI systems can be a significant barrier, particularly for smallholder farmers. In countries like India, pilot programs offering subsidies for AI-based sowing systems have shown considerable success (Jha 2019). Extending such initiatives to more regions can foster broader AI adoption.

Capacity Building and Education

To facilitate the widespread adoption of AI, governments must invest in education and capacity-building programs for farmers and agribusiness workers. Training programs that teach the use of AI tools and digital technologies in farming will enhance the skill sets required to use these systems effectively (Paltrinieri et al. 2022). Additionally, promoting AI research and development through public-private partnerships can further boost innovation in agribusiness.

Future Directions for AI in Agribusiness

AI for Climate-Resilient Agriculture

One of the most promising areas for AI in agribusiness is in addressing the challenges of climate change. AI can help predict weather patterns, monitor soil health, and optimize irrigation systems, making farms more resilient to extreme climate conditions (Antle and Capalbo 2023). Research initiatives should focus on expanding the use of AI in climate adaptation strategies, particularly for vulnerable regions like sub-Saharan Africa and South Asia.

AI and Sustainable Supply Chains

AI can play a critical role in promoting sustainability across the agribusiness supply chain. By optimizing logistics, reducing waste, and improving resource utilization, AI technologies can help agribusinesses meet the growing demand for sustainable and environmentally responsible practices (Chavas and Roumasset 2021). Future research should explore the potential of AI to create fully sustainable, zero-waste supply chains in agribusiness.

Blockchain and AI Integration

The combination of blockchain and AI holds significant potential for improving transparency and traceability in the global food supply chain. Blockchain technology, when combined with AI-based predictive analytics, can provide real-time insights into the origin, quality, and safety of agricultural products, thus enhancing food security (Tian 2017). Future efforts should explore how these technologies can be further integrated to address food fraud and safety concerns at scale.

Conclusion

This chapter has explored the transformative impact of artificial intelligence (AI) on supply chain optimization within the agribusiness sector, focusing on AI-driven advancements in demand forecasting, inventory management, logistics, and traceability through blockchain integration. By examining key algorithms such as linear programming, reinforcement learning, and neural networks chapter illustrated how AI technologies enhance efficiency, reduce costs, and mitigate risks across various stages of the supply chain. The integration of blockchain further amplifies AI's benefits, promoting transparency and traceability essential to meeting consumer and regulatory demands. The integration of Artificial Intelligence (AI) into agribusiness supply chains marks a transformative shift in the sector, with far-reaching implications for efficiency, sustainability, and global competitiveness (Anand et al. 2024). AI's capacity to analyze vast amounts of real-time data, forecast demand, optimize logistics, and manage resources more effectively has fundamentally reshaped how agricultural supply chains operate. By reducing operational inefficiencies, minimizing waste, and enhancing decision-making, AI enables agribusinesses to navigate the inherent complexities of agriculture such as seasonality, perishability, and market volatility with greater precision. The ability to integrate AI with emerging technologies, such as blockchain, has further amplified its impact by improving transparency, traceability, and trust across supply chains. This digital transformation ensures not only higher productivity but also strengthens food security and reduces environmental impacts through smarter resource management (Chowdhury et al. 2023). Importantly, AI's application is not confined to large-scale operations but holds immense potential for smallholder farmers, particularly in developing regions, where access to real-time information can empower local communities and foster more inclusive agricultural systems. Looking ahead, the future of AI in agribusiness is promising, as ongoing advancements in machine learning, predictive analytics, and sensor technologies will continue to unlock new possibilities for innovation. However, to fully realize the potential of AI, there must be a concerted effort by governments, businesses, and the scientific community to promote its adoption, ensure equitable access, and support sustainable practices. In doing so, AI will play a central role in building a more resilient, efficient, and sustainable global agricultural system, capable of meeting the growing demands of the 21st century. Future research and development should prioritize AI models that can adapt dynamically to the evolving conditions in agriculture, including climate variability and shifting consumer demands. Advancements in integrating AI with the Internet of Things (IoT) and robotics will play a crucial role in achieving a sustainable and adaptable agribusiness supply chain. However, realizing the full potential of AI requires collaborative efforts among governments, businesses, and researchers to promote equitable access, ensure ethical usage, and support sustainable practices across the sector. Through these advancements, AI will be instrumental in building a robust, inclusive and sustainable agricultural system that can meet the demands of a growing global population in the years to come.

References

Abolghasemi, M., E. Beh, G. Tarr and R. Gerlach. 2020. Demand forecasting in supply chain: The impact of demand volatility in the presence of promotion. Comput. Ind. Eng. 142: 106380.

Aderibigbe, A.O., E.C. Ani, P.E. Ohenhen, N.C. Ohalete and D.O. Daraojimba. 2023. Enhancing energy efficiency with AI: A review of machine learning models in electricity demand forecasting. Eng. Sci. Technol. J. 4(6): 341-356.

Ahmed, S. and A. Khan. 2024. The impact of artificial intelligence on sustainable agriculture: Developing advanced computing models for environmental and economic benefits. J. Adv. Comput. Syst. 4(6): 1-8.

Akhigbe, B.I., K. Munir, O. Akinade, L. Akanbi and L.O. Oyedele. 2021. IoT technologies for livestock management: A review of present status, opportunities, and future trends. Big Data Cogn. Comput. 5(1): 10.

Akintokunbo, O.O. and B.E. Arimie. 2021. Supply chain management: A game changer in the oil and gas industry in Nigeria: A review of literature. Int. J. Supply Chain Logist. 5(3): 54-68.

Ali, S.S., H. Barman, R. Kaur, H. Tomaskova and S.K. Roy. 2021. Multi-product multi-echelon measurements of perishable supply chain: Fuzzy non-linear programming approach. Mathematics. 9(17): 2093.

Aljohani, A. 2023. Predictive analytics and machine learning for real-time supply chain risk mitigation and agility. Sustainability. 15(20): 15088.

Anand, R., S.S. Nalla, S. Nasreen, M. Chowdhury, Madhuban, B.S., A. Pandey et al. 2024. Fruit counting and analysis using artificial intelligence approaches. pp. 93-114. *In:* S.S. Chouhan, A. Saxena, U.P. Singh and S. Jain [eds.]. Artificial Intelligence Techniques in Smart Agriculture. Springer Nature, Singapore.

Antle, J.M. and S.M. Capalbo. 2023. Equity and access in AI adoption for smallholder agribusiness. Agric. Policy Dev. 16(1): 101-115.

Belhadi, A., S. Kamble, S. Fosso Wamba and M.M. Queiroz. 2022. Building supply-chain resilience: An artificial intelligence-based technique to mitigate the impact of COVID-19 on supply chains. J. Bus. Res. 147: 620-629.

Berdik, D., S. Otoum, N. Schmidt, D. Porter and Y. Jararweh. 2021. A survey on blockchain for information systems management and security. Inf. Process. Manag. 58(1): 102397.

Bhatia, M. and G.M.J. Bhat. 2020. Agriculture supply chain management – An operational perspective. Braz. J. Oper. Prod. Manag. 17(4): 1-18.

Castaldo, S. and J. Del Pino. 2022. Data-driven insights on agribusiness efficiency: A review of AI-driven analytics applications. Agribus. Rev. 35(3): 321-338.

Chan, F.T.S., M.K. Tiwari and D. Vishnu. 2020. Applications of metaheuristic algorithms for enhancing agri-supply chain network design. J. Ind. Eng. Manag. 13(1): 67-78.

Chavas, J.P. and J.A. Roumasset. 2024. Strategic innovation for resilient food supply chains: AI and sustainability in agriculture. Food Policy J. 89: 245-258.

Chowdhury, M., A. Kushwah, A.N. Satpute, S.K. Singh and A.K. Patil. 2023a. A comprehensive review on potential application of nanomaterials in the field of agricultural engineering. J. Biosyst. Eng. 48(4): 457-477.

Chowdhury, M., R. Anand, T. Dhar, R. Kurmi, R.K. Sahni and A. Kushwah. 2024a. Digital insights into plant health: Exploring vegetation indices through computer vision. pp. 7-30. *In:* S.S. Chouhan, U.P. Singh and S. Jain [eds.]. Applications of Computer Vision and Drone Technology in Agriculture 4.0. Springer Nature, Singapore.

Dong, L. and K. Gai. 2023. Advancements in blockchain-based agricultural supply chain management systems: A systematic review and research agenda. Comput. Ind. Eng. 189: 106118.

Fan, Z., B. Ghaddar, X. Wang, L. Xing, Y. Zhang and Z. Zhou. 2024. Artificial intelligence for operations research: Revolutionizing the operations research process. arXiv preprint arXiv:2401.03244.

FAO. 2022. Artificial intelligence in agriculture: A global perspective. Rome: FAO.

Feizabadi, J. 2022. Machine learning demand forecasting and supply chain performance. Int. J. Logist. Res. Appl. 25(2): 119-142.

Ghimire, R. and K. Panthi. 2023. Blockchain technology in agribusiness: Applications and future perspectives. J. Agric. Inform. 14(2): 87-103.

Gholian-Jouybari, F., S.M. Hosseini-Motlagh and A. Cheraghalipour. 2023. Heuristic and metaheuristic approaches for sustainable agri-supply chain optimization. Int. J. Agric. Econ. 35(2): 174-188.

Goyal, S. and J. Patel. 2022. Enhancing agriculture through AI-driven supply chain management: Opportunities and challenges. J. Agric. Sci. Technol. 26(4): 556-570.

Hu, J. and Y. Zhang. 2021. Smart farming: Applications and developments in artificial intelligence and data analytics. Int. J. Smart Agric. Technol. 5(1): 12-23.

Islam, N., M.A. Ganaie and A. Majid. 2023. Blockchain and AI integration for transparent and efficient supply chains in agribusiness. Agric. Econ. Policy. 55(2): 130-147.

Jackson, I., M. Jesus Saenz and D. Ivanov. 2024. From natural language to simulations: Applying AI to automate simulation modelling of logistics systems. Int. J. Prod. Res. 62(4): 1434-1457.

Jha, B.K. 2019. Conservation agriculture in Eastern India: A way forward for rice fallow management. Conserv. Agric. Clim. Resilient Farm. Doubl. Farmers' Income. 246 p.

Jonnalagadda, S. 2021, March. Predictive analytics in agriculture using geospatial mapping. *In:* 2021 IEEE Integrated STEM Education Conference (ISEC) (pp. 206-206). IEEE.

Kalwar, D., O. Shelke and H. Khadilkar. 2023. Using general value functions to learn domain-backed inventory management policies. arXiv preprint arXiv: 2311.02125.

Kamilaris, A., A. Fonts and F.X. Prenafeta-Boldύ. 2019. The rise of blockchain technology in agriculture and food supply chains. Trends Food Sci. Technol. 91: 640-652.

Klerkx, L., E. Jakku and P. Labarthe. 2019. A review of social science on digital agriculture, smart farming, and agriculture 4.0: New contributions and a future research agenda. NJAS - Wageningen J. Life Sci. 90-91: 100315.

Kong, Y., S. Zheng, Y. Li and X. Zhang. 2022. Artificial intelligence in agriculture: A review and future directions. J. AI Res. Appl. 47(7): 119-133.

Kumar, R. and S. Patel. 2023. AI and blockchain in agribusiness: Innovations for the future. Agric. Syst. 102: 78-91.

Kumari, N., D. Chaudhary, H. Kaur and A.L. Yadav. 2023, June. Artificial intelligence in supply chain optimization. *In:* 2023 International Conference on IoT, Communication and Automation Technology (ICICAT) (pp. 1-6). IEEE.

Kuźnar, M. and A. Lorenc. 2021. A method of predicting wear and damage of pantograph sliding strips based on artificial neural networks. Mater. 15(1): 98.

Lamb, H.J., B.J. Hayes, L.T. Nguyen and E.M. Ross. 2020. The future of livestock management: A review of real-time portable sequencing applied to livestock. Genes. 11(12): 1478.

Lamb, H.J., B.J. Hayes, L.T. Nguyen and E.M. Ross. 2024. Advances in AI for sustainable livestock management. Genes. 12(1): 101-115.

Laskar, A.A., K. Kumar, P. Roy, A.S. Mazumder and B. Das. 2024. Exploring the role of smart systems in farm machinery for soil fertility and crop productivity. Int. J. Res. Appl. Sci. Eng. Technol. 12(12): 2063-2075.

Leong, Y.M., E.H. Lim, N.F.B. Subri and N.B.A. Jalil. 2023, September. Transforming agriculture: Navigating the challenges and embracing the opportunities of artificial intelligence of things. *In:* 2023 IEEE International Conference on Agrosystem Engineering, Technology and Applications (AGRETA) (pp. 142-147). IEEE.

Liu, Y., T. Zhang and C. Wang. 2023. Improving supply chain management in agribusiness: The role of AI and machine learning. Int. J. Prod. Econ. 243: 108306.

Maohua, W. 2001. Possible adoption of precision agriculture for developing countries at the threshold of the new millennium. Comput. Electron. Agric. 30(1): 45-50.

Mohamed-Iliasse, B., B. Schroeder and C. Lodemann. 2020. Reinforcement learning applications in agribusiness supply chain optimization. Agric. Syst. Res. 18(1): 45-56.

Momenitabar, S., A.M. Fathollahi-Fard and A. Ahmadi. 2022. Applying particle swarm optimization for inventory and scheduling in agriculture. J. Comput. Agric. Econ. 14(3): 289-300.

Muzylyov, D., N. Shramenko and V. Shramenko. 2020. Integrated business-criterion to choose a rational supply chain for perishable agricultural goods at automobile transportations. Int. J. Bus. Perform. Manag. 21(1-2): 166-183.

Nie, X. and J. Yang. 2023. AI-enhanced supply chain optimization in the agricultural sector: A case study. J. Agric. Econ. Policy. 12(3): 214-228.

Pacella, M. and G. Papadia. 2021. Evaluation of deep learning with long short-term memory networks for time series forecasting in supply chain management. Procedia CIRP. 99: 604-609.

Pal, S. 2023. Advancements in AI-enhanced just-in-time inventory: Elevating demand forecasting accuracy. Int. J. Res. Appl. Sci. Eng. Technol. 11: 282-289.

Pandey, S., A. Gehlot, V. Pandey, N. Kathuria, G. Chhabra and P.K. Malik. 2023, April. Subsuming AI, IoT, and Big Data in smart farm practices. *In:* 2023 IEEE Devices for Integrated Circuit (DevIC) (pp. 229-232). IEEE.

Patil, S., N. Aklade and A. Uikey. 2023. Revolutionizing vegetable value chains: A comprehensive review of digital technologies and their impact on agricultural transformation. Curr. J. Appl. Sci. Technol. 42(47): 54-65.

Perez, A. and N. Singh. 2022. Leveraging AI for supply chain efficiency in agribusiness. Food Policy Rev. 38(4): 254-272.

Pradhan, N.C., M.A. Naik, M. Chowdhury, A. Kushwah, K.R. Asha, T. Dhar et al. 2024. Robotic seeding or sowing system in smart agriculture. pp. 495-520. *In:* K. Pandey, N.L. Kushwaha, C.B. Pande and K.G. Singh [eds.]. Artificial Intelligence and Smart Agriculture: Technology and Applications. Springer Nature, Singapore.

Rattan, P. 2023. Cultivating agricultural evolution: Revolutionizing farming through the power of AI and technology. Rev. Artif. Intell. Educ. 4(00): e010.

Satpute, A.N., K.P. Gavhane, S. Kaur, A. Jha, N.C. Pradhan and M. Chowdhury. 2024. Integration of AI and IoT in soilless cultivation to power sustainable agricultural revolution. pp. 387-411. *In:* K. Pandey, N.L. Kushwaha, C.B. Pande and K.G. Singh [eds.]. Artificial Intelligence and Smart Agriculture: Technology and Applications. Springer Nature, Singapore.

Schroeder, B. and C. Lodemann. 2021. Machine learning for sustainable yield prediction in agriculture. Environ. Syst. Sci. 45(2): 210-225.

Singh, P.K. 2023. Digital transformation in supply chain management: Artificial intelligence (AI) and machine learning (ML) as catalysts for value creation. Int. J. Supply Chain Manag. 12(6): 57-63.

Singh, R.K., M.H. Rahmani, M. Weyn and R. Berkvens. 2022. Joint communication and sensing: A proof of concept and datasets for greenhouse monitoring using LoRaWAN. Sensors. 22(4): 1326.

Smith, J. and X. Li. 2023. Machine learning applications in agricultural supply chains. J. Agric. Econ. 45(2): 123-145.

Storer, M., P. Hyland, M. Ferrer, R. Santa and A. Griffiths. 2014. Strategic supply chain management factors influencing agribusiness innovation utilization. Int. J. Logist. Manag. 25(3): 487-521.

Taşkıner, T. and B. Bilgen. 2021. Optimization models for harvest and production planning in agri-food supply chain: A systematic review. Logist. 5(3): 52.

Terrada, L., M. El Khaili and H. Ouajji. 2022. Demand forecasting model using deep learning methods for supply chain management 4.0. Int. J. Adv. Comput. Sci. Appl. 13(5): 704-711.

Tian, F. 2017. A supply chain traceability system for food safety based on HACCP, blockchain & Internet of Things. *In:* 2017 Int. Conf. Serv. Syst. Serv. Manag. (pp. 1-6). IEEE.

Tirkolaee, E.B., I. Mahdavi and G.-W. Weber. 2021. AI techniques for dynamic pricing and logistics optimization in agriculture. Comput. Ind. Eng. 157: 107229.

Udokwu, C.J., C. Ukaegbu and E. Adedigba. 2023. Support vector machines for agribusiness risk assessment: A case study. J. Agric. Inform. 13(4): 88-96.

Vishnu, D., K. Baskar and P. Ganesh. 2021. Optimization in agri-supply chains: Applications of simulated annealing in network design. J. Optim. Agric. Sci. 6(2): 99-114.

Wang, F., R. Zhang and X. Liu. 2024. Artificial intelligence and big data analytics in supply chain management: A systematic review. J. Bus. Logist. 45(1): 59-76.

Waqas, M., U.W. Humphries, A. Wangwongchai, P. Dechpichai and S. Ahmad. 2023. Potential of artificial intelligence-based techniques for rainfall forecasting in Thailand: A comprehensive review. Water 15(16): 2979.

Weinberg, E., S. Sorenson and T. James. 2023. Heuristic approaches for sustainable agricultural supply chains. Agric. Syst. Manag. 22(3): 345-360.

World Bank. 2023. AI and Agribusiness Supply Chains: Opportunities and Challenges. Washington, DC: World Bank.

Wüst, K. and A. Gervais. 2018. Do you need a blockchain? *In:* 2018 Crypto Valley Conf. Blockchain Technol. (CVCBT), pp. 45-54. IEEE.

Yang, X., L. Zhang and J. Liu. 2024. Enhancing agricultural supply chains with AI-driven predictive analytics. J. Agric. Policy Manag. 16(3): 67-85.

Zhang, L., Q. Li and J. Liu. 2023. AI-driven solutions for agri-supply chain management: Current status and future prospects. Agric. Syst. 213: 103630.

Zhao, X. and Y. Zheng. 2023. Data-driven innovations in agri-supply chain management: A review of AI methodologies. Agric. Econ. Rev. 22(4): 204-219.

Zhou, T. and X. Sun. 2023. Blockchain technology and AI in agribusiness: A comprehensive review of applications and future research directions. J. Agric. Inform. 14(1): 45-62.

Zilberman, D., L. Lu and T. Reardon. 2022. Applications of LSTM in agricultural demand forecasting. J. Agric. Forecast. 58(1): 100-115.

CHAPTER

11

Sustainable Farming Practices Enhanced by Artificial Intelligence

Abdul Basit[1*], Mishal Kainat[2], Kashif Ali Khan[2], Aqsa Benish[1], Umar Farooq[3], Faiz Ul Hassan[4], Muhammad Amjad Bashir[5], Abdur Rehim[6], Sikandar Ibrahim Khan[2], Mir Sujaul Islam[4] and Ijaz Ahmad[1]

[1] Department of Agronomy, Engro Fertilizers Pvt. Ltd, Pakistan
[2] Department of Soil Science, University of Layyah, Pakistan
[3] Department of Plant Breeding and Genetics, College of Agriculture, University of Layyah, Pakistan
[4] Institute for Environment and Development (LESTARI), Universiti Kebangsaan Malaysia (UKM), Bangi 43600, Selangor, Malaysia
[5] Sichuan Provincial Key Laboratory of Philosophy and Social Sciences for Monitoring and Evaluation of Rural Land Utilization, Chengdu Normal University, Chengdu, China
[6] Department of Soil Science, Bahauddin Zakariya University, Multan, Pakistan

Introduction

Agricultural sustainability is at risk due to challenges such as soil degradation caused by intensive agricultural practices such as over-farming, monoculture and excessive use of chemical fertilizers (Chen et al. 2024). These practices lead to soil nutrient depletion, organic matter loss, and increased erosion, which in turn reduces crop yields and damages soil carbon sequestration capabilities, exacerbating climate change (Chowdhury et al. 2024). Water shortage is another key issue in modern agriculture. Global agriculture consumes about 70% of freshwater, largely due to inefficient irrigation and water-intensive crops in arid regions (Jha et al. 2023; Tariq et al. 2020). This unsustainable use of water depletes aquifers and reduces water availability, especially in areas already facing water scarcity, leading to increased competition for water between agriculture, industry and domestic use (Zhang et al. 2020; Patra et al. 2023).

Pest resistance to herbicides and pesticides is undermining the sustainability of agriculture. The overuse and misuse of these chemicals has led to the evolution of

*Corresponding author: sheikhu.ab@gmail.com

DOI: 10.1201/9781003536932-11

resistant pest populations, increased chemical inputs and higher production costs (Warra and Prasad 2020). This creates serious environmental risks, including water pollution and reduced populations of beneficial insects. Climate change is further exacerbating these challenges by changing weather patterns, increasing the frequency and severity of extreme events, and changing the distribution of pests and diseases (Skendžić et al. 2021; said et al. 2024). These changes can lead to unpredictable crop failures, threatening food security and requiring more resilient agricultural practices. However, adapting to climate change requires significant investments in research, technology and infrastructure, which are often beyond the capabilities of smallholder farmers in developing regions (Kuhl 2020).

Unsustainable farming practices have huge environmental and economic costs. Recent scientific research suggests that soil degradation could lead to billions of dollars in annual losses in global crop yields (Mahata and Sharma 2021; Gawande et al. 2023). Water shortages caused by unsustainable agricultural practices can cost the economy up to $500 billion annually due to reduced agricultural productivity and increased competition for water (Zhang et al. 2020; Jha et al. 2024). Additionally, managing pest resistance requires increased costs to develop and apply new chemical treatments, placing an economic burden on farmers and consumers. Sustainable agriculture (SA) is a holistic approach to agriculture that seeks to balance environmental health, economic profitability, and social equity (Chowdhury et al. 2023). At the core of its scientific principles are the maintenance of soil health, the protection of biodiversity and the efficient use of resources (Shahane and Shivay 2021).

Soil health is critical as it underpins crop productivity and resilience, and practices such as crop rotation, cover cropping and organic amendments can enhance soil structure, fertility and microbial activity, thereby reducing reliance on chemical inputs and mitigating soil degradation (Singh et al. 2024b; Chowdhury et al. 2025). Biodiversity conservation is crucial to resilient agriculture, promoting the development of a variety of beneficial organisms through SA methods such as agroforestry and polyculture. SA also focuses on resource efficiency by optimizing the use of water, energy and nutrients. Precision farming techniques minimize waste and address issues such as water scarcity and overuse of chemicals (Gawande et al. 2023).

Artificial intelligence (AI) is increasingly being integrated into agriculture to address sustainability challenges and improve decision-making. In agricultural science, AI is critical for data analysis, pattern recognition and predictive modeling (Ayoub Shaikh, Rasool, and Rasheed Lone 2022). By analyzing large data sets from disparate sources such as sensors and satellite imagery, AI can provide insights into soil conditions, crop health and environmental factors to facilitate informed decision-making. AI's ability to recognize patterns can enable early detection of problems such as pest infestations and nutrient deficiencies, allowing timely intervention to prevent crop loss and reduce the need for chemical treatments (Waheed et al. 2022).

In SA, AI provides data-driven recommendations for best agricultural practices, increasing productivity and reducing the environmental and economic costs associated with unsustainable practices. This integration paves the way for a more resilient and sustainable agricultural future (Vishnoi and Goel 2024). This chapter critically analyzes current challenges to SA, including soil degradation, water scarcity, pest resistance and the impacts of climate change. It assesses the environmental and economic costs of unsustainable agricultural practices and explores the potential of SA and AI as integrated solutions. The goal is to use scientific principles and advanced technologies to achieve a more resilient and SA future.

Losses in Agriculture

Unbalanced agricultural practices (UA) has significant environmental and economic impacts, affecting soil health, water resources, pest management and climate resilience (Faye and Braun 2022). These challenges not only endanger agricultural productivity but also place a heavy economic burden on farmers, economies and ecosystems around the world. Urgent action and a shift to more SA practices are essential to address these issues (Zurek et al. 2022).

Agricultural soil degradation imposes huge environmental costs. Soil erosion causes significant productivity losses, requiring more than \$44 billion in chemical fertilizers annually to restore soil fertility (Montgomery 2021). Additionally, nutrient depletion from soil degradation costs the global economy more than \$10 billion annually, highlighting the serious impact of unsustainable agricultural practices on soil health (Brevik 2022). Water scarcity is a major environmental problem due to over-extraction and inefficient irrigation methods. It is estimated that poor irrigation systems waste \$150 billion globally each year and lead to more than 70% of agricultural water loss (Abdelhafez et al. 2020). Agricultural droughts, especially in the United States, cause an estimated \$10 billion in economic losses yearly, exacerbating the water crisis (Fleming-Muñoz et al. 2023).

Pest resistance poses significant challenges, leading to increased pesticide use and costs. As pests develop resistance, conventional pesticides become 30-35% less effective, costing more than \$10 billion in pesticide costs annually (Lykogianni et al. 2021). Despite these efforts, resistant pests continue to cause severe damage to crops, with losses totaling more than \$20 billion annually (Zaller 2020). Climate change is exacerbating environmental challenges, with extreme weather events such as hurricanes and floods causing \$12 billion in damage annually (Zaller 2020). Global agricultural production is projected to decline by 17% by 2050, potentially causing economic losses of more than \$200 billion (Mehrabi 2023).

The economic costs of UA are enormous, primarily due to soil erosion that reduces crop yields by 0.03% per year, causing losses of more than \$40 billion (Prada and Dante 2005). Soil degradation and pest resistance require greater use of chemical fertilizers and pesticides, increasing agricultural input costs by 10-20% and adding \$25 billion to agricultural spending annually (Weller et al. 2014). Inefficient agricultural practices result in huge health costs and environmental damage. In the United States, more than \$5 billion is spent each year to address health problems caused by the use of pesticides and chemical fertilizers (Osteen and Szmedra 1989). Additionally, biodiversity loss and ecosystem destruction cost the world more than \$300 trillion (Kumar et al. 2024).

Artificial Intelligence in Sustainable Agriculture

The integration of AI into crop management began in 1985 with McKinion and Lemmon's book "Expert Systems in Agriculture" (Elbasi et al. 2023). This lays the foundation for the development of agricultural smart systems, using AI to enhance the resilience and productivity of agricultural processes, and food systems (Pradhan et al. 2024). Agricultural research made significant progress in the 1990s and 2000s with the widespread adoption of machine-learning algorithms (Sharma et al. 2021). Use techniques such as neural networks, decision trees, and support vector machines to analyze complex agricultural data (Satpute et al. 2024). In addition, remote sensing technologies such as satellite imagery and drones facilitate remote monitoring of crops. AI algorithms process this data to assess crop health, identify disease, and optimize irrigation, ultimately making crop management practices more efficient and precise (Kowalska and Ashraf 2021).

The 2010s saw the rise of AI-powered precision agriculture, integrating data from sensors, GPS, and weather forecasts to optimize planting, fertilization, and pest control (Anand et al. 2024). Deep learning models improved crop disease detection, while AI-driven robotics and drones made planting and monitoring more efficient. Predictive analytics helped forecast yield, market prices, and climate impacts for better planning. Currently, AI is integrated with IoT to collect real-time data, processed for precise crop management decisions. It also processes big data for historical and actionable insights and combines with blockchain for supply chain traceability and transparency, shaping SA as described in Fig. 1 (Kumari et al. 2024).

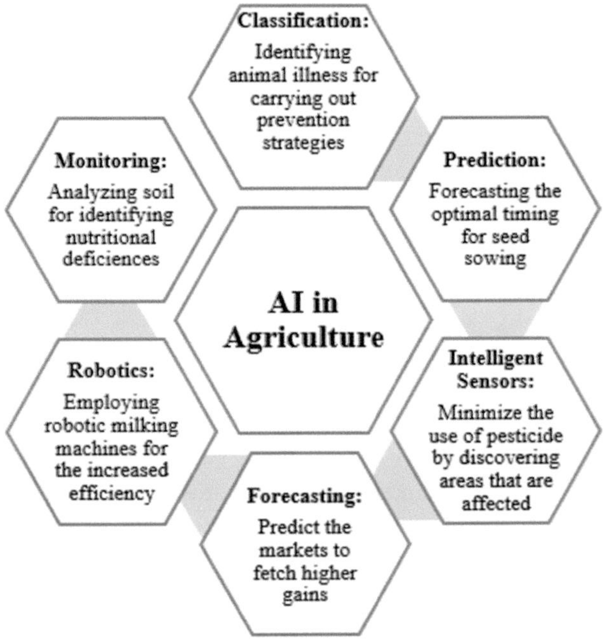

Fig. 1: Applications of AI in agriculture.

Artificial Intelligence for Soil Health

Using different plant species and agricultural management techniques creates diverse habitats for soil organisms and contributes to soil health. This diverse plant species mitigates root disease and supports soil biomes critical for nutrient uptake. Reduced tillage and strategic soil disturbance techniques can aerate the soil and prevent compaction, limiting impacts on soil organisms (Cardoso et al. 2013). To minimize soil disturbance, it is recommended to plant different species with unique root structures. Due to their chemical makeup, the benefits of deep-rooted plants extend beyond the topsoil, providing a diverse habitat for soil organisms. Organic matter from plants sustains soil biological communities. Soil amendments such as humates, clays and compost are used in agriculture to change soil properties and improve biological habitats. For example, adding compost can enrich the soil matrix, while adding lime can improve the symbiosis between rhizobia and legumes (Pukalchik et al. 2019).

Integrating AI into agricultural practices for different crop varieties can significantly improve soil health by optimizing habitat creation for soil organisms. AI analyzes plant species to identify those that are effective in reducing the spread of root diseases and promoting healthy soil biomes, which are critical for nutrient uptake (Dias et al. 2015). AI-driven tools identify optimal plant combinations and management techniques to create optimal habitats for soil organisms. Guided by AI, reduced tillage and strategic soil disturbance techniques can enhance soil aeration and prevent compaction, supporting soil health (Fahad et al. 2022; Sunil Kumar et al. 2024).

AI assesses soil conditions and recommends suitable plant varieties with unique root structures to minimize soil disturbance, improve water and air movement, and reduce compaction. For example, deep-rooted plants identified through AI provide benefits beyond the topsoil because of their chemical composition and interactions with soil organisms (Raghavan et al. 1990). In a recent study, the implementation of AI-driven crop selection and management techniques reduced soil compaction by 30% and increased soil aeration by 25% compared to traditional methods. Additionally, using AI to guide plant variety selection increased microbial diversity within soil biomes by 40%, indicating significant improvements in soil health and nutrient uptake (Du et al. 2018, Malkani et al. 2024). These measurable results highlight the transformative potential of AI in optimizing agricultural practices to enhance soil health (Zhang et al. 2021).

AI significantly enhances the application of organic matter and soil amendments. AI can identify plants that store more carbon in the soil by analyzing molecular structure, nutrient requirements and how different plants respond to soil disturbance. This ensures optimal integration of organic matter into the soil biome. AI recommends using appropriate soil amendments such as humates, clays, gypsum, lime and biological resources to modify soil habitat, improve soil pH, strengthen soil structure and provide nutrients. For example, various research conducted and AI-driven analysis identified specific plant varieties that increased soil carbon storage by 35% compared to traditional selections (Katsuumi et al. 2024). Additionally, AI-assisted compost application increased microbial richness in targeted soil zones by 45%, while adding lime increased rhizobia-legume symbiosis by 30% (Saberi Kamarposhti et al. 2024; Mithöfer 2002). In a controlled study, AI suggestions for soil improvement improved soil pH by 20% and soil structure by 25% (Singh et al. 2024a). As Table 1 summarized role of AI in improving soil health through optimized plant selection, reduced tillage, and effective use of organic matter and soil amendments. It highlights measurable impacts and references key studies supporting the findings.

Table 1: Applications of artificial intelligence for soil health optimization

AI-Enabled Technique	Objective	Impact on Soil Health	References
Optimized Plant Species Selection	Identifying plant species effective in reducing root diseases and supporting soil biomes.	• Increased microbial diversity by 40%. • Improved nutrient uptake and root disease mitigation.	Dias et al. 2015; Du et al. 2018
Guided Crop Combinations	Designing optimal plant combinations for diverse habitats.	• Enhanced soil biological communities. • Enriched habitat for nutrient cycling and organism support.	Fahad et al. 2022
Reduced Tillage and Soil Disturbance	Analyzing soil conditions for minimal disruption techniques.	• Reduced soil compaction by 30%. • Improved soil aeration by 25%.	Cardoso et al. 2013; Du et al. 2018
Deep-Rooted Plant Integration	Recommending deep-rooted plant varieties based on chemical composition.	• Benefits extended beyond topsoil. • Enhanced water and nutrient movement.	Raghavan et al. 1990
AI-Assisted Organic Matter Application	Optimizing compost and organic inputs for soil matrix enrichment.	• Increased microbial richness by 45%. • Boosted soil carbon storage by 35%.	Pukalchik et al. 2019; Katsuumi et al. 2024
Targeted Soil Amendments	Recommending specific amendments (e.g., lime, gypsum) based on AI analysis.	• Improved soil pH by 20%. • Strengthened soil structure by 25%. • Enhanced rhizobia-legume symbiosis by 30%.	Mithöfer 2002; Saberi Kamarposhti et al. 2024

Artificial Intelligence for Pest and Disease Management

AI has become a powerful tool in pest and disease management, providing advanced technologies for monitoring, predicting and controlling agricultural threats. Image recognition technology is a key scientific method that enables early detection and identification of pests and diseases by analyzing crop images and comparing them with an extensive database of known pathogens (Qin et al. 2016). Recent advances in deep learning have significantly improved the identification and classification of agricultural pests and diseases. Researchers developed a new object detection framework using Cascade RCNN (Region with CNN) to address challenges such as large image size, diverse pest species, and limited sample datasets. This method utilizes data augmentation techniques to overcome the problem of limited and imbalanced data sets. In addition, a sliding window cropping method is also utilized to improve feature learning while retaining the original image size (Cai and Vasconcelos 2021).

To enhance the model's accuracy, attention mechanisms were integrated with Feature Pyramid Networks (FPN) to focus on critical features within sample images (Cao et al. 2023). The model exhibited strong performance, with key metrics such as an Average Precision of 84.16%, Recall of 65.23%, precision of 67.79%, and an F1 Score of 82.34%. These results highlight the model's effectiveness in addressing the challenges posed by the wide variety of pest types and large image sizes in agricultural environments (Zhou and Zhang 2022; Guo et al. 2023a).

In addition, genome sequencing helps to understand the genetic makeup of pests and pathogens, helping to track the evolution of resistance and develop targeted control strategies. AI-driven epidemiological models predict disease spread by analyzing environmental factors, crop conditions and pest behavior, allowing farmers to implement preventive measures (Bisht et al. 2019). Genome editing technologies have advanced significantly over the past 20 years, allowing for precise modification of genome sequences and targeted gene inactivation. The CRISPR/Cas9 system is a revolutionary tool in the field of functional genomics, especially in pest research. This programmable genome editing tool is critical for understanding gene function, tracking the evolution of resistance, and developing improved pest management strategies (Cui et al. 2017). CRISPR/Cas9 enables precise genomic changes and offers the potential for more effective and sustainable pest control methods (Singh et al. 2022).

Recent scientific research shows that AI-based pest management methods, such as AI pheromone traps, can effectively monitor and control agricultural pest populations. These traps release chemical attractants and are integrated with AI systems that analyze captured data to determine the optimal time and location for intervention. This approach reduces reliance on broad-spectrum pesticides and minimizes environmental impact (Rosado et al. 2022). Similarly, AI optimizes the deployment of biocontrol agents such as beneficial insects by analyzing environmental conditions and pest dynamics, thereby increasing their effectiveness (Hanif et al. 2022).

With the help of AI, precision spraying has made significant progress. AI algorithms analyze data from sensors, drones and satellite imagery to determine the precise amount and location of pesticides to apply. This targeted approach reduces chemical use, cost and risk of pest resistance (Talaviya et al. 2020). Studies show that precision spraying driven by AI can produce results that are equivalent to or better than traditional methods while minimizing environmental and economic costs.

AI-Driven Plant Breeding and Genomics

SA is the agricultural practice of producing food, fiber and shelter potentially increasing crop yield and production in an eco-friendly environment and maintaining economic viability (Basit et al. 2023). It involves gaining the maximum outputs by giving minimum inputs to sustain the overall farmer's economics. AI is a system that uses human intelligence to carry out numerous tasks like learning, decision-making, and problem-solving. AI has various fields including robotics, machine learning, and neural networks to help humans in assisting different tasks. The dilation of established agricultural efficient methods and uniform culture creates acute natural catastrophes globally (Redhu et al. 2022).

Scientists oppose unsurpassed obstructions exigencies to analyze in the ecological label the output of ruler production network (primary production) in terms of eco-friendliness. The production Enhancement and technological innovations should be acquired by researchers such as AI transcriptome analysis, and next-generation phonemics have already helped to study genotypes and their emergence with phenotype particularly for critical characteristics by parallel genome sequencing and transcriptomes. This is the age of AI with groundbreaking expansion in Omics studies blending of genomes and phenomes in genomes wide association studies and anticipating of phenotypes from genotype in DNA selection has ameliorated crops significantly (Felt 2008).

AI is a captivating futuristic system yielding infinite windows regarding agricultural implementations. It expands boundaries E-breeding future AI generations planned to stimulate modern varieties of intellectually inspired networks and compositional skilled and AI-based results efficient and effective than ever and we are already familiar with AI, information Analysis, mega Data Being used in corporate writing that records analyze and interpret data (van Rijmenam 2019). The use of AI in the breeding realm is due to technological advancements data deluge can be examined vastly and faultlessly uncovering unexpected correlations. AI enables plant scientists and breeders to organize unorganized data then convert it into genetics planning and contemplate to use of beneficial technology to infrastructure procedures in crop breeding (Harfouche et al. 2019).

AI-driven genomics and crop modeling have revamped agricultural breeding and management genetic mutations affiliated with marketable crops are skillfully detected by scientists consequently. AI has effectively boosted yield and limited ecological effects (Glover and Stone 2018). AI-based Genome-wide association studies have helped to recognize genetic polymorphs that are connected to distinct characteristics by evaluating the genome of Flora (Guo et al. 2023b). AI-based algorithms for phenotypic analysis can scrutinize photo datasets of Flora and recognize arrangements and interdependence of plant phenotypes and genetic records (Márquez and Korhan 2023).

AI-driven farming has helped to establish the amount of fertilizer required for healthy plant growth all three of these AI-based techniques have aided in making plants unresponsive to diseases, and drought consequently, enhancing the nutritional value of plants (Katsuumi et al. 2024). Accurate forecasting of plant characteristics and recognition of genetic mutants, limiting the use of fertilization, and boosting yield for quality crops all is possible by the use of these AI-powered techniques. AI-driven gene-based crop modeling is universally pertinent as it has enabled researchers to enhance crop breeding, amplify crop regulation traditions, pathologies detections, recognition of mutations and analyze data sets (Desai et al. 2022).

AI has upgraded the productivity and efficacy of conventional breeding modalities conspicuously by pinpointing genetic alterations that are responsible for appealing traits breeders can discreetly breed plants with those characteristics to cultivate high-achieving and precision crops immune to disease, pests, eco-friendly, and nutrition (Weller et al. 2014; Harfouche et al. 2019). AI-driven agriculture utilizes perfect data and machinery to increase crop management procedures using yield-elevating fertilizations and hydration executing modern technologies analyzing ecological and genetic information to diminish surplus and augment resilient productivity and adaptive methods in agriculture (McArthur and McCord 2017; Ye et al. 2020).

Data Analysis and Decision Support System in Agriculture

AI plays a key role in modern agriculture by providing advanced data analysis and decision support systems to improve resource efficiency, sustainability and productivity. AI uses scenario analysis, optimization models, and simulations to analyze a wide range of data sets and support decision-making. AI uses scenario analysis to predict potential outcomes and associated risks based on multiple agricultural scenarios (Nathaniel et al. 2020). It also uses optimization models such as linear programming and genetic algorithms to efficiently allocate resources such as water and fertilizer. In addition, AI-driven simulation models evaluate various agricultural technologies, providing insights into best practices under different conditions (Kumari et al. 2024).

AI prioritizes user-centered design when developing Decision Support Systems (DSS) to understand the specific needs of farmers and agricultural stakeholders. This involves conducting a thorough needs assessment to customize decision support tools to address the unique challenges faced by users (Baldin et al. 2021). Designing the user interface (UI) is critical to ensuring accessibility and navigability for a broad range of users (Shneiderman and Plaisant 2010). AI also enables seamless system integration and interoperability, allowing DSS to connect with existing agricultural management systems and data sources, thereby improving overall efficiency (Li et al. 2018).

Data collection and management are critical to AI in agriculture. AI ensures timely, relevant and accurate data by integrating information from various sources such as weather stations, soil sensors, agricultural models and historical databases (Fuentes-Peñailillo et al. 2024). Maintain high-quality data through rigorous cleaning and validation processes to eliminate errors and anomalies. AI-driven techniques such as preprocessing, exploratory data analysis (EDA), and data reduction using methods such as principal component analysis (PCA) convert raw data into a usable format. This allows farmers and agricultural professionals to focus on key factors affecting crop production and environmental health (Vuppalapati 2021).

AI-driven agricultural practices prioritize sustainability and environmental health by promoting resource optimization and reducing chemical use. AI tools enable cost-benefit and market analysis to support informed decisions that balance profitability and environmental stewardship. One practical application of AI in agriculture is the implementation of sensor-based water pumping systems. The system utilizes soil moisture sensors to measure soil moisture and adjust water flow accordingly (Abdelhafez et al. 2020). The sensors communicate with the microcontroller, which serves as the central device for controlling the system. Based on the input from the sensor, the microcontroller activates or deactivates the pumping motor through a relay and displays the soil status on the liquid crystal display (Winston and Osikibo 2022). The technology aims to optimize water use by continuously monitoring temperature, soil moisture and water levels, and transmits data via SMS or GPRS technology for remote monitoring. AI also plays a vital role in enhancing the automation of irrigation systems, with artificial neural network (ANN) controllers providing advanced solutions to irrigation control problems, ensuring efficient water management and conservation (Lakhiar et al. 2024).

Figure 2 describes that AI presents multiple benefits in agriculture, it is important to evaluate and actively handle the moral issues and possible hazards associated with AI systems (Ali et al. 2024). Inadequate design and configuration of intelligent systems can cause damage and unforeseen repercussions in the field of digital agriculture. Examples of ethical concerns in digital agriculture encompass the invasion of farmers' privacy, adverse effects on animal welfare caused by robotic technologies, and the lack of accountability for complications that arise from the use of AI tools. Digital agriculture encompasses the utilization of cutting-edge technology that have revolutionized the agricultural industry, enhancing the intelligence and efficiency of farm operations. Automated techniques, including AI, IoT, and the collecting and analysis of farm data, can be employed to do this (Subeesh and Mehta 2021). AI-based solutions in farming applications capture and utilize significant quantities of data in data-driven services and decision support systems (Nathaniel et al. 2020).

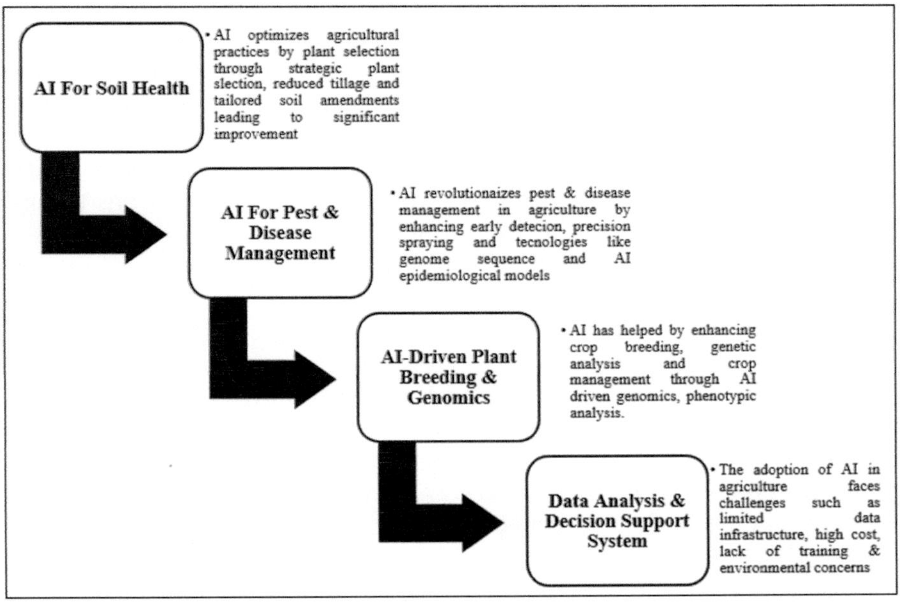

Fig. 2: Flow chart of AI-assisted sustainable agriculture.

Limitations

The use of AI in agriculture relies heavily on the availability and quality of data. AI-driven agricultural practices may be hampered in areas with limited data infrastructure and advanced technologies. Furthermore, the large initial investments required to implement AI in agriculture may not be feasible for smallholders or farmers in developing regions, which may exacerbate existing agricultural productivity inequalities. Farmers and agricultural professionals may face challenges in using AI effectively due to the complexity of AI algorithms and the need for specialized training. A lack of such training may hinder the successful integration of AI into agricultural practices. While AI provides precision and efficiency, overreliance on these technologies may lead to the reduction of traditional agricultural knowledge and skills.

The study acknowledges the environmental concerns posed by the increased use of agricultural technology, including electronic waste and energy consumption by AI systems. Further research is needed to understand the long-term impacts of AI-driven practices on soil health, biodiversity, and overall ecosystem sustainability. These limitations emphasize the importance of considering the technical and socioeconomic context when promoting AI in SA.

Future Aspects

The application of AI in agriculture shows great potential to overcome current limitations and promote sustainable development. The focus is on improving data collection and integration methods through remote sensing, satellite imagery and IoT devices. These advances will enable AI algorithms to provide precise recommendations based on local conditions, even in remote areas. The combination of the CRISPR/Cas9 system and AI will further revolutionize crop improvement through precise genome editing. Future efforts should prioritize making AI technologies more accessible and affordable for smallholder farmers. This can be achieved by developing cost-effective AI solutions and creating collaborative platforms for knowledge sharing. Investments in education and training programs are critical to equip farmers and agricultural professionals with the skills needed to effectively utilize AI tools. Simplifying AI technology for non-experts through intuitive user interfaces and applications is also critical.

AI has the potential to significantly improve resource efficiency and minimize the environmental impact of SA. Future research could look into combining AI with technologies such as blockchain to create transparent supply chains. Additionally, the development of AI-driven decision support systems can help predict the long-term effects of agricultural practices on soil health and biodiversity. Integrating EDA and PCA will provide deeper insights into complex agricultural data, while ANN can improve the accuracy of predictions and optimization. As AI technology continues to evolve, it is critical to consider its energy efficiency and environmental footprint as it becomes possible to incorporate renewable energy into AI-driven agricultural systems. The future of AI in agriculture depends on developing technologies that are inclusive, accessible, and environmentally sustainable. By addressing current challenges and exploring innovative solutions, AI can revolutionize agriculture, increase productivity, and contribute to the long-term health of agricultural ecosystems.

Conclusion

The application of AI in SA is now delivering significant real-world benefits. This article highlights the benefits of AI technology, including enabling more resource-efficient agriculture by analyzing a wide range of data sets to enhance decision-making in precision agriculture. Additionally, AI-recommended practices improve soil health by promoting biodiversity and optimizing the use of organic matter. In addition, AI-driven pest management strategies can effectively reduce environmental impacts while maintaining agricultural productivity. Finally, advances in AI-based plant breeding are revolutionizing crop improvement methods and increasing resistance to climate change and biological threats.

The convergence of AI and SA is transforming modern agriculture by solving a range of challenges. AI has evolved from expert systems in the 1980s to advanced deep learning models and IoT platforms, impacting areas such as crop management, soil health, pest control and plant breeding. In crop management, AI increases productivity and resilience through remote sensing and predictive analytics, enabling data-driven decision-making to optimize irrigation and pest control while minimizing environmental impact. When it comes to soil health, AI promotes practices that enhance microbial diversity and nutrient uptake, contributing to long-term fertility. In pest management, AI tools can achieve early detection and precise pesticide application, significantly reducing pesticide use and promoting ecological balance. Finally, the role of AI in plant breeding facilitates the identification of desirable traits to develop high-yielding, disease-resistant crops that support SA practices.

AI is critical to increasing agricultural productivity while ensuring environmental sustainability. It effectively addresses key challenges in agricultural value chains, fostering resilient food systems that can adapt to a growing global population and a changing climate. Future research should aim to combine AI with emerging technologies such as blockchain and quantum computing to improve the transparency, scalability and precision of SA.

References

Abdelhafez, A.A., S.M. Metwalley and H.H. Abbas. 2020. Irrigation: Water resources, types and common problems in Egypt. pp. 15-34. *In:* E.S. Omran and A. Negm [eds.]. Technological and Modern Irrigation Environment in Egypt. Springer Water, Springer, Cham.

Ali, G., M.M. Mijwil, B.A. Buruga, M. Abotaleb and I. Adamopoulos. 2024. A survey on artificial intelligence in cybersecurity for smart agriculture: State-of-the-art, cyber threats, artificial intelligence applications, and ethical concerns. Mesopot. J. Comput. Sci. 2024: 53-103.

Anand, R., S.S. Nalla, S. Nasreen, M. Chowdhury, Madhuban, B.S., A. Pandey et al. 2024. Fruit Counting and analysis using artificial intelligence approaches. pp. 93-114. *In:* S.S. Chouhan, A. Saxena, U.P. Singh and S. Jain [eds.]. Artificial Intelligence Techniques in Smart Agriculture. Springer Nature, Singapore.

Baldin, M.T. Breunig, R. Cue, A. De Vries, M. Doornink, J. Drevenak and V.E. Cabrera. 2021. Integrated decision support systems (IDSS) for dairy farming: A discussion on how to improve their sustained adoption. Animals. 11(7): 2025.

Basit, A., K.A. Khan, M. Kainat, M.M. Akram and M. Murtaza. 2024. Biochar and minerals impact on plant defense mechanism. J. Agric. Livest. Farm. 1(1): 1-7. https://doi.org/10.61577/jalf.2023.100002.

Bisht, D.S., V. Bhatia and R. Bhattacharya. 2019. Improving plant-resistance to insect-pests and pathogens: The new opportunities through targeted genome editing. *In:* Semin. Cell Dev. Biol. 96: 65-76. Academic Press.

Brevik, E.C. 2022. Agricultural land degradation in the United States of America. *In:* Impact Agric. Soil Degrad. I: Perspect. Africa, Asia, Am. Oceania (pp. 363-391). Cham: Springer International Publishing.

Cai, Z. and N. Vasconcelos. 2019. Cascade R-CNN: High quality object detection and instance segmentation. IEEE Trans. Pattern Anal. Mach. Intell. 43(5): 1483-1498.

Cao, Y., Z. Zhao, Y. Huang, X. Lin, S. Luo, B. Xiang and H. Yang. 2023. Case instance segmentation of small farmland based on Mask R-CNN of feature pyramid network with double attention mechanism in high resolution satellite images. Comput. Electron. Agric. 212: 108073.

Cardoso, E.J.B.N., R.L.F. Vasconcellos, D. Bini, M.Y.H. Miyauchi, C.A.D. Santos, P.R.L. Alves and M.A. Nogueira. 2013. Soil health: Looking for suitable indicators. What should be considered to assess the effects of use and management on soil health? Sci. Agric. 70: 274-289.

Chen, Y., P. Yu, L. Wang, Y. Chen and E.H.W. Chan. 2024. The impact of rice-crayfish field on socio-ecological system in traditional farming areas: Implications for sustainable agricultural landscape transformation. J. Clean. Prod. 434: 139625.

Chowdhury, M., E.V. Thomas, A. Jha, A. Kushwah, R. Kurmi, T.K. Khura et al. 2023. An automatic pressure control system for precise spray pattern analysis on spray patternator. Comput. Electron. Agric. 214: 108287.

Chowdhury, M., R. Anand, T. Dhar, R. Kurmi, R.K. Sahni and A. Kushwah. 2024. Digital insights into plant health: Exploring vegetation indices through computer vision. pp. 7-30. *In:* S.S. Chouhan, U.P. Singh and S. Jain [eds.]. Applications of Computer Vision and Drone Technology in Agriculture 4.0. Springer Nature, Singapore.

Chowdhury, M., T.K. Khura, R.A. Parray, P.K. Upadhyay, H.L. Kushwaha, A. Lama and A. Kushwah. 2025. Multi-vegetation indices based handheld device for precise nitrogen assessment and prescription in direct seeded rice. Comput. Electron. Agric. 229: 109886.

Clarke, B.J., F.E. Otto and R.G. Jones. 2021. Inventories of extreme weather events and impacts: Implications for loss and damage from and adaptation to climate extremes. Clim. Risk Manag. 32: 100285.

Cui, Y., J.L. Sun and L. Yu. 2017. Application of the CRISPR gene-editing technique in insect functional genome studies – A review. Entomol. Exp. Appl. 162(2): 124-132.

De Prada, J.D. 2005. Economics of Soil Erosion and Wetland Degradation: A Case Study from Argentina. Univ. Connecticut.

Desai, R.H., N. Zandvakili and S.K. Bohlander. 2022. Dissecting the genetic and non-genetic heterogeneity of acute myeloid leukemia using next-generation sequencing and in vivo models. Cancers, 14(9): 2182.

Dias, T., A. Dukes and P.M. Antunes. 2015. Accounting for soil biotic effects on soil health and crop productivity in the design of crop rotations. J. Sci. Food Agric. 95(3): 447-454.

Du, Y.D., W.Q. Niu, X.B. Gu, Q. Zhang, B.J. Cui and Y. Zhao. 2018. Crop yield and water use efficiency under aerated irrigation: A meta-analysis. Agric. Water Manag. 210: 158-164.

Elbasi, E., N. Mostafa, Z. Al Arnaout, A.I. Zreikat, E. Cina, G. Varghese and C. Zaki. 2022. Artificial intelligence technology in the agricultural sector: A systematic literature review. IEEE Access. 11: 171-202.

Fahad, S., S.B. Chavan, A.R. Chichaghare, A.R. Uthappa, M. Kumar, V. Kakade and P. Poczai. 2022. Agroforestry systems for soil health improvement and maintenance. Sustainability, 14(22): 14877.

Faye, J.B. and Y.A. Braun. 2022. Soil and human health: Understanding agricultural and socio-environmental risk and resilience in the age of climate change. Health & Place, 77: 102799.

Felt, U. 2008. Science and Governance. Ipoc Press.

Fleming-Muñoz, D.A., S. Whitten and G.D. Bonnett. 2023. The economics of drought: A review of impacts and costs. Aust. J. Agric. Resour. Econ. 67(4): 501-523.

Fuentes-Peñailillo, F., K. Gutter. R. Vega and G.C. Silva. 2024. Transformative technologies in digital agriculture: Leveraging Internet of Things, remote sensing, and artificial intelligence for smart crop management. J. Sens. Actuator Netw. 13(4): 39.

Gawande, V., D.R.K. Saikanth, B.S. Sumithra, S.A. Aravind, G.N. Swamy, M. Chowdhury and B.V. Singh. 2023. Potential of precision farming technologies for eco-friendly agriculture. Int. J. Plant Soil Sci. 35(19): 101-112.

Glover, D. and G.D. Stone. 2018. Heirloom rice in Ifugao: An 'anti-commodity' in the process of commodification. J. Peasant Stud. 45(4): 776-804.

Guo, D., Z. Wu, J. Feng and T. Zou. 2023a. Multi-scale semantic enhancement network for object detection. Sci. Rep. 13(1): 7178.

Guo, K., M. Wu, Z. Soo, Y. Yang, Y. Zhang, Q. Zhang and J. Lu. 2023b. Artificial intelligence-driven biomedical genomics. Knowl.-Based Syst. 279(4): 110937.

Hanif, M.K., S.Z. Khan and M. Bibi. 2022. Applications of artificial intelligence in pest management. *In:* AI & Smart Agri Apps (pp. 277-300). Auerbach Publications.

Harfouche, A.L., D.A. Jacobson, D. Kainer, J.C. Romero, A.H. Harfouche, G.S. Mugnozza and A. Altman. 2019. Accelerating climate resilient plant breeding by applying next-generation artificial intelligence. Trends Biotechnol. 37(11): 1217-1235.

Hassan, M., A. Kowalska and H. Ashraf. 2023. Advances in deep learning algorithms for agricultural monitoring and management. Appl. Res. Artif. Intell. Cloud Comput. 6(1): 68-88.

Jha, A., M. Chowdhury and A.N. Satpute. 2023. Surface water quality forecasting using machine learning approach. pp. 293-315. *In:* C.B. Pande, M. Kumar and N.L. Kushwaha [eds.]. Surface and Groundwater Resources Development and Management in Semi-arid Region: Strategies and Solutions for Sustainable Water Management. Cham: Springer International Publishing.

Jha, A., S. Garg, S. Satpute, J.P. Singh, D. Singh, U.G Phutela et al. 2024. Isotherm and kinetic modelling for assessing the effectiveness of economical adsorbents in sewage water treatment. Indian J. Chem. Technol. 31: 438-450.

Katsuumi, A., Y. Cong and S. Inazumi. 2024. AI-driven prediction and mapping of soil liquefaction risks for enhancing earthquake resilience in smart cities. Smart Cities 7(4): 1836-1856.

Kuhl, L. 2020. Technology transfer and adoption for smallholder climate change adaptation: Opportunities and challenges. Clim. Dev. 12(4): 353-368.

Kumar, A., A. Shukla and S. Kailkhura. 2024. Biodiversity loss and its economic costs: A global perspective. Res. J. Recent Sci. 13(1): 31-37.

Kumari, S., S. Tiwari, K. Naithani and P. Subbiah. 2024. Internet of Things for smart agriculture: A survey. *In:* AI Appl. Bus. Med. Agric. Sustain. (pp. 100-121). IGI Global.

Lakhiar, I.A., H. Yan, C. Zhang, G. Wang, B. He, B. Hao and M. Rakibuzzaman. 2024. A review of precision irrigation water-saving technology under changing climate for enhancing water use efficiency, crop yield, and environmental footprints. Agric. 14(7): 1141.

Li, H., J. Li, Y. Shen, X. Zhang and Y. Lei. 2018. Web-based irrigation decision support system with limited inputs for farmers. Agric. Water Manag. 210: 279-285.

Lykogianni, M., E. Bempelou, F. Karamaouna and K.A. Aliferis. 2021. Do pesticides promote or hinder sustainability in agriculture? The challenge of sustainable use of pesticides in modern agriculture. Sci. Total Environ. 795: 148625.

Mahata, S. and V.N. Sharma. 2021. The global problem of land degradation: A review. Natl. Geogr. J. India. 67: 216-231.

Malkani, P., I. Mani, P.K. Sahoo, R.A. Parray, S.S. Swain, K.R. Asha et al. 2024. Development and evaluation of a sensor enabled smart herbicide applicator for inter row crops. Smart Agric. Technol. 9: 100649.

Márquez, F.P.G. and O. Korhan. 2023. Digital Twin Tech.-Fundamentals and Appl. BoD – Books on Demand.

McArthur, J.W. and G.C. McCord. 2017. Fertilizing growth: Agricultural inputs and their effects in economic development. J. Dev. Econ. 217: 133-152.

Mehrabi, Z. 2023. Likely decline in the number of farms globally by the middle of the century. Nat. Sustain. 6(8): 949-954.

Mithöfer, A. 2002. Suppression of plant defence in rhizobia–legume symbiosis. Trends Plant Sci. 7(10): 440-444.

Montgomery, D.R. 2021. Soil health and the revolutionary potential of Conservation Agriculture. *In:* Rethinking Food & Agriculture (pp. 219-229). Woodhead Publishing.

Narra, N., P. Nevavuori, P. Linna and T. Lipping. 2020. A data driven approach to decision support in farming. *In:* Inf. Modelling & Knowl. Bases XXXI (pp. 175-185). IOS Press.

Osteen, C.D. and P.I. Szmedra. 1989. Agricultural pesticide use trends and policy issues. USDA, Econ. Res. Ser., Agric. Econ. Rep. No. 622, Washington, DC.

Patra, K., C.M. Parihar, H.S. Nayak, B. Rana, D.R. Sena, A. Anand et al. 2023. Appraisal of complementarity of subsurface drip fertigation and conservation agriculture for physiological performance and water economy of maize. Agric. Water Manag., 283: 108308.

Pradhan, N.C., M.A. Naik, M. Chowdhury, A. Kushwah, K.R. Asha, T. Dhar et al. 2024. Robotic seeding or sowing system in smart agriculture. pp. 495-520. *In:* K. Pandey, N.L. Kushwaha, C.B. Pande and K.G. Singh [eds.]. Artificial Intelligence and Smart Agriculture: Technology and Applications. Springer Nature, Singapore.

Pukalchik, M., K. Kydralieva, O. Yakimenko, E. Fedoseeva and V. Terekhova. 2019. Outlining the potential role of humic products in modifying biological properties of the soil: A review. Front. Environ. Sci. 7: 80.

Qin, F., D. Liu, B. Sun, L. Ruan, Z. Ma and H. Wang. 2016. Identification of alfalfa leaf diseases using image recognition technology. PloS One 11(12): e0168274.

Raghavan, G.S.V., P. Alvo and E. McKyes. 1990. Soil compaction in agriculture: A view toward managing the problem. Adv. Soil Sci.: Soil Degrad. 11: 1-36.

Redhu, N.S., Z. Thakur, S. Yashveer and P. Mor. 2022. Artificial intelligence: A way forward for agricultural sciences. *In:* Bioinform. Agric. (pp. 641-668). Academic Press.

Rosado, L., P. Faria, J. Gonçalves, E. Silva, A. Vasconcelos, C. Braga and C. Carlos. 2022. Eyesontraps: AI-powered mobile-based solution for pest monitoring in viticulture. Sustainability. 14(15): 9729.

Saberi Kamarposhti, M., N.K. Why, M. Yadollahi, H. Kamyab, J. Cheng and M. Khorami. 2024. Cultivating a sustainable future in the artificial intelligence era: A comprehensive assessment of greenhouse gas emissions and removals in agriculture. Environ. Res. 250: 118528.

Said, M.A., F.U. Hassan, M.S. Islam, M.E.B. Toriman, A. Rehim, M.A. Bashir and R. Nawaz. 2024. Environmental sustainability. *In:* Global Challenges for the Environment and Climate Change (pp. 1-15). IGI Global.

Satpute, A.N., K.P. Gavhane, S. Kaur, A. Jha, N.C. Pradhan and M. Chowdhury. 2024. Integration of AI and IoT in soilless cultivation to power sustainable agricultural revolution. pp. 387-411. *In:* K. Pandey, N.L. Kushwaha, C.B. Pande and K.G. Singh [eds.]. Artificial Intelligence and Smart Agriculture: Technology and Applications. Springer Nature, Singapore.

Shahane, A.A. and Y.S. Shivay. 2021. Soil health and its improvement through novel agronomic and innovative approaches. Front. Agron. 3: 680456.

Shaikh, T.A., T. Rasool and F.R. Lone. 2022. Towards leveraging the role of machine learning and artificial intelligence in precision agriculture and smart farming. Comput. Electron. Agric. 198: 107119.

Sharma, A., A. Jain, P. Gupta and V. Chowdary. 2020. Machine learning applications for precision agriculture: A comprehensive review. IEEE Access. 9: 4843-4873.

Shneiderman, B. and C. Plaisant. 2010. Designing the User Interface: Strategies for Effective Human-computer Interaction. Pearson Education India.

Singh, G., N. Kaur, R. Khanna, R. Kaur, S. Gudi, R. Kaur and G.S. Mangat. 2024a. 2Gs and plant architecture: Breaking grain yield ceiling through breeding approaches for next wave of revolution in rice (*Oryza sativa* L.). Crit. Rev. Biotechnol. 44(1): 139-162.

Singh, N.K., K. Sachan, M. BP, N. Panotra and D. Katiyar. 2024b. Building soil health and fertility through organic amendments and practices: A review. Asian J. Soil Sci. Plant Nutr. 10(1): 175-197.

Singh, S., S. Rahangdale, S. Pandita, G. Saxena, S.K. Upadhyay, G. Mishra and P.C. Verma. 2022. CRISPR/Cas9 for insect pests management: A comprehensive review of advances and applications. Agriculture. 12(11): 1896.

Skendžić, S., M. Zovko, I.P. Živković, V. Lešić and D. Lemić. 2021. The impact of climate change on agricultural insect pests. Insects. 12(5): 440.

Subeesh, A. and C.R. Mehta. 2021. Automation and digitization of agriculture using artificial intelligence and internet of things. Artif. Intell. Agric. 5: 278-291.

Sunil Kumar, T., H.M. Virdia, K.G. Patel, S. Ragi, M. Chowdhury, P. Kumar et al. 2024. Effect of summer legume residue incorporation and fertilizer regimes on rice growth, yield, and nutrient uptake. Front. Sustain. Food Syst. 8: 1467201.

Talaviya, T., D. Shah, N. Patel, H. Yagnik and M. Shah. 2020. Implementation of artificial intelligence in agriculture for optimisation of irrigation and application of pesticides and herbicides. Artif. Intell. Agric. 4: 58-73.

Tariq, M.A.U.R., R.R. Damnics, Z. Rajabi, M.L.U.R. Shahid and N. Muttil. 2020. Identification of major inefficient water consumption areas considering water consumption, efficiencies, and footprints in Australia. Appl. Sci. 10(18): 6156.

Van Rijmenam, M. 2019. Org. Tomorrow: How AI, Blockchain, and Analytics Turn Your Business into a Data Org. Routledge.

Vishnoi, S. and R.K. Goel. 2024. Climate smart agriculture for sustainable productivity and healthy landscapes. Environ. Sci. Policy. 151: 103600.

Vuppalapati, C. 2021. ML & AI for Agri-Economics: Prognostic Data Analytics for Small-Scale Farmers Worldwide (Vol. 314). Springer Nature.

Waheed, H., N. Zafar, W. Akram, A. Manzoor, A. Gani and S.U. Islam. 2022. Deep learning-based disease, pest pattern and nutritional deficiency detection system for "Zingiberaceae" crop. Agriculture. 12(6): 742.

Warra, A.A. and M.N.V. Prasad. 2020. African perspective of chemical usage in agriculture and horticulture—Their impact on human health and environment. *In:* Agrochem. Detect. Treat. Remediat. (pp. 401-436). Butterworth-Heinemann.

Weller, S., A. Culbreath, L. Gianessi, L. Godfrey, J. Jachetta, J. Norsworthy and J. Madsen. 2014. The contributions of pesticides to pest management in meeting the global need for food production by 2050. Counc. Agric. Sci. Technol. 55: 28.

Winston, G.I. and L.T. Osikibo. 2022. Design and construction of an IMPROVED automatic irrigation system. Am. J. Eng. Res. (AJER).

Ye, L., X. Zhao, E. Bao, J. Li, Z. Zou and K. Cao. 2020. Bio-organic fertilizer with reduced rates of chemical fertilization improves soil fertility and enhances tomato yield and quality. Sci. Rep. 10(1): 177.

Zaller, J.G. 2020. Where are the solutions to the pesticide problem? Daily Poison: Pesticides – An Underestimated Danger. 223-295.

Zhang, D., M.S. Sial, N. Ahmad, A.J. Filipe, P.A. Thu, M. Zia-Ud-Din and A.B. Caleiro. 2020. Water scarcity and sustainability in an emerging economy: A management perspective for future. Sustainability. 13(1): 144.

Zhang, P., Z. Guo, S. Ullah, G. Melagraki, A. Afantitis and I. Lynch. 2021. Nanotechnology and artificial intelligence to enable sustainable and precision agriculture. Nat. Plants. 7(7): 864-876.

Zhou, X. and L. Zhang. 2022. SA-FPN: An effective feature pyramid network for crowded human detection. Appl. Intell. 52(11): 12556-12568.

Zurek, M., A. Hebinck and O. Selomane. 2022. Climate change and the urgency to transform food systems. Science. 376(6600): 1416-1421.

CHAPTER

12

Enhancing Food Safety and Quality Assurance through Artificial Intelligence

Subrata Mandal[1], Patil Rajvardhan Kiran[1*], Silpa Mandal[2] and Pramod Aradwad[3]

[1] Division of Agricultural Engineering, Indian Agricultural Research Institute, New Delhi - 110012, India

[2] School of Agriculture Science, Sandip University, Sijoul, Bihar - 847235, India

[3] Department of Systems Process Engineering, Leibniz Institute for Agricultural Engineering and Bioeconomy eV (ATB), Max-Eyth-Allee 100, 14469 Potsdam, Germany

Introduction

Hygienic food production and quality assurance play crucial roles in both food and nutrition security, as well as the economic health of a nation. The World Health Organization (WHO) reports that around 600 million people fall ill annually due to contaminated food, highlighting significant deficiencies in food safety practices (WHO 2020). Traditional methods for assessing food safety and quality often involve destructive techniques, such as sampling followed by lab analysis, and adherence to standard operating procedures and codex regulations. However, these approaches can be inefficient and may yield inaccurate results due to human error and also time consuming (Palakurti 2022).

In recent years, the food industry has come under heightened scrutiny regarding safety and quality standards, fueled by growing consumer awareness, stricter regulations, and the complexities of global food supply chains (Chhetri 2024, Awuchi 2023). These combined pressures have created an urgent demand for innovative approaches to guarantee that food products are both safe for consumption and meet consumer quality expectations. To maintain food safety and ensure high standards, the industry faces several significant challenges. Recently, artificial intelligence (AI) and machine learning (ML) have shown great potential in addressing these challenges with promising solutions (Qian et al. 2023; Satpute et al. 2024).

*Corresponding author: rajvardhanp4545@gmail.com

DOI: 10.1201/9781003536932-12

AI refers to the intelligence programmed into machines by humans, enabling them to perform tasks that mimic human cognition. Its true potential is demonstrated not only by its ability to optimize existing processes, enhance automation, and improve information and transformation outcomes, but also by its capacity to recognize patterns, anticipate future needs, and engage with people effectively (Wamba-Taguimdje 2020; Nychas 2016). Machine learning and AI rely on mathematical algorithms, computations, and predictions. However, they can't be compared to the complexity of the human brain (Pradhan et al. 2024). Even so, the rapid evolution of this technology's impact on food industry innovation remains difficult to fully understand (Lee 2019; Shah 2023). These technologies allow for the analysis of massive data generated across the entire food supply chain, from farm to table, offering insights that were previously out of reach. By utilizing AI, food producers, processors, and retailers can detect potential risks, streamline operations, and ensure regulatory compliance more efficiently than conventional methods (Friedlander and Zoellner 2020; Guddaraddi et al. 2023; Jha et al. 2023).

The global food supply chain is complex, with multiple stakeholders and unique risks at every stage, from farming to retail. Traditional risk management methods, like manual inspections and periodic testing, may fall short in today's fast-paced environment. AI, however, allows for continuous monitoring and real-time decisions, greatly improving proactive risk management (Kushwaha et al. 2024). Moreover, the integration of AI in food safety and quality assurance can lead to increased efficiency and reduced waste (Kudashkina et al. 2022; Suthar et al. 2024). For instance, predictive analytics can forecast potential issues before they arise, allowing for timely interventions that prevent spoilage or contamination. This capability not only protects consumers but also contributes to sustainability by minimizing food waste throughout the supply chain (Vegesna et al. 2024).

As the food industry adopts digital transformation, AI's role is growing increasingly important. Advanced algorithms can analyze data from sources like sensors, IoT devices, and historical records to identify patterns and potential safety issues (Anand et al. 2024). Additionally, AI-powered automation can optimize processes from ingredient sourcing to production and distribution, improving overall quality management (Chen and Yu 2021).

This chapter aims to explore how artificial intelligence is revolutionizing food safety and quality standards. It focuses on the use of AI technologies, including machine learning, computer vision, and data analytics, to enhance safety protocols, prevent contamination, and streamline quality control across the supply chain. By tackling challenges in real-time monitoring and prediction, the chapter emphasizes the potential of AI to create more efficient, reliable, and scalable systems for ensuring food safety and quality assurance.

Challenges in Food Safety and Quality Assurance

Ensuring food safety and quality assurance remains a significant challenge for both developing and developed countries, as it directly impacts public health and economic stability. Several key challenges arise in this domain:

Foodborne Diseases

The occurrence of foodborne illnesses due to contamination by pathogens like bacteria, viruses, and parasites continues to be a major concern. Outbreaks of diseases such as E. coli, Salmonella, and Listeria highlight vulnerabilities in the food supply chain, from production to consumption.

Globalization of the Food Supply

The increasing globalization of the food supply chain complicates monitoring and regulation. Foods are often grown, processed, and distributed across multiple countries, creating gaps in regulatory oversight, making it difficult to track contamination sources or enforce consistent safety standards.

Adulteration and Fraud

The deliberate contamination or adulteration of food products for economic gain is a serious issue. Common examples include mixing substandard ingredients or mislabelling products, which threatens consumer trust and health. Food fraud is difficult to detect and prevent due to the complex nature of global supply chains.

Lack of Infrastructure and Technology

In many regions, especially developing countries, inadequate infrastructure, such as poor storage facilities, transportation, and lack of access to proper sanitation, hampers efforts to maintain food safety. Additionally, limited access to modern technology for food testing and monitoring exacerbates the problem.

Climate Change and Environmental Factors

Climate change impacts agricultural practices, leading to altered pest patterns and food production cycles. Rising temperatures and unpredictable weather conditions increase the risk of food spoilage and contamination, putting more pressure on food safety systems.

Regulatory Inconsistencies

Variations in food safety regulations across regions and countries result in inconsistent quality standards, leading to gaps in safety assurance.

Addressing these challenges requires robust policies, enhanced regulatory frameworks, and the adoption of advanced technologies to ensure that food safety and quality assurance systems are effective across the entire supply chain.

Importance of AI in Food Safety and Quality Control

AI is increasingly preferred over traditional approaches in food safety and quality control due to its ability to provide faster, more accurate, and cost-effective solutions. Traditional methods rely heavily on manual inspections, random sampling, and laboratory-based testing, which are often time-consuming, labor-intensive, and prone to human error. In contrast, AI systems leverage advanced data analytics, machine learning, and automation to enhance detection accuracy, minimize human intervention, and improve decision-making in real time. Artificial Intelligence (AI) plays a critical role in food safety and quality control, helping to ensure that food products meet safety standards and are free from contamination or defects (Kumar et al. 2021).

AI-powered computer vision systems are increasingly used in food production lines for real-time inspection of food products. These systems can detect physical defects, contamination, or irregularities in products with high accuracy and speed. For example: AI-driven cameras and sensors can identify foreign objects (like plastic or metal) in food products, reducing the risk of contamination. Image recognition algorithms can detect visible blemishes or color variations in fruits, vegetables, or packaged goods, ensuring that only high-quality products reach consumers.

AI can analyze large datasets, such as historical contamination events, environmental data, and supply chain information, to predict potential contamination risks before they occur. This is particularly important for monitoring and preventing contamination from pathogens like Salmonella or Listeria in processing plants, and from chemicals like pesticides. By applying machine learning algorithms to this data, AI can identify patterns and provide early warnings, helping companies take preventive actions and avoid costly recalls or foodborne illness outbreaks.

AI improves food traceability by overseeing the entire supply chain, from production to consumption, playing a crucial role in maintaining food safety. AI systems can track and trace food products in real time, providing detailed records of where the food came from, how it was processed, and how it was stored. This ensures that if there is a contamination issue, companies can quickly identify and isolate the affected batch, minimizing the risk of widespread recalls. AI algorithms can also assess storage conditions (e.g., temperature, humidity) to prevent spoilage or contamination during transport and storage (Pandey and Pandey 2023).

Machine learning algorithms can improve food quality control by analyzing large datasets from sensors and quality assessments (Malkani et al. 2024). For instance, AI can evaluate sensor data related to texture, color, or taste to ensure that products consistently meet quality standards. In processing plants, AI can analyze sensor data to monitor critical factors like cooking times, temperatures, or chemical levels in real time, ensuring that the final products meet safety and quality criteria.

AI-driven robots and automation systems are being deployed in food processing to ensure consistent and hygienic handling of food products. Automation reduces human contact with food, which lowers the risk of contamination, particularly in high-risk environments like meat processing plants. AI can ensure:

- Precise measurements and ingredient mixing, maintaining product consistency and safety.
- Real-time monitoring of processes, detecting anomalies or errors, and adjusting production parameters to ensure safety and quality.

Application of AI in Food Safety

AI-driven solutions are rapidly transforming the approach to tackling complex food safety challenges. By incorporating advanced technologies such as machine learning, computer vision, and IoT sensors, AI enables precise detection of contaminants, real-time monitoring of food quality, and enhanced traceability across supply chains (Fig. 1). These innovations not only minimize risks but also improve regulatory compliance and operational efficiency. As food safety demands grow due to increasing globalization and consumer awareness, AI offers a promising pathway to create safer, more transparent, and sustainable food systems.

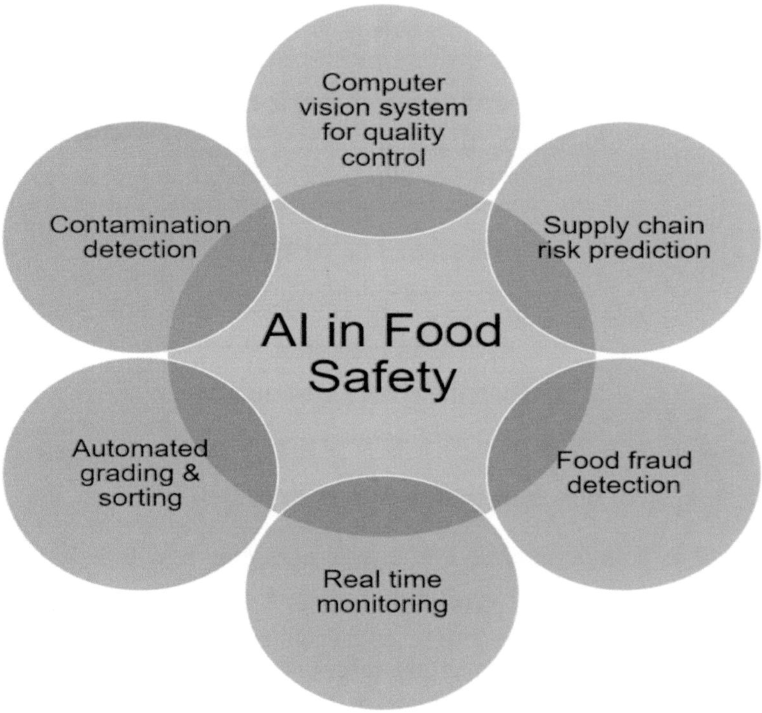

Fig. 1: AI in food safety.

Predictive Analytics

Artificial Intelligence (AI) has become a game-changer in the field of predictive analytics for food safety. By leveraging large datasets, machine learning algorithms, and real-time monitoring, AI can predict potential risks, improve decision-making, and prevent food safety incidents before they occur. Here are some key applications of AI in predictive analytics for food safety:

Early Detection of Contamination and Pathogens

AI can analyze historical data and environmental factors (e.g., temperature, humidity) to predict the likelihood of contamination by pathogens like E. coli, Salmonella, or Listeria in specific food products. Sensors integrated with AI systems can monitor food production environments in real-time to detect any early signs of contamination or bacterial growth (Fig. 2) (Rahul et al. 2023).

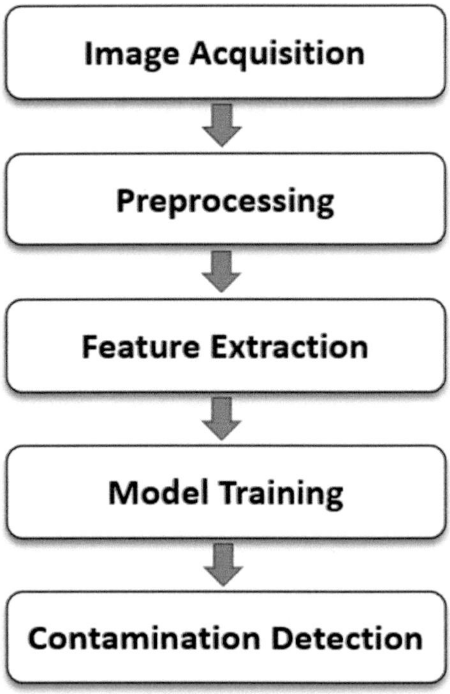

Fig. 2: Contamination detection method by CNN.

Predictive Maintenance of Food Processing Equipment

AI can be used to monitor and predict when critical food processing equipment (e.g., refrigeration units, cooking systems) might fail, allowing for proactive maintenance and preventing contamination due to improper processing. Predictive analytics can help optimize food production processes by identifying potential inefficiencies or deviations from standard operating procedures that could impact food safety.

Outbreak Prediction and Prevention

AI models can analyze vast amounts of data from multiple sources (e.g., historical health data, weather patterns, food production reports) to predict the likelihood of foodborne illness outbreaks and help authorities take preventive measures. AI can assist in the early identification of foodborne illness outbreaks by analyzing social media, health reports, and other data streams for signs of emerging health threats.

Predictive Food Safety Inspections

AI can analyze data from past food safety inspections and incidents to predict which facilities or food products are most at risk, allowing regulatory agencies to prioritize high-risk inspections. AI-powered computer vision and machine learning models can be used to automate routine inspection tasks, such as detecting cleanliness, compliance with safety standards, and proper food handling practices in real-time.

Food Allergen Risk Prediction

AI can analyze food ingredient data and production environments to predict potential cross-contamination with allergens, helping manufacturers ensure allergen-free production and reduce the risk of recalls. AI can help consumers manage food allergies by analyzing ingredient labels and predicting potential allergen risks in packaged foods or restaurant meals based on historical data.

Personalized Food Safety Recommendations for Consumers

AI can provide consumers with personalized recommendations on food safety practices (e.g., proper storage, handling, cooking) based on their purchasing habits, dietary preferences, and location. AI-powered mobile apps can analyze food labels, ingredients, and expiration dates to offer real-time food safety guidance to consumers (Abdel Qader 2023).

Automated Inspection

AI-powered automated inspection systems are revolutionizing food safety by enhancing accuracy, efficiency, and scalability in detecting quality issues and contaminants in food products. These systems minimize human error and provide real-time analysis, making them a valuable tool for food manufacturers, retailers, and regulators. Here are some key applications of AI in automated inspection for food safety:

Foreign Object Detection

AI-powered computer vision systems can detect foreign materials (such as plastic, metal, wood, or glass) in food products during production and packaging. These systems can identify even small or camouflaged objects in real-time, preventing contamination from reaching consumers. AI can enhance the analysis of X-ray or ultrasound images to detect hidden foreign objects embedded within food items, providing a non-invasive and thorough inspection (Patel et al. 2023).

Automated Grading and Sorting

AI algorithms are widely used to grade food items based on size, shape, and color. For example, fruits, nuts, and vegetables can be sorted automatically to ensure uniform quality. This enhances sorting speed and accuracy compared to traditional manual methods. AI systems can analyze visual data to assess ripeness, freshness, or maturity, allowing only products at their peak quality to be packaged and sold (Ranjan et al. 2023).

Label Inspection and Compliance

AI-based systems inspect food labels to verify the accuracy and presence of essential information, such as expiration dates, ingredient lists, and allergen warnings. AI ensures that mislabeled products are flagged before they reach the market, preventing potential safety issues. AI can detect packaging defects such as broken seals, misalignment, or tears, ensuring the integrity of the product is maintained during distribution and storage.

Food Fraud Detection

AI can analyze patterns in data related to food origins, supply chain transactions, and ingredient authenticity to predict and detect fraudulent practices such as adulteration, mislabeling, or counterfeiting. AI models can detect anomalies in ingredient compositions that could indicate tampering or substitution with inferior products.

Meat and Poultry Inspection

AI-based vision systems can assess the quality of meat and poultry by analyzing features like marbling, color, fat content, and texture. These systems ensure consistent grading of meat products based on predefined quality standards. AI, combined with hyperspectral imaging or ultraviolet light analysis, can detect microbial contamination on the surface of meat, identifying pathogens like Salmonella or E. coli before the products leave the facility.

Seafood Inspection

AI systems can evaluate seafood freshness by analyzing visual cues like eye clarity, gill color, and texture. These systems ensure only fresh, high-quality seafood reaches the consumer, reducing the risk of spoilage-related illnesses. AI is used to grade and sort seafood such as shrimp and fish based on size, weight, and quality attributes. This speeds up processing while maintaining quality standards (Adetunji et al. 2022).

Thermal Imaging for Temperature Control

AI-driven thermal cameras monitor temperature during cooking, pasteurization, or cooling processes, ensuring that food is processed within safe temperature ranges. AI can detect anomalies and trigger alerts if temperature deviations occur. AI is used to monitor the temperature of food products throughout the supply chain, ensuring that perishable items are stored and transported at optimal temperatures to prevent spoilage or contamination.

Automation of Regulatory Compliance

AI-driven inspection systems can ensure that food processing facilities meet regulatory standards in real-time by monitoring food safety compliance throughout the production process. AI can automate inspections for critical control points (HACCP) and provide documentation for audits. AI can generate automatic reports based on inspections, making it easier to track compliance, identify potential issues, and implement corrective actions.

Real Time Monitoring

AI has significantly enhanced real-time monitoring in food safety, helping food producers, manufacturers, and regulators ensure product quality and reduce contamination risks. By leveraging AI-driven sensors, cameras, and data analytics, real-time monitoring systems can detect potential hazards, ensure compliance, and provide instant feedback to prevent foodborne illnesses. Here are key applications of AI in real-time monitoring for food safety:

Temperature and Humidity Monitoring

AI systems monitor temperature and humidity conditions in real-time during the storage and transportation of perishable food products, ensuring they remain within safe limits. If temperatures deviate from safe ranges, the AI system can trigger alerts or corrective actions to prevent spoilage or contamination. AI-driven thermal cameras can monitor the temperature of food during cooking, pasteurization, or sterilization processes. This ensures that food is heated to the correct temperature for the required duration to eliminate harmful microorganisms.

Real-Time Quality Control

AI models are trained to identify visual defects such as discoloration, bruises, deformities, and other quality defects in fruits, vegetables, and processed foods. AI can ensure uniformity and compliance with quality standards by inspecting at high speeds in real time (Chowdhury et al. 2024). AI combined with techniques like hyperspectral or thermal imaging can assess internal properties like ripeness, moisture content, or potential spoilage in food products without causing damage, providing non-invasive quality control (da Costa et al. 2022).

Real-Time Traceability and Integration of Blockchain Technology

AI systems can track food products in real-time throughout the supply chain, from farm to table. This traceability ensures that every step of the supply chain is monitored for compliance with food safety standards, providing a full history of each product's journey. AI can integrate with blockchain technology to provide real-time, immutable records of each transaction in the food supply chain. This enhances traceability and ensures food safety by verifying the authenticity of products and identifying sources of contamination during outbreaks (Pandey and Pandey 2023).

Real-Time Food Quality and Freshness Indicators

AI-enabled smart packaging with integrated sensors can predict food quality and freshness levels by monitoring environmental conditions, such as temperature and humidity, throughout the supply chain. AI can use image processing to assess the visual quality of food items (e.g., color, texture) and predict spoilage or degradation before it becomes apparent to the naked eye.

Employee Hygiene and Sanitation Monitoring

AI-driven cameras and sensors can monitor employee hygiene practices in real-time, ensuring compliance with food safety protocols such as hand-washing, wearing protective clothing, and following proper sanitization procedures. AI systems can track cleaning schedules and monitor sanitation efforts in food processing facilities. By analyzing environmental data and comparing it to cleaning protocols, AI ensures that surfaces, equipment, and workspaces meet cleanliness standards.

Table 1: Application of AI in food safety and quality control

Sample	Objectives	AI Methods	Outcomes	References
Apples	Detection of bruises	Convolutional Neural Network (CNN) with Hyperspectral Imaging	Detected internal bruises with 90% accuracy, reducing defective apples in the market and improving sorting efficiency.	Gai et al. 2022
Poultry	Contaminant detection (e.g., Salmonella)	Support Vector Machine (SVM)	Reduced detection time of Salmonella, improving food safety and allowing faster intervention in poultry processing.	Hortinela et al. 2020
Milk	Prediction of spoilage and shelf life	Artificial Neural Network (ANN)	Predicted spoilage with 92% accuracy, improving management of milk shelf life and reducing wastage in the supply chain.	Yang and Wei 2021
Fish	Freshness assessment	Deep Learning (DL)	Enabled real-time freshness assessment, leading to improved handling and less spoilage during storage and transport.	Yasin et al. 2023
Tomatoes	Ripeness classification	Machine Vision and CNN	Achieved 89% accuracy in ripeness classification, helping reduce labor costs in manual sorting and enhancing consistency.	Alam Siddiquee et al. 2020
Packaged Foods	Foreign object detection	CNN with X-ray Imaging	Detected foreign objects like plastic and metal with 94% accuracy, lowering contamination risks in packaged foods.	Kim et al. 2021
Ground Beef	E. coli detection	Random Forest	Improved safety by detecting E. coli contamination, reducing the risk of recalls and public health incidents.	Al et al. 2024
Wheat	Quality assessment (moisture content)	SVM and ANN	Achieved accurate predictions of moisture content, improving post-harvest storage conditions and minimizing the risk of spoilage.	Abdollahpour et al. 2020
Shrimp	Chemical contamination detection	Fuzzy Logic	Reduced chemical contamination detection time by 25%, enhancing food safety in shrimp export and processing industries.	Adetunji et al. 2022

Potatoes	Disease detection	CNN and Decision Trees	Detected blight and other diseases with 91% accuracy, allowing early treatment and reducing crop losses.	Kumar and Patel 2023
Salmon	Shelf-life prediction	CNN	Improved shelf-life prediction, enabling better inventory management and reducing food waste by up to 15%.	Wu et al. 2022
Strawberries	Quality control and sorting	Image Processing and CNN	Enhanced sorting efficiency by accurately classifying strawberries based on size, shape, and color, reducing manual labor requirements.	Choi et al. 2021
Bread	Texture analysis	K-Nearest Neighbors (KNN)	Detected texture defects in bread production with 85% accuracy, ensuring consistent quality and reducing waste.	Kavitha et al. 2022
Snack Foods	Packaging defect detection	Machine Vision with CNN	Identified packaging defects like seal failures with 91% accuracy, improving packaging quality and reducing customer complaints.	Minghui et al. 2023
Cheese	Prediction of fat and protein content	ANN	Improved predictions of fat and protein content, ensuring consistent product quality and regulatory compliance.	Karapetkovska-Hristova et al. 2018
Peanuts	Allergen contamination detection	Decision Tree Algorithm	Detected cross-contamination risks of allergens with 83% accuracy, helping prevent allergic reactions and ensuring food safety.	Luparelli et al. 2023
Lettuce	Microbial contamination prediction	Random Forest	Predicted contamination risk during handling, reducing the likelihood of foodborne illness outbreaks in leafy greens.	Santos et al. 2024
Rice	Automated grading and sorting	CNN	Enhanced grading and sorting operations, achieving faster processing times and improved consistency in rice quality.	Ranjan et al. 2023

Real-Time Food Spoilage Detection

Smart packaging equipped with AI-enabled sensors can monitor the freshness and quality of food products in real-time. Sensors embedded in the packaging can detect changes in temperature, gas emissions (e.g., ethylene for ripeness or spoilage), or moisture levels, signalling when a product is near spoilage. AI systems can analyze gas emissions from food products (e.g., carbon dioxide, ammonia) to detect spoilage or chemical reactions that indicate the food is deteriorating. Real-time monitoring helps prevent spoiled products from reaching consumers.

AI in real-time monitoring for food safety brings advanced capabilities to the food industry, enabling timely detection and prevention of contamination, spoilage, and other risks. From supply chain monitoring to equipment maintenance and employee hygiene, AI-driven real-time monitoring ensures safer food production processes, better compliance with regulatory standards, and improved product quality. This proactive approach to food safety is helping to reduce recalls, minimize foodborne illnesses, and promote sustainability throughout the food industry. Different applications of AI in different food samples are discussed in Table 1.

Applications of AI in Quality Assurance

Sensory evaluation

Overview of AI-Driven Sensory Evaluation Systems for Quality Assessment

In recent years, the integration of artificial intelligence (AI) in sensory evaluation systems has revolutionized the way industries, especially in food and agriculture, assess the quality of their products (Pateiro et al. 2020). Traditional sensory evaluation methods rely on human perception to analyze taste, smell, texture, and visual aspects, which can be subjective and inconsistent (Zhang et al. 2020). AI-driven sensory evaluation systems, on the other hand, offer a more objective, scalable, and efficient approach to quality assessment. This article provides an overview of how AI is applied to sensory evaluation systems, its advantages, and some challenges associated with the technology (Fernández-López et al. 2019).

The Need for AI-Driven Sensory Evaluation Systems

Human sensory panels have been the standard for quality assessment in industries like food, beverages, cosmetics, and even textiles. While human perception is rich and complex, it is also inherently subjective, prone to variability, fatigue, and biases. Standardizing quality across large batches of products becomes difficult when subjective opinions vary (Fernández-López et al. 2019).

AI-based systems, combined with advanced sensory hardware (such as electronic noses, electronic tongues, and machine vision systems), can objectively analyze products and provide consistent results (Pateiro et al. 2020). These systems can be programmed to recognize patterns, interpret data from multiple sources, and learn from feedback, making them invaluable for quality control.

Key Technologies in AI-Driven Sensory Evaluation

Several AI technologies form the backbone of modern sensory evaluation systems. These include machine learning (ML), deep learning (DL), computer vision, and sensor fusion technologies. Each of these technologies plays a unique role in evaluating different sensory attributes.

Machine Learning (ML) and Deep Learning (DL): ML algorithms allow sensory systems to recognize patterns in data and classify products based on quality parameters. Deep learning, a subset of ML, is particularly useful for complex sensory tasks like recognizing intricate taste profiles or identifying visual imperfections in food products (Yang et al. 2020; Kiran et al. 2024).

Computer Vision: Computer vision technology enables the assessment of the visual characteristics of products, such as shape, color, and texture. AI algorithms process images captured by cameras and can instantly detect flaws like bruises on fruits, discoloration in vegetables, or defects in textiles (Zhang et al. 2021).

Electronic Noses and Electronic Tongues: Electronic tongues (e-tongues) and electronic noses (e-noses) are sensory technologies designed to replicate the functions of the human sense of smell and taste. Combined with AI, these devices can differentiate between various chemical compounds, identify off-flavors or undesirable odors, and ensure flavour consistency.

Sensor Fusion: Sensor fusion refers to the integration of data from multiple sensors (e.g., visual, olfactory, and gustatory sensors) to provide a comprehensive assessment of product quality. AI-driven systems can analyze data from various sensory inputs to offer a more holistic evaluation (Fernández et al. 2019).

Role of Machine Learning Algorithms in Interpreting Sensory Data

Machine learning (ML) algorithms play a transformative role in interpreting sensory data, offering robust solutions for analyzing complex and multi-dimensional data sets that arise from human sensory inputs or sensory devices like electronic noses, tongues, and vision systems. These algorithms enable the extraction of valuable patterns from noisy data, leading to accurate, objective, and scalable sensory evaluations (Pateiro et al. 2020).

In sensory science, traditional methods rely heavily on human panels, which can be subjective and inconsistent. Machine learning provides a more objective approach by training models to recognize patterns in data collected from sensory devices (Fernández et al. 2019). For instance, in food quality control, ML algorithms analyze data from electronic noses and tongues to assess taste, aroma, and texture, making it possible to distinguish between slight variations in flavor or freshness (Zhang et al. 2020).

Several ML techniques are commonly used to interpret sensory data. Supervised learning algorithms, such as decision trees, support vector machines (SVM), and random forests, are widely used for classification tasks, where products are categorized based on sensory attributes. Unsupervised learning methods, such as clustering, can identify natural groupings in data, helping discover hidden patterns in consumer preferences or product defects (Li et al. 2022). Deep learning, particularly convolutional neural networks (CNNs), is particularly effective in image analysis, where sensory data in the form of images are processed to detect flaws or quality differences in products like fruits and textiles (Kiran et al. 2024a). By utilizing these algorithms, industries can achieve more consistent quality control, reduced reliance on human evaluators, and faster processing of sensory data. The integration of ML algorithms in sensory data interpretation is a key step toward the future of automated, AI-driven quality assessments (Fernández-López et al. 2019).

Applications of AI-Driven Sensory Evaluation Systems

AI-driven sensory evaluation systems are being increasingly applied across different industries. Some of the most prominent applications include:

Food Industry: AI systems are being used to assess the quality of food products by analyzing sensory attributes like flavor, aroma, texture, and appearance. For instance, AI-powered computer vision can inspect the ripeness of fruits, while electronic tongues can evaluate the bitterness or sweetness of beverages. The food industry also uses AI to predict shelf-life, enhance product development, and ensure product safety (Zhang et al. 2020).

Wine and Beverage Industry: E-noses and e-tongues, paired with AI, have been applied to evaluate the quality of wines, beers, and other beverages. These systems can help identify subtle flavor profiles and ensure consistency between batches (Domínguez et al. 2021).

Agriculture: AI-driven sensory systems are being used to monitor the quality of crops and livestock. These systems can detect diseases, assess crop ripeness, and even evaluate the quality of milk or meat products (Zhang et al. 2021).

Benefits of AI-Driven Sensory Evaluation Systems

AI-driven sensory evaluation systems offer several advantages over traditional human-based assessments:

Consistency and Objectivity: AI systems provide consistent and objective results, reducing human error and variability. This leads to more reliable quality control and ensures that products meet the desired standards.

Scalability: AI-driven systems can process large volumes of data at high speeds, making them suitable for industries where rapid quality assessment is necessary. This scalability allows industries to maintain high production rates without compromising on quality.

Cost Efficiency: While the initial setup of AI-driven sensory systems may be expensive, they can reduce long-term costs by minimizing the need for human sensory panels and reducing product wastage due to inaccurate quality assessment.

Continuous Learning and Improvement: AI systems can learn from new data and improve over time. This makes them adaptive to changing quality parameters and evolving consumer preferences.

Reduction in Sensory Fatigue: Human sensory evaluators can experience fatigue, which can affect the accuracy of assessments. AI systems, however, are immune to fatigue and can operate continuously without a drop in performance.

Supply Chain Management

The global food industry, characterized by its complexity and scale, is increasingly adopting advanced technologies such as Artificial Intelligence (AI) and blockchain to address challenges related to traceability, transparency, and operational efficiency. These technologies are particularly valuable for enhancing the safety, quality, and authenticity of food and its by-products. In this context, AI and blockchain are playing crucial roles in transforming how food supply chains operate, from farm to fork (Galvez et al. 2018). This article provides an overview of AI-driven supply chain solutions for traceability and transparency, the application of blockchain in verifying food product authenticity and origin, and the benefits of AI in optimizing supply chain processes.

AI-Driven Supply Chain Solutions for Traceability and Transparency in the Food Industry

Traceability and transparency are essential in the food industry, where food safety, quality, and compliance with regulations are paramount. Consumers are increasingly demanding to know the origins of their food and how it was produced, particularly in the wake of concerns about food fraud, contamination, and sustainability. AI-driven supply chain solutions provide a powerful means to address these concerns, enabling businesses to ensure product safety and build consumer trust (Fig. 3) (Feng et al. 2020).

Traceability refers to the ability to track a product through the different stages of the supply chain, from raw materials to the final consumer. AI technologies such as machine learning, computer vision, and predictive analytics are used to gather and analyze large volumes of data at each stage of the food supply chain, ensuring real-time visibility and control over the movement of food products (Kiran et al. 2024b).

For example, in the context of fresh produce, AI-driven systems can integrate data from sensors, satellite imagery, and weather forecasts to monitor crop conditions, predict yields, and track harvesting processes. This level of insight enables food producers to better manage quality and reduce the likelihood of food safety issues such as contamination or spoilage. Furthermore, AI systems can analyze historical data on supply chain disruptions and provide early warnings of potential issues, such as transport delays or temperature fluctuations in storage, which could affect the quality of food products.

Computer vision technologies and natural language processing (NLP) are used to analyze labeling and packaging information, ensuring adherence to food safety standards and regulations. This enables companies to promptly detect non-compliant batches and remove them from the supply chain before they reach consumers.

Additionally, AI enhances transparency by offering a detailed view of the food supply chain. By combining data from suppliers, manufacturers, distributors, and retailers, AI can create a digital replica of the entire supply chain for better oversight. This real-time model allows stakeholders to see where products are at any given

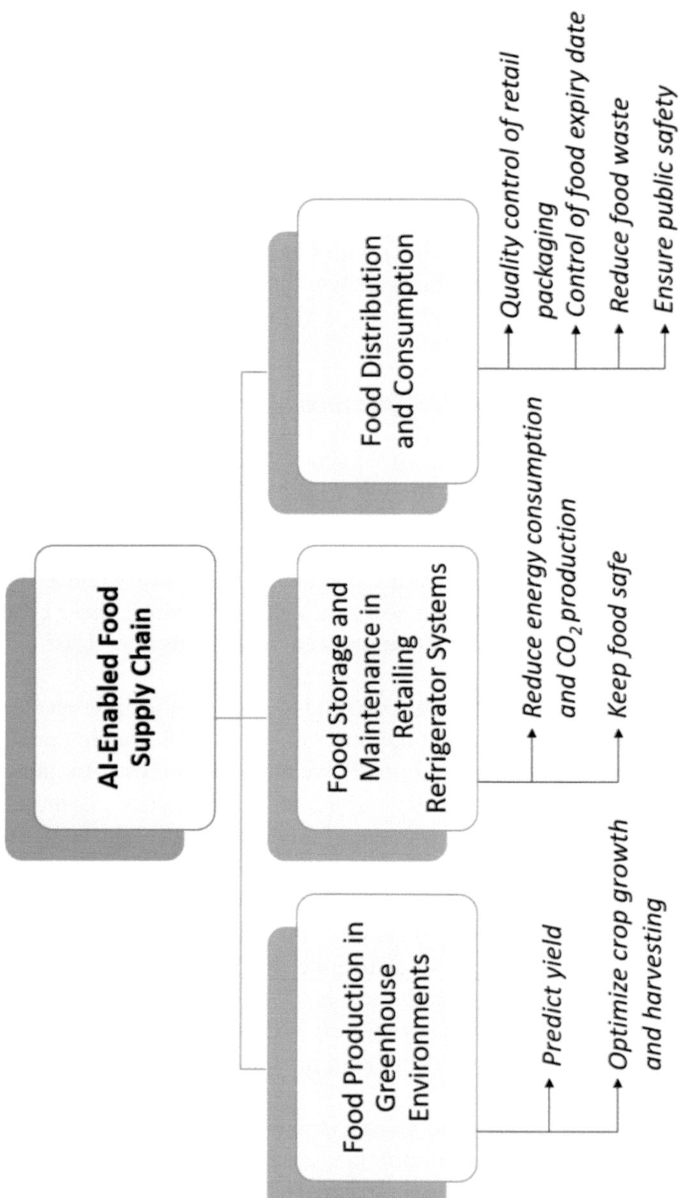

Fig. 3: AI-enabled food supply chain.

moment, how they are being stored, and any changes in their conditions (Kumar 2020). This information can then be shared with consumers via digital platforms, empowering them to make informed decisions about the food they purchase.

Application of Blockchain Technology in Verifying Product Authenticity and Origin

Blockchain technology is rapidly gaining traction in the food industry for its ability to verify the authenticity and origin of products. Food fraud, such as the mislabelling of organic produce or the substitution of premium ingredients with cheaper alternatives, is a significant concern for both consumers and businesses. Blockchain provides a decentralized and tamper-proof system that records every transaction and movement of food products throughout the supply chain (Regattieri et al. 2021).

One of the key applications of blockchain is the verification of product authenticity. By using blockchain, food producers can create an immutable record of the journey a food product takes from the farm to the consumer. Each step in the process, such as harvesting, processing, packaging, and distribution, is recorded on the blockchain, creating a digital history that cannot be altered. This makes it easy for businesses and consumers to verify whether a product is genuine and trace its origin back to the source.

For instance, a coffee company that sources beans from multiple countries can use blockchain to document where the beans were grown, how they were processed, and when they were shipped. By scanning a QR code on the packaging, consumers can access the blockchain record and verify that the coffee is indeed organic, fair trade-certified, or sustainably sourced, as claimed by the manufacturer (Galvez et al. 2018).

Blockchain is being utilized to improve food safety by creating a transparent and traceable record of food handling throughout the supply chain. In the case of a foodborne illness outbreak, blockchain technology can accurately identify the source of contamination, allowing for quicker and more precise recalls. This helps prevent further spread of contamination while reducing the financial losses faced by food producers and retailers.

In the seafood industry, for example, blockchain is being used to combat illegal fishing practices and ensure that seafood products are sustainably sourced. Companies can record each catch on the blockchain, along with information about the fishing vessel, location, and method used. This allows consumers and regulators to verify that the seafood they purchase was caught in compliance with environmental and legal standards.

The combination of AI and blockchain further enhances the ability to verify product authenticity and origin. While blockchain provides a secure, immutable record of transactions, AI can analyze the data stored on the blockchain to detect patterns and anomalies, such as discrepancies in product sourcing or suspicious behavior from suppliers. This creates a more robust system for ensuring the integrity of the food supply chain (Tripathi et al. 2019).

Benefits of AI in Optimizing Food Supply Chain Processes

Beyond traceability and transparency, AI is transforming the food supply chain by optimizing key processes, improving efficiency, and reducing costs. Some of the main areas where AI has a significant impact include demand forecasting, inventory management, transportation logistics, and warehouse automation.

Demand forecasting is one of the most critical areas where AI offers substantial benefits. By analyzing historical sales data, market trends, weather patterns, and even social media activity, AI systems can accurately predict consumer demand for various food products. This allows food producers and retailers to adjust their production schedules, optimize inventory levels, and reduce the risk of overstocking or stockouts (Regattieri et al. 2021). In the highly perishable food industry, where waste is a significant issue, AI-driven demand forecasting helps minimize food spoilage by ensuring that supply matches demand more closely.

AI is also transforming inventory management in food warehouses and distribution centers. Through the use of AI-powered robots and autonomous systems, companies can automate tasks such as sorting, packing, and inventory tracking (Tripathi et al. 2019). These systems use computer vision and machine learning algorithms to detect product quality, monitor expiration dates, and ensure that food is stored and rotated according to best practices. This automation reduces human error, improves efficiency, and ensures that products reach consumers in the best possible condition (Kiran et al. 2024b).

In transportation logistics, AI helps optimize delivery routes and schedules based on factors such as traffic conditions, fuel costs, and delivery deadlines. By using predictive analytics, AI systems can also anticipate potential disruptions, such as adverse weather or road closures, and adjust routes accordingly. This ensures that food products are delivered on time and in optimal condition, reducing the likelihood of spoilage and minimizing transportation costs (Feng et al. 2020).

Moreover, AI is playing a pivotal role in food processing and quality control. For example, AI-powered cameras and sensors can inspect food products on the production line, detecting defects such as bruising, discoloration, or contamination that human inspectors might miss. Machine learning algorithms can then analyze this data in real time, enabling companies to make quick decisions about whether to accept or reject a batch of products. This level of precision helps maintain product quality and reduces the risk of unsafe or substandard food reaching consumers (Galvez et al. 2018).

Sustainability is another critical area where AI is making a difference in food supply chains. By optimizing processes such as production, transportation, and storage, AI helps reduce food waste, lower energy consumption, and minimize the environmental impact of supply chain operations (Kumar 2020). For example, AI can analyze energy usage data in cold storage facilities and recommend adjustments to minimize energy waste while maintaining optimal storage conditions for perishable products.

Shelf-life Prediction

Introduction to AI Models for Predicting Product Shelf Life

Shelf-life prediction in the food industry is crucial for ensuring food safety, quality control, and minimizing food waste. Traditionally, the shelf life of food products has been determined through time-consuming empirical methods, including sensory evaluations, chemical analyses, and microbial testing. These approaches, while effective, are resource-intensive and often lack precision due to the variability in environmental factors, storage conditions, and inherent product characteristics (Galanakis et al. 2019).

With advancements in technology, Artificial Intelligence (AI) models have emerged as powerful tools for predicting the shelf life of food products. These models leverage vast datasets and sophisticated algorithms to analyze various parameters that impact food degradation, allowing for more accurate and faster predictions (Charlebois et al. 2020). The integration of AI-driven models enables real-time shelf-life monitoring and prediction, which can enhance food safety, optimize inventory management, and reduce waste throughout the supply chain (Galanakis et al. 2019).

AI models rely on techniques such as machine learning (ML) and deep learning (DL) to create predictive models. Machine learning algorithms such as decision trees, support vector machines (SVM), and artificial neural networks (ANNs) are employed to study historical data and identify patterns that correlate with shelf-life degradation. Deep learning models, especially convolutional neural networks (CNNs), are used in more complex cases where sensory data (e.g., images of food products) or multi-modal inputs (e.g., environmental conditions, packaging type) need to be analyzed for shelf-life estimation.

Moreover, Internet of Things (IoT) devices and sensors are often integrated into these AI systems to collect real-time data on variables like temperature, humidity, and light exposure. This data is then fed into the AI models, allowing for continuous monitoring and dynamic adjustment of shelf-life predictions based on changing storage or transportation conditions (Charlebois et al. 2020). The food industry stands to benefit significantly from the adoption of AI models for shelf-life prediction, as these systems offer a more dynamic, flexible, and accurate approach compared to traditional methods.

Factors Influencing Shelf-Life Prediction and Accuracy of Models

Accurate shelf-life prediction using AI models depends on a wide range of factors, both intrinsic and extrinsic to the food products. Understanding these factors is essential for developing robust models that can provide reliable predictions.

Intrinsic Factors

Food Composition: The chemical and physical properties of the food itself, such as its moisture content, fat composition, pH, and microbial load, play a crucial role in determining its shelf life. AI models need to take into account the specific composition of each product to accurately predict its shelf life. For example, high-fat foods are more prone to oxidative rancidity, while high-moisture products may be more susceptible to microbial spoilage.

Product Formulation: Changes in the recipe or formulation of a product, such as the use of preservatives or antioxidants, can significantly impact its shelf life. AI models must account for these variations to accurately predict how different formulations will behave over time.

Packaging: The type of packaging used for food products can greatly influence their shelf life by controlling factors such as oxygen exposure, moisture retention, and light penetration. AI models often incorporate packaging information into their predictions to assess how different packaging materials and techniques (e.g., vacuum sealing, modified atmosphere packaging) affect shelf stability.

Extrinsic Factors

Storage Conditions: Temperature, humidity, and light are critical factors that influence the rate of food deterioration. AI models rely on data collected from sensors placed in storage facilities or transportation vehicles to monitor these variables in real-time. By incorporating this data, AI systems can predict how fluctuating storage conditions affect the remaining shelf life of products (Kiran et al. 2024d).

Transportation Variables: The way food is handled and transported can have a significant impact on its shelf life. Variations in temperature during transportation, exposure to contaminants, and delays can all accelerate spoilage. AI models that include transportation data can make real-time adjustments to shelf-life predictions, providing more accurate and dynamic estimations (García-Gimeno et al. 2019).

Supply Chain Dynamics: The length and complexity of the supply chain affect shelf life. Food products that undergo extended transportation times or pass through multiple intermediaries are more likely to experience shelf-life degradation. AI models must account for these factors, making them integral to an end-to-end view of product life.

Model Factors

Data Quality and Availability: The accuracy of AI predictions is directly tied to the quality and volume of data available. Reliable sensors and comprehensive historical data are necessary for building precise models. If the data is incomplete or inconsistent, the model's predictions may be less accurate (Kiran et al. 2024c).

Algorithm Selection: Different machine learning algorithms are suited to different types of data. For instance, neural networks are well-suited for complex, non-linear relationships between variables, while decision trees might be better for simpler, hierarchical decisions. Selecting the right algorithm for the specific food product and environment is crucial for maximizing accuracy.

Model Training: AI models need to be trained on large datasets to identify the factors that most affect shelf life. A well-trained model will recognize patterns and make accurate predictions. However, a model that has been trained on insufficient or biased data may struggle to generalize to new scenarios, leading to inaccurate predictions (Kiran et al. 2024c).

Case Examples Illustrating the Use of AI in Shelf-Life Prediction

Several companies and research institutions have successfully implemented AI-driven shelf-life prediction models across various food sectors. These case examples highlight how AI can be used to enhance shelf-life prediction and address key challenges in the food industry.

Fresh Produce Monitoring Using AI: The perishable nature of fresh fruits and vegetables makes shelf-life prediction a crucial aspect of reducing food waste and ensuring product quality. A major supermarket chain in Europe implemented an AI-driven model that uses data from IoT sensors to monitor the temperature and humidity levels in its warehouses (Verdouw et al. 2021). By integrating real-time data from storage and transportation with predictive algorithms, the AI model was able to provide dynamic shelf-life estimates for fresh produce. As a result, the supermarket reduced food spoilage by 30% and improved its inventory management, ensuring that products were sold before reaching their expiration dates (García-Gimeno et al. 2019). Additionally, image recognition models were used to detect spoilage in fruits and vegetables. AI models trained to recognize visual cues of degradation, such as color changes or texture alterations, were integrated into quality control systems to automatically discard produce that had exceeded its predicted shelf life (Kiran et al. 2024c).

Meat Shelf-Life Prediction Using AI and IoT: In the meat industry, spoilage is a significant concern due to the risk of bacterial contamination, particularly from pathogens such as E. coli and Salmonella. A leading meat processing company developed an AI-powered system that combined real-time data from IoT sensors in cold storage facilities with historical data on microbial growth rates under varying conditions (Charlebois et al. 2020). The AI model was able to predict spoilage more accurately than traditional methods, adjusting the estimated shelf life based on temperature fluctuations during storage and transportation. By using this AI model, the company improved its inventory turnover rate and reduced the need for costly recalls, as it could now identify potential risks earlier and prevent contaminated products from reaching consumers (Karim and Karim 2021).

Dairy Product Shelf-Life Management: Dairy products, particularly milk and yogurt, have relatively short shelf lives and are highly sensitive to temperature changes. An AI model developed by a dairy company used historical sales data, environmental data (temperature, humidity), and microbial growth data to predict the shelf life of its products. The model was able to predict the remaining shelf life of dairy products at different stages of the supply chain, from production to retail shelves. The system was also linked to an automated inventory management platform, which adjusted orders and distribution schedules based on real-time shelf-life predictions. This led to a significant reduction in product waste and ensured that consumers received fresher products (Karim et al. 2021).

Advantages and Disadvantages of AI in Food Industry Application

Advantages

Personalization and Consumer Engagement

- *Customized Nutrition Plans:* Apps can use AI to recommend personalized meal plans based on individual health metrics and dietary restrictions. Companies like Noom leverage AI for weight loss and health management.

- *Dynamic Menu Design:* Restaurants can use AI to adjust menus in real-time based on what ingredients are in stock or popular among customers, enhancing responsiveness to consumer demand.

Sustainability Initiatives

- *Resource Optimization:* AI can analyze data to optimize water usage in agriculture, leading to more sustainable farming practices. For example, precision irrigation systems can adjust water supply based on soil moisture data.
- *Waste Management:* AI can track inventory levels and expiration dates, suggesting redistribution or discounting of products nearing their expiration date, reducing waste.

Food Safety Enhancements

- *Traceability:* Blockchain and AI can work together to trace the supply chain of food products, allowing quick identification of contamination sources. This is vital in recalls, such as when a pathogen is found in spinach.
- *Pathogen Detection:* AI can be used in combination with sensors and imaging technologies to detect pathogens on food surfaces, improving safety before products reach consumers.

Research and Development

- *Innovative Product Creation:* AI can analyze market data to identify gaps in the market, helping companies innovate new products that meet emerging consumer needs. For instance, AI can suggest new flavors or ingredients that are gaining popularity.
- *Flavor Pairing:* AI can analyze chemical compositions of various ingredients to suggest unique flavor pairings. This technique is utilized by chefs to create novel dishes.

Consumer Experience

- *Chatbots for Customer Service:* AI-driven chatbots can handle customer inquiries 24/7, improving service efficiency. For example, restaurants can use chatbots for reservation management and order-taking.
- *Enhanced Ordering Systems:* AI can streamline online food ordering processes by providing personalized suggestions based on past orders and preferences.

Disadvantages

Job Displacement

- *Skill Gap:* Automation may lead to the displacement of workers who perform routine tasks. The transition to an AI-driven environment requires upskilling existing workers, which can be a significant challenge.

High Initial Investment

- *Cost Barriers:* Small and medium-sized enterprises (SMEs) might face challenges in affording the technology and infrastructure required for implementing AI solutions, which could increase the disparity between larger corporations and smaller businesses.

Data Privacy and Security

- *Sensitive Data Handling:* The collection of consumer data for personalization and analysis raises concerns about data breaches and unauthorized access. Companies must ensure robust cybersecurity measures are in place to protect consumer information.

Complexity and Integration Issues

- *System Compatibility:* Integrating AI with existing systems can be complicated and resource-intensive. For example, a restaurant trying to incorporate AI into its inventory system might face challenges aligning new technologies with legacy systems.
- *Need for Skilled Personnel:* Organizations require skilled data scientists and AI experts to interpret data and manage AI systems, which can be difficult to recruit.

Bias and Ethical Concerns

- *Algorithmic Bias:* AI systems can inadvertently perpetuate existing biases if trained on biased datasets, leading to unfair treatment in marketing or product development. For example, targeting certain demographics more aggressively based on historical data could lead to discriminatory practices.
- *Transparency Issues:* Consumers may be sceptical about how AI-driven recommendations are generated, leading to mistrust. Transparency in algorithms and decision-making processes is crucial for building trust.

Conclusions and Future Directions

The use of AI in food safety and quality assurance has the potential to transform the industry by enhancing contaminant detection, quality monitoring, and supply chain transparency. Technologies like machine learning, computer vision, and IoT sensors provide real-time insights that minimize human error and ensure compliance with regulations. However, challenges remain, including the need for extensive datasets, high costs, and difficulties in system integration. Future research should prioritize scalable AI solutions for smaller businesses, improved algorithms for complex food systems, and better data-sharing frameworks. Collaboration between AI, blockchain, and automation could further revolutionize global food safety in the future.

References

Abdel Qader, A. 2023. Utilizing digital technologies to ensure food safety. Int. J. Mod. Agric. Environ. 3(2): 17-29.

Abdollahpour, S., A. Kosari-Moghaddam and M. Bannayan. 2020. Prediction of wheat moisture content at harvest time through ANN and SVR modeling techniques. Inf. Process. Agric. 7(4): 500-510.

Adetunji, C.O., O.A. Anani, O.T. Olugbemi, D.I. Hefft, N. Wilson and A.S. Olayinka. 2022. Toward the design of an intelligent system for enhancing salt water shrimp production using fuzzy logic. *In:* AI, Edge and IoT-Based Smart Agriculture (pp. 533-541). Academic Press.

Al, S., F.U. Ciloglu, A. Akcay and A. Koluman. 2024. Machine learning models for prediction of Escherichia coli O157:H7 growth in raw ground beef at different storage temperatures. Meat Sci. 210: 109421.

Alam Siddiquee, K.N.E., M.S. Islam, M.Y.U. Dowla, K.M. Rezaul and V. Grout. 2020. Detection, quantification and classification of ripened tomatoes: A comparative analysis of image processing and machine learning. IET Image Process. 14(11): 2442-2456.

Anand, R., S.S. Nalla, S. Nasreen, M. Chowdhury, Madhusudan, B.S., A. Pandey et al. 2024. Fruit counting and analysis using artificial intelligence approaches. pp. 93-114. *In:* S.S. Chouhan, A. Saxena, U.P. Singh and S. Jain [eds.]. Artificial Intelligence Techniques in Smart Agriculture. Springer Nature, Singapore.

Awuchi, C.G. 2023. HACCP, quality, and food safety management in food and agricultural systems. Cogent Food Agric. 9(1): 2176280.

Charlebois, S. and J. Music. 2020. The influence of digital technologies and AI on food supply chains: A review. J. Food Sci. Technol. 58(7): 2189-2201.

Chen, T.C. and S.Y. Yu. 2021. The review of food safety inspection system based on artificial intelligence, image processing, and robotic. Food Sci. Technol. 42: e35421.

Chhetri, K.B. 2024. Applications of artificial intelligence and machine learning in food quality control and safety assessment. Food Eng. Rev. 16(1): 1-21.

Choi, J.Y., K. Seo, J.S. Cho and K.D. Moon. 2021. Applying convolutional neural networks to assess the external quality of strawberries. J. Food Compos. Anal. 102: 104071.

Chowdhury, M., R. Anand, T. Dhar, R. Kurmi, R.K. Sahni and A. Kushwah. 2024. Digital insights into plant health: Exploring vegetation indices through computer vision. pp. 7-30. *In*: S.S. Chouhan, U.P. Singh and S. Jain [eds.]. Applications of Computer Vision and Drone Technology in Agriculture 4.0. Springer Nature, Singapore.

da Costa, T.P., J. Gillespie, X. Cama-Moncunill, S. Ward, J. Condell, R. Ramanathan and F. Murphy. 2022. A systematic review of real-time monitoring technologies and its potential application to reduce food loss and waste: Key elements of food supply chains and IoT technologies. Sustainability. 15(1): 614.

Domínguez, R., M. Pateiro and P. González. 2021. Application of AI-driven sensory evaluation systems for quality assessment of wine. J. Wine Res. 32(1): 64-78.

Feng, H., X. Wang, Y. Duan, J. Zhang and X. Zhang. 2020. Applying blockchain technology to improve agri-food traceability: A review of development methods, benefits, and challenges. J. Clean. Prod. 260: 121031.

Fernández, R. 2019. Machine learning in sensory evaluation: A review. Food Control. 105: 55-67.

Fernández-López, J.A. and S. Cofrades. 2019. AI-based sensory systems in meat and fish product quality assessment. Trends Food Sci. Technol. 84: 52-59.

Food Safety, World Health Organization. 2020. https://www.who.int/health-topics/food-safety#tab=tab_1

Friedlander, A. and C. Zoellner. 2020. Artificial intelligence opportunities to improve food safety at retail. Food Prot. Trends. 40(4): 264-267.

Gai, Z., L. Sun, H. Bai, X. Li, J. Wang and S. Bai. 2022. Convolutional neural network for apple bruise detection based on hyperspectral. Spectrochim. Acta A Mol. Biomol. Spectrosc. 279: 121432.

Galanakis, C.M. 2019. Food Waste Recovery: Processing Technologies and Industrial Techniques. Academic Press.

Galvez, J.F., J.C. Mejuto and J. Simal-Gandara. 2018. Future challenges on the use of blockchain for food traceability analysis. Trends Anal. Chem. 107: 222-232.

García-Gimeno, R.M. et al. 2019. Application of machine learning algorithms for shelf-life prediction of fresh-cut products. J. Food Eng. 250: 1-10.

Guddaraddi, A., A. Singh, G. Amrutha, D.R.K. Saikanth, R. Kurmi, G. Singh et al. 2023. Sustainable biofuel production from agricultural residues an eco-friendly approach: A review. Int. J. Environ. Clim. Chang. 13(10): 2905-2914.

Hortinela, C.C., J.R. Balbin, J.C. Fausto, J.E.C. Espanillo and J.K.P. Padilla. 2020. Detection of staleness in raw chicken meat due to Salmonella spp. and Escherichia coli bacteria using metal oxide gas sensor with support vector machine. *In:* 2020 IEEE 12th Int. Conf. Humanoid, Nanotechnol., Inf. Technol., Commun. Control, Environ. Manag. (HNICEM) (pp. 1-5). IEEE.

Jha, A., M. Chowdhury and A.N. Satpute. 2023. Surface water quality forecasting using machine learning approach. pp. 293-315. *In:* C.B. Pande, M. Kumar and N.L. Kushwaha [eds.]. Surface and Groundwater Resources Development and Management in Semi-arid Region: Strategies and Solutions for Sustainable Water Management. Cham: Springer International Publishing.

Karapetkovska-Hristova, V., B. Makarijoski, M.A. Ahmad, G. Ntsomboh-Ntsefong, E. Mousa, B. Trajkovska and G. Dimitrovska. 2018. Study of fat content dynamics in dairy products with the simulations of artificial neural network (ANN) model. IOSR J. Environ. Sci. Toxicol. Food Technol. 12(3): 68-74.

Karim, S. and A. Karim. 2021. The impact of AI on supply chain and shelf-life prediction in perishable goods: A case study. Int. J. Food Sci. Technol. 56(9): 4934-4942.

Kavitha, R., D. Nandini, D.S. Guru and G. Parvathi. 2022. Texture features in prediction of bread edibility. *In:* 2022 Int. Conf. Eng. Emerg. Technol. (ICEET) (pp. 1-6). IEEE.

Kim, K., H. Kim, J. Chun, M. Kang, M. Hong and B. Min. 2021. Real-time anomaly detection in packaged food X-ray images using supervised learning. Comput. Mater. Contin. 67(2): 2547-2568.

Kiran, P.R., G. Avinash, M. Ray, S. Nigam and R.A. Parray. 2024a. Deep learning models for detection and classification of spongy tissue disorder in mango using X-ray images. J. Food Meas. Charact. 1-13.

Kiran, P.R., P. Aradwad, T.V. Arunkumar and R.A. Parray. 2024b. Enhancing mango quality control: A novel approach to spongy tissue inspection through image clustering and machine learning models via X-ray imaging. J. Food Process Eng. 47(6): e14664.

Kiran, P.R., M. Yeasin, P. Aradwad, T.V. Arunkumar and R.A. Parray. 2024c. Advancing mango quality assurance: Non-destructive detection of spongy tissue using visible near-infrared spectroscopy and machine learning classification. Future Postharvest Food. 1(2): 237-251.

Kiran, P.R., P. Aradwad, T.V. Arunkumar, M. Sahoo, I. Mani et al. 2024d. A comprehensive review on recent advances in postharvest treatment, storage, and quality evaluation of onion (Allium cepa): Current status, and challenges. Future Postharvest Food. 1(1): 124-157.

Kudashkina, K., M.G. Corradini, P. Thirunathan, R.Y. Yada and E.D. Fraser. 2022. Artificial intelligence technology in food safety: A behavioral approach. Trends Food Sci. Technol. 123: 376-381.

Kumar, A. 2020. Blockchain technology in the food industry: Review of the status quo and future trends. Food Control. 120: 107447.

Kumar, A. and V.K. Patel. 2023. Classification and identification of disease in potato leaf using hierarchical-based deep learning convolutional neural network. Multimed. Tools Appl. 82(20): 31101-31127.

Kumar, I., J. Rawat, N. Mohd and S. Husain. 2021. Opportunities of artificial intelligence and machine learning in the food industry. J. Food Qual. 2021(1): 4535567.

Kushwah, A., P.K. Sharma, H.L. Kushwaha, R.H. Nag, G. Carpenter, M.K. Choudhary et al. 2024. Economic evaluation of precise intelligent cauliflower harvester: A comparative study with manual harvesting. J. Sci. Res. Rep. 30(1): 33-42.

Lee, J., T. Suh, D. Roy and M. Baucus. 2019. Emerging technology and business model innovation: The case of artificial intelligence. J. Open Innov. Technol. Mark. Complex. 5(3): 44.

Li, M. and Z. Li. 2022. The role of AI in sensory evaluation and product development. Sensors. 22(1): 65-81.

Luparelli, A., I. Losito, E. De Angelis, R. Pilolli and L. Monaci. 2023. Multi-target detection of nuts and peanuts as hidden allergens in bakery products through bottom-up proteomics and high-resolution mass spectrometry. Foods. 12(4): 726.

Malkani, P., I. Mani, P.K. Sahoo, R.A. Parray, S.S. Swain, K.R. Asha et al. 2024. Development and evaluation of a sensor enabled smart herbicide applicator for inter row crops. Smart Agric. Technol. 9: 100649.

Minghui, G.E., Z.H. Jun and L.U. Huijuan. 2023. Research progress of food packaging defect detection based on machine vision. Food Mach. 39(9): 95-102.

Nychas, G.J.E., E.Z. Panagou and F. Mohareb. 2016. Novel approaches for food safety management and communication. Curr. Opin. Food Sci. 12: 13-20.

Palakurti, N.R. 2022. AI applications in food safety and quality control. ESP J. Eng. Technol. Adv. 2(3): 48-61.

Pandey, D. and K. Pandey. 2023. Exploiting artificial intelligence of things (AIoT) for product safety and quality control in supply chain management. *In:* Int. Conf. Entrep. Innov. Leadersh (pp. 415-423). Springer Nature Singapore.

Pateiro, M. 2020. AI and machine learning algorithms in interpreting sensory data for quality assessment. Trends Food Sci. Technol. 97: 91-100.

Patel, K.K., S.K. Goyal and Y.K. Patel. 2023. Image processing for food safety and quality. Novel Technol. Food Sci. 451-478.

Qian, C., S.I. Murphy, R.H. Orsi and M. Wiedmann. 2023. How can AI help improve food safety? Annu. Rev. Food Sci. Technol. 14(1): 517-538.

Rahul, K., R.K. Banyal and N. Arora. 2023. Machine learning and its application in food safety. *In:* Int. Conf. Adv. Data-Driven Comput. Intell. Syst. (pp. 131-142). Springer Nature, Singapore.

Ranjan, S., A. Sinha and S. Ranjan. 2023. Employing image processing and deep learning in gradation and classification of paddy grain. *In:* Artif. Intell. Societal Issues (pp. 85-111). Springer Int. Publ.

Regattieri, A., M. Gamberi and R. Manzini. 2021. Artificial intelligence applications in food supply chain management. Trends Food Sci. Technol. 114: 38-50.

Santos, J.L.O., G.M.G.B. Bueno, É.L. de Moraes Flores, L.E. Ogava, F. de Souza Dias and O.D. Leite. 2024. Support vector machine in the evaluation elementomic of arugula (Eruca sativa) and lettuce (Lactuca sativa) grown in soils from a decommissioned mining area. J. Food Compos. Anal. 106702.

Shah, H., H. Kumar and A. Akgül. 2023. Artificial intelligence and deep learning-based system for agri-food quality and safety detection. Wasit J. Comput. Math. Sci. 2(2): 26-32.

Suthar, A.P., S.H. Sindhi, J.B. Kathiriya, A.K. Sharma, V.K. Singh and K.R. Bhedi. 2024. Use of artificial intelligence (AI) in ensuring quality and safety of food of animal origin: A review. Int. J. Vet. Sci. Anim. Husb. 240-247.

Tripathi, S. 2019. Role of artificial intelligence in food industry. J. Food Sci. Technol. 56(9): 4087-4093.

Vegesna, V.V., S.M. Prabhakaran and P. Whig. 2024. AI for food safety: Leveraging artificial intelligence to ensure a safe and reliable food supply. *In:* Nutr. Controv. Adv. Autoimmune Dis. (pp. 501-524). IGI Global.

Verdouw, C.N. 2021. Internet of things for food and agriculture: Data-driven supply chains. J. Food Eng. 294: 110408.

Wamba-Taguimdje, S.L., S.F. Wamba, J.R.K. Kamdjoug and C.E.T. Wanko. 2020. Influence of artificial intelligence (AI) on firm performance: The business value of AI-based transformation projects. Bus. Process Manag. J. 26(7): 1893-1924.

Wu, T., J. Lu, J. Zou, N. Chen and L. Yang. 2022. Accurate prediction of salmon freshness under temperature fluctuations using the convolutional neural network long short-term memory model. J. Food Eng. 334: 111171.

Yang, J. and W. Liang. 2020. AI-driven innovations in electronic tongues and electronic noses for food quality control. Food Chem. 310: 125823.

Yang, Y. and L. Wei. 2021. Application of E-nose technology combined with artificial neural network to predict total bacterial count in milk. J. Dairy Sci. 104(10): 10558-10565.

Yasin, E.T., I.A. Ozkan and M. Koklu. 2023. Detection of fish freshness using artificial intelligence methods. Eur. Food Res. Technol. 249(8): 1979-1990.

Zhang, D., X. Liu and Y. Jiang. 2020. Applications of artificial intelligence in food quality evaluation: A review. Food Rev. Int. 36(6): 477-495.

Zhang, Y. and G. Tian. 2021. Applications of machine learning algorithms in sensory science. J. Food Sci. 86(4): 1450-1460.

Index

G

H

I

L

M

N

P

Q

R

S

For Product Safety Concerns and Information please contact our
EU representative GPSR@taylorandfrancis.com Taylor & Francis
Verlag GmbH, Kaufingerstraße 24, 80331 München, Germany